KU-489-736

SPIDER COMMUNICATION

SPIDER COMMUNICATION
Mechanisms and Ecological Significance

Edited by Peter N. Witt
and Jerome S. Rovner

Princeton University Press
Princeton, New Jersey
1982

Published by Princeton University Press,
41 William Street, Princeton, New Jersey
In the United Kingdom: Princeton University Press,
Guildford, Surrey

Library of Congress Cataloging in Publication Data
will be found on the last printed page of this book

This book has been composed in Linotron Trump

Clothbound editions of Princeton University Press
books are printed on acid-free paper, and binding
materials are chosen for strength and durability

Printed in the United States of America by
Princeton University Press, Princeton, New Jersey

Designed by Laury A. Egan

This book is dedicated to all the women and men, past and present, who have contributed to our knowledge of the taxonomy of spiders. Their efforts have enabled us to know what kinds of spiders we are investigating when we study communication in these animals.

CONTENTS

SPIDER COMMUNICATION

Chapter 1

INTRODUCTION: COMMUNICATION IN SPIDERS

Peter N. Witt
North Carolina Mental Health Research
Anderson Hall
Dorothea Dix Hospital
Raleigh, North Carolina 27611

Most of the authors of this book were invited by me to come together as part of the International Meeting sponsored by the American Arachnological Society in the summer of 1978 in Gainesville, Florida. They agreed to discuss their work as it contributed to our knowledge about communication in spiders. I conceived the plan for the book at the symposium and thus became its senior editor. At a later date I asked Jerome Rovner to join me in the editorial work, and we bear together the responsibility for the present list of contributors and the present shape. The final version developed after the symposium through additional reviews of the relevant literature and inclusion of current laboratory and field work. All along we tried to preserve the attraction of the immediateness of the reports from the authors' own laboratories and combine them with a more general review of the field. The text aims at presenting the present state of knowledge and is a compromise between completeness and readability. Sometimes a chapter fits only the widest definition of "communication," e.g., Riechert and Łuczak's "Spider Foraging: Behavioral Responses to Prey." Because it contains valuable information on spider behavior and ecology which is not to be found summarized elsewhere, and because it rounds off the other chapters by applying some of the sensory physiology and other knowledge to predator-prey interactions, the editors decided to include it. After all, the subject matter of that chapter is an important part of the spider's relationship with its environment. A special effort is made to address readers beyond arachnologists, so that they can sample and compare how much (or little) is known about one aspect of this one group of animals at the present time.

To many readers opening this volume, the title will seem strange, if not unworthy of serious and lengthy treatment. On the face of it, spiders do not deserve reputations for communicability. Rather, they are popularly regarded as solitary and silent predators, who neither seek, nor are likely to obtain, partners in any gentle exchange. A moment's reflection will revise that judgment, but it is probably the first task of an introduction to offer some explanation of why the volume is thicker than one might anticipate. The events of courtship would probably be granted at once to be a matter of communication. Even the "news" which prey provides, however regretfully it does so, could be construed as communication. Indeed, this latter, very broad interpretation of the term "communication" is the one I have chosen to use in order to include material in this volume which is relevant to a consideration of mechanisms underlying communication *sensu strictu*. Thus, even where the research dealt with a system for prey detection, the findings of such a study could also be of interest to those seeking to understand the use of the same channel for signaling by a male spider when courting a female.

For the reader who pictures only the solitary web-builder, there will be surprises in this book, especially when communal spiders are considered. But the topic is capable of even wider expansion; so much so that in preparing this book one became convinced that to cover the problem thoroughly, it will never be thick enough. Scarcity of knowledge, rather than lack of subject matter worthy of discussion, has limited the size of the present work. In the search for knowledge, this book is a way station where we review how far we have come in order to determine where we want to go.

Invertebrates have recently become particularly interesting to neuroscientists. Chase (1979) points out that papers in invertebrate neurobiology constituted the third largest of 41 topical categories at the 1978 meeting, which represented the whole Society of Neuroscience's interest. He discusses the advantages of relatively fewer neuronal elements for identification in invertebrates, together with greater technical accessibility. This accounts for the fact that physiological descriptions of behavioral control are more complete for invertebrate models than they are for vertebrate models. If one assumes evolutionary continuity from animals to man, it follows that studies of identification of the role of particular neurons in invertebrates become relevant to the human situation. This in turn permits more enlightened speculation on the mind-body problem (see Chase, 1979).

Spiders, one of several groups making up the class Arachnida, are invertebrates which show a number of peculiarities. An example of the special body structure of spiders is shown in Figure 1.1. Spiders show a number of precoded behavior patterns, frequently called fixed action patterns. In several places in this book such behavior

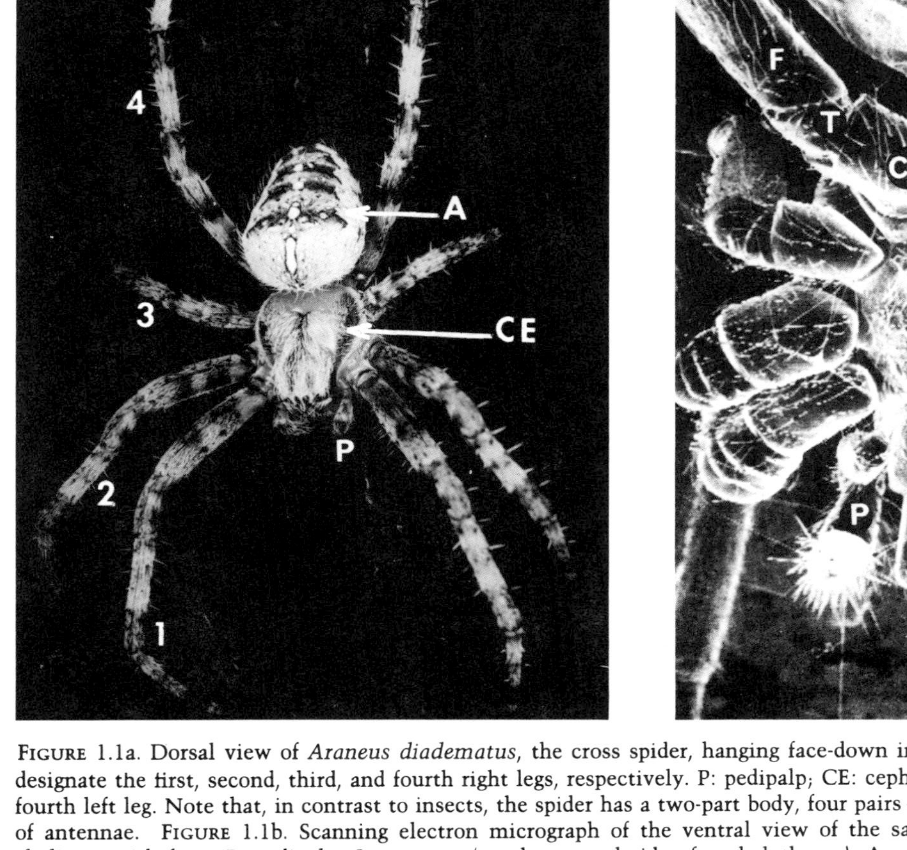

FIGURE 1.1a. Dorsal view of *Araneus diadematus*, the cross spider, hanging face-down in its orb-web. Numbers 1, 2, 3, and 4 designate the first, second, third, and fourth right legs, respectively. P: pedipalp; CE: cephalothorax; A: abdomen; TA: tarsus of fourth left leg. Note that, in contrast to insects, the spider has a two-part body, four pairs of legs, and a pair of pedipalps instead of antennae. FIGURE 1.1b. Scanning electron micrograph of the ventral view of the same spider (face-down as in 1.1a). C: chelicera with fang; P: pedipalp; S: sternum (on the ventral side of cephalothorax); A: abdomen; CO: coxa; T: trochanter; F: femur of fourth leg.

patterns as courtship, web-building, and feeding will be described as highly ritualized and species-specific. The neuronal substrates of such behavior probably will be elucidated one day. The following chapters review much of the present knowledge as an early step in this process. Table 1.1 lists the families of spiders containing one or more species whose names will appear in these chapters.

Why do we focus on communication? Communication is important for an animal's survival. Animals change their behavior as a consequence of the information they receive. In social animals all communal life is based on a communication network, which car-

TABLE 1.1. Families of spiders included in this book.

Web-weavers	*Wanderers*
Agelenidae - funnel weavers	Anyphaenidae
Amaurobiidae	Archaeidae
Araneidae - "ordinary" orb-weavers	Clubionidae - sac spiders
Argyronetidae - water spiders	Ctenidae
Atypidae - purse-web spiders	Dysderidae
Barychelidae	Gnaphosidae (Drassidae)
Ctenizidae - trapdoor spiders	Lycosidae - wolf spiders
Dictynidae	Mimetidae - pirate spiders
Diguetidae	Oxyopidae - lynx spiders
Dinopidae - ogre-faced spiders	Philodromidae
Dipluridae - funnel-web tarantulas (mygalomorphs)	Pisauridae - nursery-web spiders
Eresidae	Salticidae - jumping spiders
Filistatidae	Scytodidae - spitting spiders
Hahniidae	Sicariidae
Hypochilidae	Sparassidae (Heteropodidae) - huntsman spiders
Leptonetidae	Theraphosidae - "ordinary" tarantulas (mygalomorphs)
Linyphiidae - sheet-web weavers	Thomisidae - crab spiders
Liphistiidae	
Loxoscelidae	
Micryphantidae (Erigonidae) - dwarf spiders	
Ochyroceratidae	
Oecobiidae	
Pholcidae - long-legged spiders	
Segestriidae	
Tetragnathidae - long-jawed orb-weavers	
Theridiidae - comb-footed spiders	
Uloboridae	
Urocteidae	

ries information among the individuals, letting each know what it has to do to assist in the survival of relatives and thereby insure its own best genetic interest. Communication between sexually reproducing animals is one prerequisite for their genes' survival.

There are many ways in which living beings communicate. Some of the ways are so characteristic for a species or genus that they can be used for defining the difference between that group and others. Humans are frequently distinguished from animals by the ability of our species to use language for communication. The more recent investigations of chimpanzees' ability to master and apply American Sign Language are efforts to resolve the controversy over the degree of sophistication, abstraction, and generalization which these apes can develop in this communication medium as compared to humans (Griffin, 1977, 1978; Premack and Woodruff, 1978).

Understanding communication between living beings requires insight into many different aspects of life. The organs which transmit and receive signals can be studied as to their physical appearance and particular function. The signal itself—its variation, nature, means of transmittal, and its information content—is another subject for investigation. These components cannot be understood as part of the communication process if the meaning of the message is not clarified: how it was encoded by the communicator and how it was decoded and interpreted by the receiver. Usually, observation and the measurement of the individual's behavior under specified conditions are used as methods for gaining some understanding of the content and meaning of a communication process.

The central position which communication plays in animals' lives can be deduced from the observation that studying communication leads to an understanding of the peculiarities of that animal and its conspecifics. The description of a simple experiment will illustrate the point.

A spider hangs face down in the center of its intricate orb web. A low-frequency tuning fork is struck. As soon as the vibrating prongs of the instrument touch a radius of the web, the spider turns and positions one front leg on the moving radius (Figure 1.2). This trial can be repeated over and over, and the results will nearly always be the same. Further reaction is more variable and depends on a number of circumstances: the duration of the signal, the response of the web to short jerks on the radius by the first legs, the spider's appetite, the number and character of preceding trials, and even general circumstances such as time of day, sun, wind, and rain.

FIGURE 1.2. A vibrating tuning fork is brought into contact with one radius at the periphery of the orb-web of an adult female *Araneus diadematus*. The animal orients toward the vibrating radius and probes with its front legs to monitor the stimulus.

What has just been described represents certain aspects of communication (broadly defined) which are characteristic of many species of spiders. The tuning fork produced a measurable signal; frequency, intensity, and variability of the signal could have been tested for the ranges in which they elicit a response. The instrument was used in place of another animal—a conspecific or prey—to test the nature, frequency, and intensity of effective signal production. The signal was transmitted through a specific channel, in this case mechanically through the silken radial thread. The channel connected the signaler with the receiver. The resulting behavior of the spider, i.e., turning in the direction of the tuning fork, provided evidence that the signal was received and decoded. The lyriform and slit organs on the legs of the spider (here *Araneus diadematus*) have been shown in other spiders to be sensitive to vibration-induced strains in the exoskeleton, as will be discussed by Barth. These receptors sent impulses (action potentials) along nerve fibers through the legs to the central nervous system. Here the message was decoded and translated into outgoing nerve-borne signals, which resulted in patterned muscular contractions that produced movements of the legs and body. All sections of this communication system worked together to produce an observable action-reaction sequence. As will be fully discussed in later chapters, the investigator can analyze the various parts of the system, and define the role each plays in the total process.

Beyond the general conception of communication which we have derived from this observation, it has taught us much about the animal in its living space. This particular spider, like many of its relatives, has made use of a specific signal quality, namely the vibration of the substrate on which it rests. Each type of signal, be it chemical, acoustic, visual or, as in this case, vibratory, has properties which make it practical for a specific environment. Vibration, for example, is independent of light and can be as effective at night as during the day. It is relatively independent of air currents, which can, on the other hand, carry chemicals with them. Nocturnal spiders, which build an invisible trap to catch visually orienting flies, were preadapted with a sensory system which is highly receptive, independent of vision. Many spiders have compensated for the absence of a suitable substrate over which a vibratory signal can be conducted, as well as for limitations in the distance of conduction in available substrates, by extending the perceptual range of the legs with a silken structure. It is not enough that the area of capture is enlarged; the fact of contact by the prey with the silk must be communicated. Suddenly the radiating shape of the web (Figure

1.2) takes on a new meaning for the observer: the web extends the perceptual range of the sense of vibration from about 15mm to more than 500 mm by forming a suitable substrate for the transmission of vibrations to the legs. The legs, in turn, improve reception by pulling the silk tight.

Knowledge of the organs for communication lets us understand other behaviors of the animal. During orb-web construction, the spider pulls and probes existing threads to gather information on the degree of completion, before new strands are laid (Peters, 1938; Reed, 1969). It straddles angles, apparently assessing their width and comparing them to an internal "plan" or template, which guides web construction. In the laboratory we have let the cross-spider build an orb-web in a closed box in complete darkness. Elaborate measurements carried out on the resulting webs showed no differences between these webs and those built by the same animal in partial light. No longer dependent on vision, and in darkness safe from visually searching predators, the spider builds an almost invisible web on which it will catch visually orienting prey.

Pertinent to the topic of communication is the fact that the web also plays a delicate and essential role in the spider's courtship behavior. A male may drum or pluck on the web of a female over long periods of time, until the aggressive attack of the female changes to acceptance of the male for copulation. It can be shown, by comparison of the webs of a few spider species (Risch, 1977), that the web built by the adult female is more species-specific than is the juvenile web. This observation holds for measures of size, fine structure, and shape. It raises the question of whether the specific resonance of the female web plays a role in species recognition for the "short-sighted" male. Blanke's (1973) experiments revealed reactions of males to wind which had blown across sexually mature females, suggesting that in finding the female web the male may be aided by an airborne signal, probably a chemical. This may then be followed by vibratory communication. Rarely is only one sensory modality involved in communication behavior; instead, one channel is usually predominant and others assist in one stage of communication. The roles may be reversed in the next stage.

It is not necessary to suppose that the spider has any conception of the world which she so nicely manages. In this respect, the condition bears some resemblance to the human condition. The world outside us, and the mental picture we have of it, is a product of what we can detect through our receptor organs and the subsequent analyses in the brain. We see colors and shades from which we guess at shapes and materials. We smell, feel, hear; and we use those inputs to recognize, categorize, understand objects and other living beings. Philosophers like Schopenhauer have gone so far as to describe the world as a product of our will and imagination.

Arthur Schopenhauer's principal work is the book *Die Welt als Wille und Vorstellung*, which appeared first in Leipzig in 1819, and was translated into English in 1883 under the title *The World as Will and Idea* (see ref. Schopenhauer). The second part of the title, which describes the first part of his philosophy, is of special interest here. The German word *Vorstellung* can also be translated as "representation, conception, mental image" (The New Cassell's Dictionary, 1958). Any of those three words appear to this writer to describe part of Schopenhauer's philosophy better than the common translation "idea." Based on Descartes, Kant, and Locke, Schopenhauer stressed the distinction between the phenomenon, or the appearance that a thing presents to the perceiving mind, and the thing as it is in itself. Through perception the mind is presumed to be aware only of the observable facts or events, i.e., of the phenomena: what lies behind them, being beyond all possible experience, is unknowable. It is impossible to dissociate conceptual thinking from the perceptual experience on which it is based. To use Schopenhauer's words (in translation): "Conceptions and abstractions which do not ultimately refer to *perception* [translator's italics] are like paths in the wood that end without leading out of it." In brief, our conception or understanding of the world is formed by the organs we have to perceive it; thus, we have a predominantly visual world. A very different world, mainly filled with touch and vibration signals, exists for the web-building spider. By discussing the spider's organs for communication, some of the contributors to this book try to introduce the reader to the "inner world" of the spider.

Such an argument is not intended to mean that the present author believes exclusively in the familiar "idea that there can be nothing in our intellect which has not entered it through our senses." We must assume that "every animal is born with expectations and anticipations" (Popper, 1974), which means that it possesses inborn knowledge. However, it is argued that observation of the perceptual repertoire of an animal (including the central nervous processing of signals) introduces some special understanding of the animal and its works, in our case of spiders, which cannot be gained otherwise.

So far the word "communication" has been used loosely; no definition has been given. We have talked about communication between living beings, thereby excluding communication processes inside one being, i.e., nervous or chemical signals which carry messages from one part of the body to another. We have excluded passive communication between an individual and its environment, i.e., the perception of and reaction to temperature changes can sometimes be called passive communication (E. O. Wilson, 1975). Internal messages and the impact of the environment will only be discussed so far as they affect communication processes between animals. There are many definitions of communication available (see Smith, 1977; Sebeok, 1977; Wilson, 1975). Each definition serves a specific purpose; they are not mutually exclusive, but rather complementary. For that reason, none has been singled out. Rather it has been left to the authors of the different chapters to choose the definition which best suits their approach to the overall theme of communication in spiders.

Special signals carry messages for communication (for exceptions see Smith, 1977, p. 13). Such signals have to be produced by an organ which is specially adapted for signal production. A number of conditions have to be fulfulled before a signal is actually sent out: the physiological stage of the signaling organism has to be right before the environment (or the partner) can elicit the signal. Both the state of the organism and the environmental releaser determine whether a signal is sent.

Some of the following chapters focus attention on the nature of the signal, i.e., they discuss airborne, chemical, and other modes of communication. In pulsed signals, for example, one may distinguish between frequency and intensity, find graded and stereotyped repertoires, and note the manner in which the beginning and end of the signal are determined by the signaler or the environment. However, very few messages depend on one channel of communication only. In humans the visual perception of the signaler's face joins with the tactile message of a handshake and the auditory message of a greeting to communicate "welcome." The quality of any one of these elements, or its absence, may radically alter the message. The problems raised by the spectrum of signals and their possible combinations require additional kinds of study, and are the concern of other chapters.

The message can only get across to another living being if that individual has an organ sensitive to the nature of the signal, a way to decode the message, and is ready to receive it. The study of receptor organs—their sensitivity and ability to discriminate—is combined with observation of the behavior of the receiver to reveal whether a message was received and what its significance was. Analysis of communication behavior requires a special line of investigation, at a level of analysis which is just as important for our understanding of communication as are the anatomy and physiology of the communicating organisms and the physics of signal production, transmission, and reception.

Another way to look at communication is to place it in the wider context of its contribution to the survival of the organism, i.e., to gauge its adaptive value (Burghardt, 1977). It is generally assumed that organisms which live communally, frequently in structured societies, use more and a greater variety of signals than do solitary animals. However, an animal which lives alone is still dependent on communication with conspecifics (and sometimes with animals of other species). One has to assume that the lonely spider of our example, in the middle of its orb-web, is at least occasionally interacting with other orb-web builders with whom she competes for prey.

Communication systems change throughout the lifetime of an animal as requirements to communicate differ. The tiny immature spiderling, its nutritional needs supplied by the yolk, probably has communication requirements different from those of the adult female, one hundred times heavier, sexually mature, and a voracious feeder (Burch, 1979). Such ontogenetic development in the communication system of spiders has been little studied, an exception being Aspey's (1975) study of ontogeny of display in a wolf spider. On the other hand, a type of communication like the visual signaling given and received by jumping spiders may vary from species to closely related species. Comparison of communication systems can assist in tracing the phylogeny of whole taxonomic groups of spiders.

Communication involves at least two, and frequently more than two animals. If there is no receiver for a message sent out by a display of an animal, communication cannot take place. There is always a mutual evolutionary advantage (Smith, 1977) involved in successful communication. Frequently it achieves a central social function. One can assume that the degree of social organization of a species can be measured by its use of intraspecific communication.

Finally, one may ask whether spiders have developed ways specific to them in which they communicate? Do their communication systems set them apart from other animals? In many, the production and daily use of silk has opened up for them a means of communication which only few other animals can rival. When some of that knowledge was reviewed several years ago, Witt (1975), focusing on the orb-web, found that a small amount of evidence went together with a large array of speculation and surmise.

Much has been discovered since that time, and some chapters in this book update our knowledge of various aspects of spider communication.

The reader may be convinced by now that there is a good case for bringing together much of the existing knowledge about this topic. By reviewing communication in spiders—the organs involved, the purposes which communication serves, the circumstances under which it takes place, and the effect it has—we will increase our understanding of the more than 30,000 species of spiders (Levi and Levi, 1968) as much as we would by a review of their size, shape, and color. One may even be convinced that several authors are necessary to bring all the pieces of knowledge together in order to make it possible to understand fully the peculiarities of spiders and the relationship of species with each other. Platnick (1971) stresses that stereotyped patterns (e.g., courtship behavior) must be considered at least as important a character for systematics as morphology. But one can still ask whether the beautifully illustrated large books, like the recent volume by Gertsch (1979) on *American Spiders,* do not sufficiently cover the subject. The material presented in our book is so different from that discussed in other "spider books" (see references) that it can serve as a complement to these without repeating any of their details. Indeed, only the ten-page review by Weygoldt (1977) on "Communication in Crustaceans and Arachnids" has tried anything similar to this book and did so, of necessity, in a very much shorter form. Ours is the first book of its kind, one which discusses a specific aspect of all spiders' lives, and so increases our knowledge of this interesting group of animals.

Acknowledgments

The author's more than thirty years of web analysis, which led to the organization of this book and the insights discussed in the Introduction, was for many years supported by the Swiss National Fund, the National Science Foundation, and many co-workers, whose names appear in the reference lists. Mrs. R. Daniels worked untiringly on the arrangement and retyping of manuscripts. Dr. C. F. Reed provided invaluable stimulation and advice.

Chapter 2

THE SIGNIFICANCE AND COMPLEXITY OF COMMUNICATION IN SPIDERS

Bertrand Krafft

Laboratoire de Biologie du Comportement
Case Officielle 140
Université de Nancy 1
54037 Nancy CEDEX, France

Introduction

"Biological communication can be defined as actions on the part of one organism that alter the probability pattern of behavior in another organism in an adaptive fashion" (E. O. Wilson, 1971). This very broad definition of communication, which goes beyond the usage of many authors restricting communication to intraspecific interactions, allows the inclusion of signal exchanges between animals of different species such as occur in commensalism.

Communication is typically an exchange of information in the form of direct interactions. Animal A sends a signal toward animal B. The latter responds by a behavior modification that constitutes a signal acting in return on animal A, which changes its behavior in consequence. However, A's signal need not be directed to a particular target or receiver. It may constitute either a form of direct interaction if the receiving animals perceive the signal when it is sent, or a form of indirect interaction if the signal persists and is perceived after a delay. The emission of an alarm pheromone by an ant corresponds to the first case, and the delimitation of a territory with the aid of pheromones deposited on the substratum corresponds to the second. The animal changes the environment by adding a chemical compound that orients or modifies the behavior of any conspecific that detects the signal. In these conditions, a thread or silky structure woven by a spider can constitute a signal for a conspecific. Taking communication, in the form of a direct or indirect exchange of information, to accord with Wilson's definition, I shall consider all these mechanisms.

Spiders have long been neglected, in comparison with vertebrates and insects, as subjects for study. Their essentially solitary way of life, our scanty knowledge until recent years of their sensory organization, and the difficulties in the taxonomy of arachnids may have caused this state of affairs. Only the spectacular aspect of the courtship display of the Salticidae and the Lycosidae has prompted thorough studies. However, many other situations require two or more spiders to adjust their activities between themselves. The distribution of individuals in a biotope, the communal life of the young, and sexual, parental, and social behaviors—all these necessitate communication. Likewise, the interspecific relations found in parasitism and commensalism cannot be explained without assuming that information is transmitted between individuals.

Communication in Spiders' Behavior

Predatory Behavior

During predation spiders often use the same sensory channel to detect prey as they use to exchange information between individuals. Among Lycosidae, even immature ones, leg-waving limits cannibalism (Aspey, 1976; Vogel, 1971, 1972). Thus the spider distinguishes between a conspecific and prey on the basis of visual information. Vibratory information plays the main role in triggering predatory behavior in the web-spinning spiders. It would be tempting to believe that anything that disturbs the web is attacked, whether it be a fly, a conspecific, or a tuning fork (Burgess and Witt, 1976). In fact, the approach of a conspecific rarely triggers predatory behavior, but rather triggers agonistic behavior. Such discrimination between prey and a conspecific is seen often, even in the solitary species. Thus, the vibratory signals emitted by males during courtship displays and the disturbance caused by a conspecific moving on the web do not trigger predatory behavior, even though they fall within the same range of frequencies. It is certain that spiders can distinguish these different signals.

The interpretation that a precise, specific stimulus is necessary to trigger the predatory behavior would be difficult to support. In fact, spiders respond to a wide range of frequencies, and the range of prey captured is also wide. In addition, the disturbances of the web caused by the prey are complex phenomena in which various frequencies are superimposed (Liesenfeld, 1956; Parry, 1965). On the contrary, males emit stereotyped signals, whose recognition by the female undoubtedly depends on central mechanisms. It could be conjectured that inhibition of predatory behavior requires a particular signal, while predation itself is elicited by a variety of more general signals. However, this hypothesis does not account for all the relevant phenomena. Among the social spiders, the various activities of individuals on the web do not trigger predatory behavior in their conspecifics (Krafft, 1975a,b); the same is true of the solitary species when the male and the female are living together (Mielle, 1978).

Since the vibrations thus provoked can vary (Krafft et al., 1978; Krafft, 1979), recognition of these different signals must depend on an overall perception that takes account of the frequency, the form, and the temporal structure of the signal, and requires a relatively complex central integration mechanism. To this should be added the possibility of learning by experience (Bays, 1962; Meyer, 1928; Walcott, 1969), which allows the predatory behavior of spiders to be adapted to, among other things, the available prey. It is important to emphasize also that spiders rarely use a single sensory channel in their intraspecific and interspecific relations. Sexual and social behaviors involve stimuli that are visual and tacto-chemical or chemical, vibratory, and tactile. Integration of the diverse kinds of information certainly increases the precision of the spider's response and allows it to avoid any confusion between its prey and a conspecific.

Overwintering and Resting Aggregations

The Clubionidae, Thomisidae, Oecobiidae, Salticidae, and Gnaphosidae sometimes form overwintering aggregations (Berland, 1932; Jennings, 1972; Kaston, 1965; Simon, 1891). It is tempting to attribute these groupings to the existence of a mutual tolerance (Łuczak, 1971); this merits further study since the various aggregations often include more than one species. Underlying these groupings is doubtless a search for favorable microclimatic conditions, and the absence of aggression could be related to a reduction in predatory behavior because of the unfavorable climatic conditions.

On the other hand, an exchange of information seems to be necessary among the wandering spiders that form resting aggregations during the summer months. Kajak and Łuczak (1961) demonstrated the existence of a grouping drive in *Marpissa radiata, Sitticus littoralis* (Salticidae), and *Xysticus ulmi* (Thomisidae). The factor responsible for these groupings might be a nonspecific stimulus provided by the silky structures already constructed, since such aggregations can include spiders of different species. This hypothesis is probable, given the major role that silk plays among spiders.

Intraspecific and Interspecific Interactions in Dense Populations of Spiders

The distribution of spiders in a biotope depends not only on the environmental conditions, but also on intraspecific and interspecific interactions that differ clearly from predatory behavior. Millot and Bourgin (1942) had already noticed that several individuals of *Stegodyphus lineatus* could be raised in an enclosure so small that their webs touched one another. When two individuals met, threat displays were made. Cloudsley-Thompson (1955) made the same observation on a group of *Amaurobius similis, Amaurobius fenestralis*, and *Amaurobius ferox*, showing that these interactions can be interspecific. However, the absence of cannibalism depended on the size of the rearing box. Aggregations of *Araneus cornutus* and of *Nephila* have also been observed. *Leucauge* sp. sometimes form temporary colonies of two to eleven individuals when the biotope is favorable (Valerio and Herrero, 1977). Such associations cannot be explained without recourse to tolerance mechanisms.

The positioning of a web determines its effectiveness and plays a major role in the selection of prey (Uetz et al., 1978). As discussed in the chapter by Riechert in this volume, to maintain its energy balance, each individual spider must select and preserve a territorial space of an extent appropriate for the available prey (Riechert, 1974b, 1976, 1978a,b; Turnbull, 1973). The communication involved in an agonistic interaction between two individuals must be considered proof that there is a form of relative tolerance linked to a possibility of discriminating between a conspecific and prey.

Intraspecific and interspecific encounters are particularly frequent among wandering spiders such as the Salticidae and the Lycosidae. The threat postures that make it possible to reduce cannibalism are well developed among the predominantly visual Salticidae (Crane, 1949b). When two males of *Cyrba algerina* meet, the individuals raise their front legs in a V-shape (Legendre, 1970). This attitude, which is generally adopted by the male when he perceives the contact pheromone that the female leaves on the substratum, may also be triggered by the approach of any object, even outside the reproductive period. It must therefore be considered a signal that is partly independent of sexual behavior.

In mixed populations of *Pardosa falcifera* and *P. sternalis*, and of *P. sternalis* and *P. zionis*, whose density may reach 150 individuals per square meter, each spider defends a "privacy sphere" 7 cm in diameter (Vogel, 1971, 1972). As soon as the privacy spheres of two spiders overlap, the individuals go through a series of agonistic interactions that assure their dispersion by bringing about the flight of one of them. Various laboratory studies on the genera *Lycosa, Schizocosa,* and *Pardosa* have shown the wide distribution of such threat displays in the Lycosidae, usually consisting of a waving of the first pair of legs. Among adult males, these interactions allow the establishment of a hierarchy partially correlated with size, and may have the function of enabling each male to occupy a territorial space favorable for reproduction (Dijkstra, 1978). The appeasement function of this leg-waving was demonstrated by the fact that it stopped pursuit between two male *Schizocosa ocreata* in 85% of the cases (Aspey, 1974). The fact that this agonistic behavior appears often among juveniles and in both sexes and can be triggered by nonsexual stimuli shows that its function is not necessarily related to reproductive behavior (Aspey, 1974, 1976; Dijkstra, 1969, 1978; Koomans et al., 1974). Since the agonistic behavior pattern is very similar from one species to another, it must be a very general behavior with a function not only of intraspecific but also of interspecific communication, allowing the biotope to be better exploited by a large number of individuals. This mode of interaction is very important, considering the impact that spiders can have on the insect fauna. Indeed, these agonistic relations affect the density of spiders and hence their predatory potential toward insects.

Parasitism and Commensalism

Interspecific associations among spiders, or between spiders and insects, are frequent (Kaston, 1965); but the exchanges of signals regulating these interactions are little known.

PARASITISM The Linyphiidae and Theridiidae sometimes establish themselves in the webs of other species (Bristowe, 1958; Kullmann, 1959, 1960; Legendre, 1960; Roberts, 1969). These spiders behave as real parasites. They compete with the host to capture the prey taken in the web and steal the prey set aside by the host. *Achaearanea tepidariorum* is only a facultative parasite; when kept alone, this spider can weave a perfect web, whereas in the webs of *Cyrtophora* it makes only a rudimentary trap. *Conopistha argyrodes,* on the other hand, is an obligate parasite, since it is incapable of weaving a trap to capture prey outside the webs of *Cyrtophora* (Kullmann, 1959). These parasitic associations imply the acquisition of a minimum of information necessary for the parasite to adjust its behavior to that of the host.

It is not yet known how the parasite finds the host's web. However, the simultaneous presence of several individuals in one web suggests that some orientation stimulus is involved. The impunity that *Argyrodes elevatus* and other parasitic Theridiidae enjoy can on no account be likened to a tolerance mechanism. In fact, if dropped on the host's web, one of these spiders is captured and devoured. In addition, a specific tolerance mechanism would be incompatible with the fact that each species of parasitic Theridiidae can establish itself in the webs of many different species of orb-spiders. The survival of the parasite on the host's web therefore depends essentially on behavioral adaptations. (See the chapter by Barth in this volume.)

COMMENSALISM Parasitic associations are based mainly on the parasite's aptitude for exploiting the information transmitted by the host's web. In the case of commensalism between solitary spiders and social spiders, between spiders and social insects, or between insects and social spiders, the interactions are certainly more complex. Several species of Salticidae live in ants' nests. *Myrmarachne foenisex* associates with *Oecophylla longinoda*, *Myrmarachne legon* with *Camponotus acvapimensis*, and *Myrmarachne elongata* with *Tetraponera anthracina* (Edmunds, 1978). The genus *Cotinusa* lives in the ant-heaps of *Tapinoma melanocephalum* (Shepard, 1972), and so do certain Linyphiidae (Bristowe, 1958); the converse association also exists. Robinson (1977a) described an association between the caterpillars of *Neopalthis* and the social spider *Anelosimus eximius*. These caterpillars devour the prey taken in the web. The lepidopteran *Batrachedra stegodyphobius* lives in the nests of *Stegodyphus* (Pocock, 1903). The nests of *Mallos gregalis* contain a coleopteran and sometimes the drassid spider *Poecilochroa convictrix* (Berland, 1913; Diguet, 1909a; Semichon, 1910; Simon, 1909). Various Hemiptera (Plokiophilidae, Anthocoridae) and a coleopteran are frequently found in the nests of *Agelena consociata*.

In this type of association, encounters between the foreign species and members of the society are inevitable. However, the spider does not attack its hosts, and its presence within the society is tolerated. Therefore a mechanism of mutual tolerance must exist. It is not known whether this depends on tactile and chemical communication, as between the coleopteran *Atemeles publicolis* and the ants of the genera *Formica* and *Myrmica*. The relations between species of spiders are carried on partly by the intermediary of silky structures. Now, in many cases of commensalism this silky element is missing, and therefore interactions similar to those between insects of different species should be found. Bristowe (1958) reported that the Linyphiidae, when coming in contact with ants, wave their first pair of legs in a movement like that of the antennae of the ants. But this is undoubtedly not the only type of signal used, and an experimental study could turn out to be fruitful, particularly in the cases of commensalism between Salticidae and ants.

Sexual Behavior

The spectacular aspect of the courtship behavior of male spiders has prompted many studies, which have often been limited, however, to a description of these behaviors and to a search for the stimuli that trigger them. In fact, the specificity of these phenomena is useful for distinguishing species and constructing phylogenetic series (Blanke, 1975a; Dondale, 1964, 1967; Dumais et al., 1973; Hollander and Dijkstra, 1974; Hollander et al., 1973; Vlijm and Dijkstra, 1966). However, the sexual behavior of spiders involves many interactions which depend on chemical, visual, vibratory, and tactile signals. The reciprocal behaviors of the male and the female synchronize the partners' reproductive activities, suppress nonsexual drives, and contribute to the barrier between species (Hollander et al., 1973). The sexual behavior begins with the male actively seeking the female. Then he must approach her without triggering her predatory behavior or flight behavior, and may have to live with her for some time, even though her feeding behavior is not always inhibited. Lastly, the male must repulse competing males.

THE COMING TOGETHER OF THE SEXES In most spider species, the male actively seeks the female. But the female may facilitate the coming together of the sexes by emitting chemical signals or by turning toward the male. The rarity of studies of the coming together of the sexes can be explained by the fact that the effect of a signal can be studied only through the behavioral response of the receiver. Now, this response may be slight. In addition, most authors have looked for the stimuli liable to trigger the courtship behavior of the male rather than for those providing some directional information or modifying the male's locomotor behavior.

The males of the Dysderidae (Berland, 1912; Cooke, 1965), and of many Clubionidae and Thomisidae (Kaston, 1936; Platnick, 1971) manifest obvious sexual activity only after a direct contact with the female. But this does not rule out the possibility of information from a distance allowing the male to orient his movements. Airborne pheromones play a major role in the coming together of the sexes in insects (Beroza, 1979); their role in spiders is discussed in the chapter by Tietjen and Rovner.

The best-known signal that enters into the detection of the female by the male is surely the sexual pheromone contained in the dragline or in the threads of the web of an adult female. Owing to such a mechanism, the male can then approach a receptive female without exciting predatory behavior (Krafft, 1975a), but this is surely not the only role of the pheromone contained in the silk. Placed on the substratum of a female, male Lycosidae often alternate between courtship behavior and exploratory behavior (Dondale and Hegdekar, 1973; Richter, 1971). The latter disappears in favor of the courtship behavior only after direct contact has been made with the female (Dijkstra, 1976; Hollander et al., 1973). In these conditions one may wonder whether the main function of the dragline is not really to trigger the courtship behavior of the male, and whether the results obtained in the laboratory do not stem from an artificial increase in the concentration of the stimulus on a substratum abundantly covered with threads spun by the females. The silk certainly plays a larger role than has been supposed in orienting the male that is seeking a female. For example, orientation by means of information contained in the dragline has recently been observed for the sedentary spider *Coelotes terrestris* (Krafft and Roland, 1979). In a T-maze, the males followed the paths traced by a female, using the dragline as information. The removal of the line with a paint brush makes it impossible for the males to orient themselves, whereas a thread drawn artificially from the spinnerets of the female and placed in the maze is followed by the males (Table 2.1). There is indeed an orientation phenomenon linked with sexual behavior, since females never follow this same path. The same kind of experiments carried out with *Tegenaria domestica* have given the same results and have shown that males do not follow the paths of other males (Leborgne et al., 1980). These diverse observations show that the female may take an active part in the coming together of the sexes when she lays down draglines. But she can also respond actively to signals from the male. The female of *Lycosa rabida* moves toward a loudspeaker reproducing the acoustical signals that accompany the male's courtship behavior (Rovner, 1967, 1975).

THE MALE'S APPROACH TO THE FEMALE When the male has located a female after a brief contact made by means of visual information

TABLE 2.1. Orientation of *Coelotes terrestris* in a choice situation (T-maze).

Sex of spider tested	*Stimulus*	*Number of experiments*	*Spider's choice of stimulus (%)* Stimulus branch	*Opposite branch*	*Chi-square significance*
Male	trail of adult female	432	70.5	29.5	$P < 0.001$
Female	trail of adult female	100	51.0	49.0	n.s.
Male	thread of adult female	127	90.5	9.5	$P < 0.001$
Male	threadless trail of adult female	108	50.9	49.1	n.s.

Trail, the trail of a conspecific obtained by letting it pass just once through one of the branches of the maze; thread, a thread taken from the spinnerets of an anesthetized spider and put in one of the branches of the maze by the experimenter; threadless trail, a trail from which the thread has been removed (from Krafft and Roland, 1979).

or by contact with her silk, he must approach her without triggering either aggression or flight. The concept of a "privacy sphere" proposed by Vogel (1972) can doubtless be extended to all spiders. The function of the male's courtship behavior would be to permit the boundaries of this individual space to be crossed, by making the male known in advance to the female. The Lycosidae and the Salticidae use visual information (Bristowe and Locket, 1926). In web-spinning spiders, the male sends complex vibratory signals by drumming on the web with his pedipalps, tugging on the threads with his legs, and vibrating his abdomen when he enters the space that corresponds to the female's prey-catching area. The tactochemical information that facilitates the coming together of the sexes also enters into the male's recognition of the female. Such information is complemented by an exchange of visual or vibratory information between the partners. The male's recognition of the female therefore cannot be dissociated from the female's recognition of the male.

The females are not alone in emitting chemical signals. The characteristic movements of a female *Latrodectus hesperus* in response to vibratory signals from a male can also be elicited by putting her on the male's web, whereas the webs of females of the same species provoke only exploratory behavior, without any sexual display. Males' webs may therefore contain a sexual stimulus, probably a pheromone, analogous to that of the female (Ross and Smith, 1979). The males of other species also may emit sexual pheromones. The males of *Cyclosa conica, Araneus cucurbitinus, Araneus sericatus*, and *Linyphia hortensis*, and of certain *Tegenaria* have special gnathocoxal glands that constitute a secondary sex characteristic (Legendre and Lopez, 1974a). Those of Dysderidae and Clubionidae have epigastric glands whose function is unknown (Lopez, 1972, 1974). In *Argyrodes* (Theridiidae) there is great sexual dimorphism: the males are characterized by a prosomatic depression surrounding two frontal ocular protuberances; here are found the openings of the clypeal glands, which may play a part in mating, for the female holds the male up in such a way that her buccal region touches these structures (Legendre and Lopez, 1974b, 1975; Meijer, 1976). All these glands may secrete pheromones.

The visual signals of the Salticidae (see the chapters by Forster and Jackson) and the Lycosidae are the best known and have been the object of many comparative studies to further the identification of species (Hollander, 1971; Hollander et al., 1973). But authors have mainly limited themselves to describing the activities of the males without trying to find out the significance of the various behavioral sequences.

Many species of spiders have stridulation organs, as discussed in the chapter by Uetz and Stratton. The web-spinning spiders have the advantage of vibratory information transmitted by the silk. The male of *Zygiella x-notata* emits a vibratory signal at a frequency of 10 to 14 Hz (Liesenfeld, 1956). The signal of the males of *Cyrtophora cicatrosa* has a frequency of 25 Hz, and that of *Cyrtophora citricola*, 8 to 12 Hz (Blanke, 1972, 1975). *Mallos gregalis* vibrates its abdomen at 10 Hz, while *Mallos trivittatus* and *Dictyna calcarata* have frequencies of 2 to 3 Hz (Jackson, 1979c). The frequency of the abdominal vibration in *Tegenaria parietina* is 15 Hz, while that of *Tegenaria atrica* is 30 Hz (Mielle, 1978).

An original technique has allowed the vibratory signals emitted by the males of several species of spiders to be analyzed (Leborgne and Krafft, 1979; Leborgne et al., 1980; Krafft, 1978; Krafft et al., 1978). These signals are not simple sinusoidal vibrations, but a succession of characteristic elements. Comparison among several species shows that the signals are species-specific (Figures 2.1, 2.2 and 2.3).

During contact between the partners, spiders also exchange tactile signals. One often sees touching, and highly complex leg-play, which certainly are important in preparing the female for copula-

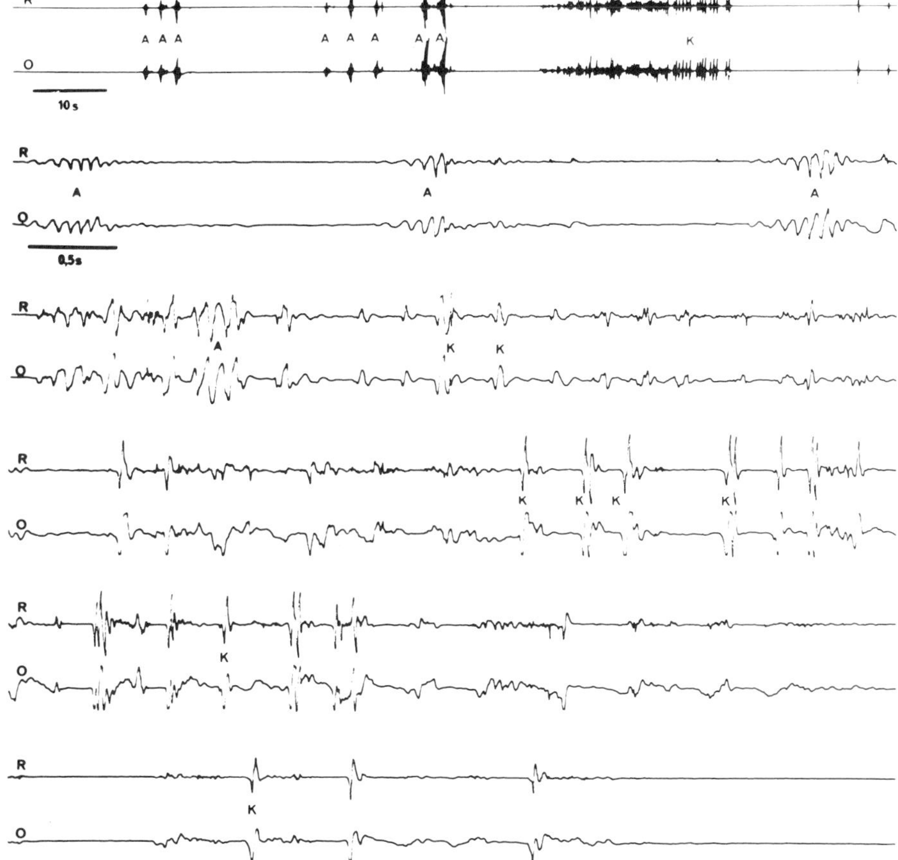

FIGURE 2.1. Signal generated by a male *Tegenaria parietina*: O: original; R: playback into the web by means of a vibrator after recording on a tape recorder; A: abdominal vibrations; K: strikes by the abdomen (from Leborgne and Krafft, 1979).

tion. In *Lycosa helluo* (Nappi, 1965), such an exchange of tactile signals seems to be essential before mating.

Description of these signals is an incomplete approach to the phenomenon. Many recent observations have shown that communication is established between the male and the female (Platnick, 1971). The female's immobility during the male's courtship display can in itself be considered a response. Sometimes the female approaches the male on the courting line. Sometimes she responds to the male's entreaties by sending vibratory signals. This is the case with females of *Filistata insidiatrix* (Berland, 1912), *Steatoda bi-*

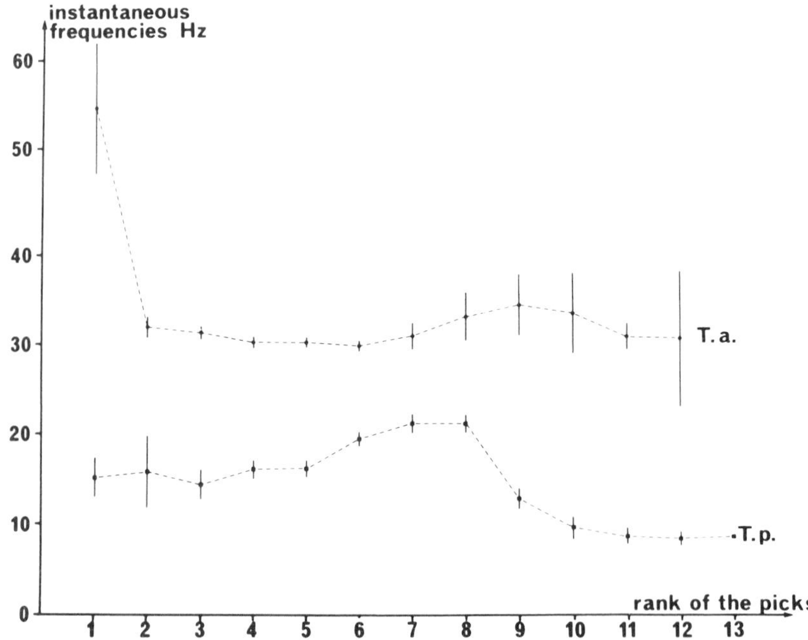

FIGURE 2.2. Changes in instantaneous frequencies of the abdominal vibrations of males of *Tegenaria atrica* and *T. parietina* during courtship behavior (from Mielle, 1978).

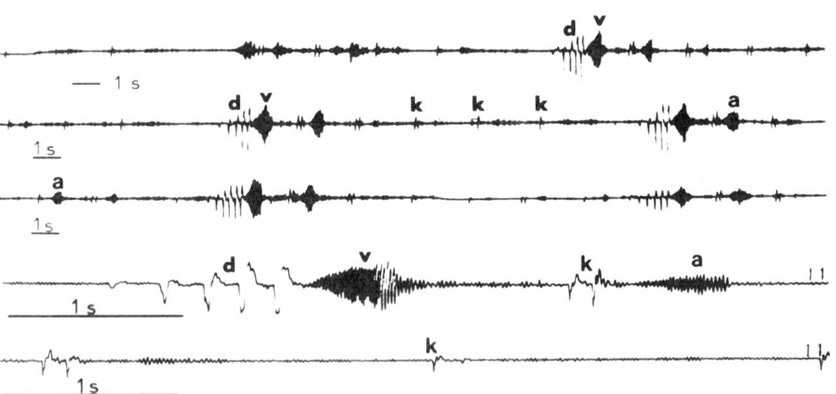

FIGURE 2.3. Courtship behavior of male *Amaurobius ferox*; d: drumming with palps, F = 4.5 Hz; k: strikes with palps, F = 0.5 Hz; v: vibrations of abdomen following d, F = 64 Hz; a: vibrations of abdomen, F = 47 Hz (adapted from Krafft, 1978).

punctata (Locket, 1926), *Theridion saxatile* (Nørgaard, 1956), and *Araneus pallidus* (Grasshoff, 1964). On a web lacking a female, the courting behavior of the male of *Araneus diadematus* rapidly ceases. Such a web is also less effective in triggering such behavior (Witt, 1975). The female thus generates information that reinforces the effect of the pheromone in the silk and maintains the male's courtship behavior, which in return prepares the female for mating (Berland, 1932). When a male of *Cyrtophora cicatrosa* ventures onto a female's web, she drums on the web at a frequency of 4 to 5 Hz before the male makes his courtship display. Only in response to this signal does the male emit his specific vibratory message (Blanke, 1975). The female then responds by drumming at a frequency of 8 to 10 Hz.

Exchange of information between the male and the female is also seen in the wandering spiders. For example, in *Lycosa rabida* the female, which responds to the male with leg-waving during the periods when the male is immobile, initiates the first contact (Rovner, 1968b). The importance of the female's responses is demonstrated by the fact that the frequency of the male's courtship display sequences in *Lycosa rabida* gradually increases in the presence of a receptive female; this response increases, in turn, the frequency of the female's leg-waving (Rovner, 1967, 1968b).

The male's courtship behavior is a mechanism that reduces the probability of predatory behavior toward him. A nonreceptive female usually responds to the male with a threat display. But other signals also enter in. With particularly aggressive females, the male of *Pardosa chelata* adopts a threat posture before beginning his courtship behavior (Hallander, 1967). When the male's information is incomplete, as is the case when a female already has a cocoon, the approaching male of the species turns away (Vlijm et al., 1970). After mating in *Lycosa rabida*, the male performs a threat display identical to that seen in encounters between males (Rovner, 1968b, 1972). Thus, the reproductive behavior of spiders involves communication by some combination of chemical, visual, vibratory, or tactile signals.

COHABITATION OF THE MALE AND THE FEMALE The use of cannibalism-limiting signals is even more evident in the species in which the male and the female live together for several days. Such an association is found in the Dictynidae (Jackson, 1977d; Locket, 1926). It can last eight days in the theridiid *Lithyphantes paykullianus* (Kullmann, 1964). In *Linyphia triangularis*, the male and female live together for two days or more without inhibiting their feeding behavior (Rovner, 1968a). The male may kill and eat another male and the female may feed on another female, but the two sexes do not attack each other. The female can discriminate among the disturbances of the web produced by an adult male, prey, or another female. The male distinguishes the female from prey and from other males, for the latter arouse special agonistic behavior. Such discrimination probably depends on vibratory information transmitted by the web. However, the possibility that tactochemical signals enter into play during contact between the two cannot be eliminated.

The male's sporadic emission of vibratory signals, produced by an abdominal vibration identical to that used during courtship behavior, probably plays an important role in *Tegenaria parietina.* In the laboratory a male and a female can be kept together on the same web for more than 10 months. When these animals are fed, the male or the female captures prey and consumes it individually. Any disturbance of the web by prey or by the female triggers the male's abdominal vibrations. They undoubtedly function as a signal intended for the female, since they are not produced when the male is on his own web, and they cease on the female's web when she is not there (Mielle, 1978).

COMPETITION BETWEEN MALES Encounters between males are naturally frequent among Lycosidae living in very dense populations. During encounters, the males make threat displays that aid in the establishment of a hierarchy and in the dispersion of individuals (Aspey, 1974, 1976; Dijkstra, 1969, 1978). These threat behaviors are either visual signals produced with the first pair of legs, or acoustic signals. The males of *Lycosa rabida* emit a threat sound consisting only of brief bursts of pulses, rather than a continuous emission—which differentiates this signal from the courtship sound, although the two are similar in frequencies. The acoustic emissions of the males of *Lycosa gulosa* may have the same functions, since the signals of one individual often provoke a response from males a meter away (Harrison, 1969).

Female spiders often attract a great many males. In *Nephila clavipes,* the males compete to occupy a central zone near the hub of the female's web. The male in the central position responds aggressively, with jerking and plucking behavior, to the approach of a male from the periphery. The emission of these vibratory signals generally results in the withdrawal of the intruder. If the latter persists, the central male moves toward him, raising his first pair of legs and may even bite him (Christenson and Goist, 1979). In *Linyphia triangularis* the male approaches the intruder with jerking movements of the abdomen and then spreads his front legs and opens his enormous chelicerae. The vibratory signal emitted is generally enough to make the intruder run away. If not, there may be a fight; the loser runs the risk of being injured or even killed (Rovner, 1968a).

These various observations show that the sexual behavior of spiders is a particularly interesting domain for the study of animal communication. Exchanges of information between the partners are often highly complex and successively or simultaneously call on signals of such different categories as pheromones, sounds, web-transmitted vibrations, visual signals, and tactile stimulations. The specificity of matings depends without any doubt on the combined effect of these signals of different types. Some receptors can respond specifically to one stimulus. This is the case with the tarsal organ of the male of *Cupiennus salei*, in which certain neurons react specifically to the odor of females (Dumpert, 1978). Such a mechanism for the exploitation of vibratory signals, however, is unthinkable. The kinds of information transmitted by the web can be of very different natures, corresponding to the arrival of prey, or to the threatening behavior of a conspecific, or to the courtship behavior of the male, or quite simply to the movement of a conspecific. Since each of these signals produces a particular response from the spider receiving it, the analysis of this information must take place centrally.

Young Spiders during the Communal Phase

The mutual tolerance shown by young spiders during their communal phase has often been attributed to the absence of development of predatory behavior before their dispersion. Nevertheless, in certain species the young are capable of feeding themselves.

The young in the second instar feed even in the cocoon, on nonviable eggs and/or first instar spiderlings. This is the case for certain Loxoscelidae, Theridiidae, Lycosidae, Gnaphosidae, Clubionidae, Thomisidae, and Agelenidae (Goldfarb, unpubl.; Schick, 1972; Valerio, 1974b). The same is perhaps true of *Araneus diadematus* (Burch, 1979). Experiments have shown that the second instar spiderlings of *Misumenops* turn toward freshly laid eggs and eat them (Schick, 1972), but do not feed on deutova or second instar spiderlings. One must therefore assume that these early instar spiderlings produce a stimulus that prevents cannibalism and that can be emitted only by living individuals, since dead deutova are eaten. It is not yet known whether this stimulus is chemical in nature or corresponds to a behavioral response of deutova and second instar spiderlings to contact with a conspecific.

The young of *Araneus diadematus* and of *Agelena labyrinthica* display predatory behavior and willingly capture living prey during their communal phase on the communal web (Krafft, 1975a; Witt, 1975). This implies the existence of a system of communication between the individuals, allowing them to discriminate between a conspecific and prey. This tolerance depends partly on vibratory information transmitted by the web and partly on information perceived during contact between individuals. The vibratory stimulus responsible for the recognition may be linked to the way of moving around on the communal web. Such a mechanism would explain the lack of specificity of this tolerance. It is in fact possible to mix young of different species belonging to the same family, or in certain conditions the young of different families. For the present we have no information about any chemical stimulus perceived by contact chemoreception, such as exists among the social spiders.

This mechanism of mutual tolerance loses much of its effectiveness when the spiders weave individual webs, but it does not totally disappear, since there are threat displays, as distinct from predatory behavior, even among adult solitary spiders. It might be thought that the young after their dispersion do not lose their faculty of distinguishing the vibrations of prey from those of a conspecific; but there is, rather, a modification of the animal's response to this signal, linked to the necessity of defending a space sufficient to ensure the subsistence of the individuals (Riechert, 1974b, 1978a,b). There seems to be a balance in the young between the predatory tendency and mutual tolerance. As long as they have no overriding need of food, the young may have a very high threshold for the release of the predatory response and a very low threshold for the perception of the stimulus for tolerance. In fact the phase of mutual tolerance among the young can be prolonged by giving them abundant food (Horel et al., 1979).

Normally the communal phase of solitary spiders is brief. It is prolonged among those species that display parental behavior and, if continued to adulthood, provides the probable evolutionary basis for social living found in colonial spiders.

The communal life of the young spiders is necessarily linked to a mechanism that limits the movement of individuals and ensures group cohesion. The mechanisms responsible are still unknown, but one may reasonably assume that the silk plays the essential role, since it is an attractive element for spiders. The silky structures woven in common may be the critical factor for the existence of these groups. But within the communal web the spiders may disperse to a greater or lesser extent, or gather together (Burch, 1979). We do not know if specific factors are involved in this phenomenon.

The interactions governing the relations between individuals during the communal phase of young spiders are still little known. It would be interesting to study the signals responsible for these phenomena, in order to compare them with the greater amount of data available about the social spiders.

Parental Behavior

Maternal behavior depends on complex interactions that prevent the female from eating her eggs or her young, and that in certain species allow a coordination of activities between the mother and her offspring (Krafft and Horel, 1979). These interactions are comparable with the behavior of the social species. The involvement of chemical stimuli has not yet been demonstrated, but it is difficult to exclude such a possibility.

Maternal behavior is a transitory phenomenon in the life of spiders. It would therefore be interesting to study the dynamics of this particular state, as has been done for certain vertebrates (Hinde, 1965; Lehrman, 1964; Rosenblatt and Lehrman, 1963) and for certain insects (Vancassel, 1977). Very few studies of this type have been carried out on spiders. However, there are regulatory mechanisms that allow a certain plasticity in their maternal behavior (Bonnet, 1940; Palmgren, 1944).

The care given to the cocoon (egg sac) depends partly on endogenous factors linked to the physiological state of the female. Laying eggs is one of the determining factors of this behavior. If a cocoon is offered to a female that has not yet laid her eggs, she refuses it or eats it (Palmgren, 1944). *Clubiona kulezynskii* and *Dolomedes plantarius* do likewise (Bonnet, 1940; Buckle, 1971). On the other hand, adult females of *Theridion saxatile* accept cocoons well before they have themselves laid eggs (Nørgaard, 1956). Maternal behavior depends also on information emanating from the cocoon itself. This is not true communication, but comes close to it to the extent that the contents of the cocoon, that is, the eggs or the young, help to determine this behavior.

The possibility exists of discrimination among various kinds of cocoons, at least for the social spider *Agelena consociata*, which suspends its cocoons like strings of beads in chambers prepared in the web, and cares for them constantly while they contain eggs or living young. If the cocoons are unfastened and placed on the floor of the galleries of the web, the spiders fasten them back up if they contain eggs or living young (Table 2.2). But cocoons that are empty or contain dried-out eggs are deserted by the spiders or even thrown out of the web. When offered a choice, the spiders prefer live cocoons to dead ones, which are in turn more attractive than cocoon-sized balls of silk made from the prey-catching web. These results show that the recognition of cocoons depends partly on their silky structure, which is probably different from that of the prey-catching web, and partly also on their contents. The nature of the information emanating from the cocoon's contents, that is, the eggs or the young, is not yet known. It could be simply a matter of the weight of the eggs or the larvae, or else a chemical stimulus. One of the stimuli that maintains maternal behavior in *Clubiona kulezynskii* may be chemical (Buckle, 1971).

TABLE 2.2. Behavior of females of *Agelena consociata* toward their cocoons.

Type of cocoon	*Cocoons reattached*	*Cocoons deserted or thrown out*
25 live cocoons	18	7
59 empty or dead cocoons	6	53

The spiders discriminate between live and empty or dead cocoons (Chi-square $P < 0.001$).

The persistence of maternal behavior depends on the presence of the cocoon. If the cocoon of *Theridion sisyphium* is surreptitiously taken away from the mother, after 72 hours she refuses to take it back (Locket, 1926). Maternal behavior in *Theridion impressum* and *Theridion pictum* ceases 3½ days and 6 days, respectively, after their cocoons have been taken from them (Hirschberg, 1969); 5 days' separation is needed in *Dolomedes plantarius* (Bonnet, 1940); and 20 hours is enough for *Lycosa tarsalis* (Palmgren, 1944).

The maternal behavior of a spider can be prolonged by replacing her cocoon with a less-developed one. If *Peucetia viridans* is made to accept a cocoon 12 days younger than her own, she opens it 12 days later than she would have opened her own. She must surely react to a stimulus emanating from the cocoon when the young are on the point of emerging (Randall, 1977). *Lycosa tarsalis* does the same (Palmgren, 1944). Artificial substitutes, which do not give off an adequate signal, are transported longer than normal cocoons by *Dolomedes plantarius* (Bonnet, 1940). Cocoon-opening behavior seems to be related to vibrations produced by the movement of the young. The opening of the cocoon by *Pirata piraticus* can be hastened by substituting an older cocoon. Finally, in normal conditions the cessation of cocoon-tending behavior in Lycosidae seems to be related to stimuli produced by the young as they climb onto the back of the female (Engelhardt, 1964). Ten hours after the beginning of hatching, when two-thirds of the young are on the female's back, she refuses cocoons offered to her (Palmgren, 1944).

These observations show that the cocoon is the source of several different stimuli that maintain and regulate the female's behavior. But the female is responsive to them only if she is in the particular psychophysiological state necessary for the appearance of maternal behavior. Indeed, when not in that state, a spider rejects or devours cocoons offered to her.

Caring for the cocoon can be followed by caring for the young in various families of spiders, including the Lycosidae, Salticidae, Agelenidae, Theridiidae, Scytodidae, and some mygalomorphs. The gathering of the young on the back in Lycosidae depends mainly on tactile stimuli related to the female's tegumentary structure (Dietlein, 1967; Engelhardt, 1964; Job, 1974; Rovner et al., 1973). The young, in turn, seem to be the source of stimulations of the female's mechanoreceptor hairs, maintaining maternal behavior (Engelhardt, 1964; Rovner et al., 1973).

When close coordination is established between the activities of individuals, a true exchange of information becomes apparent. Young Lycosidae spin a dragline when they climb down from the female's back to drink. As soon as she gives any sign of moving, changes in the tension in these lines stimulate the young to climb back on to their mother (Higashi and Rovner, 1975). Such transmission of information through the silk is particularly well developed in the sedentary web-spinning spiders. When the female of *Theridion saxatile* captures an item of prey, she gives jerking tugs on the thread of the web. These vibratory signals make all the young run for shelter. Once the prey has been paralyzed, the female gives slow pulls on the thread, thus attracting the young (Hirschberg, 1969; Nørgaard, 1956). The female of *Coelotes terrestris* also attracts her young toward captured prey (Tretzel, 1961). In the Theridiidae, the feeding of the young by regurgitation makes use of tactile stimulations. The young actively request food by stimulating the female's palps, legs, and mouth region; and the female can interrupt their feeding at any time with movements of her legs (Hirschberg, 1969).

Predatory behavior is rarely suppressed in females who are tending their young, which must therefore be protected from her attacks. Job (1974) attributed the tolerance enjoyed by the young of *Aulonia albimana* to a rise in the threshold for the release of predatory behavior. The female would thus not respond to the presence of her young. *Cyrtophora moluccensis* tolerates all those that are less than 8 mm long (Lubin, 1974). But this explanation is inadequate for species that feed their young. *Theridion impressum* reacts to their movements by tugging on the threads of the web. *Coelotes terrestris* recognizes her young from the way they move (Tretzel, 1961, 1963), but any encounter between mother and young after an unusual vibratory stimulation or the fall of prey into the web results in a detailed inspection of the young with the mother's pedipalps. The movements of the legs of the young, the vibrations or oscillations of their bodies, and their posture seem to be the main elements of signals protecting them from attack by the female. Furthermore, the maternal behavior of *Coelotes terrestris* can be prolonged in the laboratory, with the mother showing no aggressiveness toward her young in spite of their large size (Horel et al., 1979). It is therefore necessary to envisage some kind of recognition of the form of the vibratory signals produced by the movements of the young, permitting the female to discriminate between the young and possible prey.

Recognition of the young through vibratory signals transmitted by the web, however, can only partly account for tolerance between the mother and her young. Maternal behavior also involves direct contacts that certainly bring into play additional stimuli inhibiting cannibalism, as is the case among the social spiders (Krafft, 1975a). These stimuli may be chemical or tactile in nature.

The mother's tolerance extends only to the young. Females of *Anelosimus studiosus* and of *Sosippus floridanus* attack any adult or subadult of the same species (Brach, 1975, 1976). Therefore the female must be able to discriminate between young and adults on the basis of one or more stimuli peculiar to the young. These stimuli are not strictly species-specific (Hirschberg, 1969; Kullmann, 1972a). The mother's tolerance of the young also depends on her own psychophysiological state. The female of *Cyrtophora citricola* tolerates the young only if she has already produced a cocoon (Blanke, 1972). Unfertilized female Lycosidae and subadults of this family do not accept young (Higashi and Rovner, 1975).

In some species the mother must also be protected from attacks by her young whose predatory behavior is already developed. This tolerance no doubt depends on a mechanism that resembles, as we shall see, the mutual tolerance of social species. Indeed, the young tolerate one another and any adult, whether it be their mother or another female. The stimulus responsible must therefore be present in every individual of the species, regardless of its age. The disappearance of this tolerance at the time of dispersion from the maternal web could be due to a lowering of the reactivity of the young to this stimulus. Parental behavior must therefore require two distinct tolerance mechanisms, one related to the psychophysiological state of the female and depending on a specific stimulus from the young, and the other protecting the young from one another and protecting the mother from the young, dependent on a social stimulus extending to all the individuals.

Cocoon-tending behavior depends on tactile stimulations related to the cocoon's structure and contents. The mother distinguishes the disturbances of the web provoked by prey from those produced by the young moving around. The young tolerate one another and tolerate their mother. This recognition is largely based on the utilization of vibratory information, and shows once more the considerable role that the silky structures play for most species of spiders (Witt, 1975). In fact, the most highly developed forms of maternal behavior are seen in the web-spinning spiders.

Social Behavior

INTRODUCTION Various authors have tried to classify the societies of Arthropoda according to the complexity of the interactions that govern them (Gervet, 1968; Le Masne, 1952; Lindauer, 1965), using definitions whose restrictiveness varies with the species studied. Among the spiders, the organization of intraspecific relations shows a gradation of such interactions, from the solitary to the social spiders, which makes it hard to identify a sharp, objective division between these two modes of life.

Here I shall use Wilson's (1971) very broad definition of society as "a group of individuals that belong to the same species and are organized in a cooperative manner." I shall not consider as a social phenomenon the communal life of young solitary spiders after their emergence from the cocoon; this is a fleeting phenomenon, and cooperation between the individuals is limited. For the species whose young prolong their communal life and cooperate to capture prey and make the nest, but whose adults live in a solitary state, I shall use the term *subsocial*. This term suggests an evolution of the social phenomenon in accordance with the most probable hypothesis about the origin of spider societies. When the adults lead a communal life and cooperate to capture prey or to make the silky edifice of the colony, I shall speak of *social* species. This subdivision differs from that usually agreed on for insects (Michener, 1969), because the functioning of societies of spiders centers more on the elaboration of a trap to capture prey than on caring for the brood. The silky structures of spiders constitute a special environment that assures the protection of individuals and transmits the information necessary for the coordination of activities, particularly when prey are being captured. All the members of the societies, including the young, benefit from the products of group hunting.

Certain of the social and subsocial species manifest territorial behavior. Although living in a colony, each individual makes use of an individual web and keeps others of his species off it. Territorial social species are found both in the Araneidae and Uloboridae, which spin orb-webs, and in the Dictynidae, which spin irregular webs. The territoriality of the orb-spiders may be attributed to the utilization of orb-webs, whose complex structure can be efficiently exploited by only a single individual at a time, since all the information converges at one point, the hub. There are no nonterritorial social orb-spiders.

Territoriality doubtless has a different origin in the species that have irregular webs. It could be related to the exploitation of the minimum space necessary, given the supply of prey. Among the territorial social species, interactions between individuals are simpler than among the nonterritorial species. Their evolution seems to have been blocked, either by the structure of their webs or by ecological conditions. In contrast, nonterritorial subsocial species are in many respects comparable to nonterritorial social species.

Whether the spiders are subsocial or social, territorial or not, cohabitation of several individuals and their cooperation imply the existence of a certain number of interactions. What are the signals that ensure interindividual tolerance and group cohesion? What is the mechanism that allows each individual to adjust its behavior to that of its conspecifics?

NONTERRITORIAL SUBSOCIAL SPIDERS In many species that display maternal behavior, the communal life of the young is prolonged, even beyond the time when the mother dies. We see then the carrying out of common tasks by the young in the maintenance of the web and the capture of prey; this was considered a social characteristic by Chauvin (1961). Communal life continues until the fourth molt in *Theridion sisyphium* (Kullmann and Zimmermann, 1974). In *Eresus niger*, cooperation among the young lasts until the sixth molt (Kullmann et al., 1972; Kullmann and Zimmermann, 1974). *Anelosimus studiosus* do not disperse until the adult instar (Brach, 1977).

Mutual tolerance among spiders must depend on vibratory information, since the disturbances produced by a moving spider do not trigger predatory behavior in the other members of the group (Brach, 1977). But that does not rule out the existence of tactile or chemical stimuli during contacts. Indeed, the spiders of related species moving around on the web provoke no particular reaction, their style of movement being no doubt similar; and yet they are attacked if there is contact. The involvement of a tactile stimulus related to the spiders' integument has been shown in *Eresus* sp. of Afghanistan (Zimmermann, unpubl.). In general, mutual tolerance is only relatively species-specific in the subsocial spiders. The young of related species can be mixed: those of *Stegodyphus lineatus* (subsocial) placed in a group of young *Stegodyphus sarasinorum* (social) do not segregate, and they cooperate in the capture of prey (Kullmann et al., 1972).

Group cohesion is not related exclusively to the presence of the mother, for the communal life of the young continues beyond her death. Since no experimental research has yet been conducted in this area, it is not certain that there is a true interattraction of a social type, as there is in the social insects. The absence of repulsion between the individuals, combined with the attraction of the silky structures built up by the female and completed by the young, may in part explain the group cohesion. However, certain solitary species manifest a grouping drive during the communal phase of their lives, as in *Agelena labyrinthica* (Krafft, 1970b). This drive must also exist in subsocial species. Eventually, psychophysiological changes seem to result in increased aggressiveness in the young and diminish their grouping drive, bringing about the dispersion of the individuals.

TERRITORIAL SUBSOCIAL SPIDERS The young of *Ixeuticus candidus* (Australian Dictynidae) spin groups of irregular individual webs near and touching the female's web. But this gregariousness is found only during the juvenile instars, and the interactions that allow these structures to be built up are not known. The frequency of individual contacts linked to this promiscuity shows, however, that a form of mutual tolerance may exist in association with territorial behavior. There is at present only one known territorial subsocial species among the orb-spiders: *Eriophora bistriata*, of Paraguay (Fowler and Diehl, 1978). During the daytime resting period, 130 to 500 spiders group together in a rudimentary nest, pressing closely against one another and forming a compact mass. In the evening they spin together the foundation lines, 10 to 30 meters long, which support the individual orb-webs. As long as the prey are not bigger than the spider, each spider makes its own web and preys individually. In the less usual case, when the prey is bigger, cooperative behavior increases; the nearest spiders, attracted by vibrations, participate in the capture of prey and eat it together. Such cooperation clearly demonstrates that territorial behavior is not incompatible with mutual tolerance. The spiders develop synchronously and disperse only when they reach adulthood.

NONTERRITORIAL SOCIAL SPIDERS *Agelena consociata* (Darchen, 1965a,b; Krafft, 1970a,b, 1971a,b, 1975a,b), *Agelena republicana* (Darchen, 1967), *Stegodyphus sarasinorum* and *Stegodyphus mimosarum* (Bradoo, 1972; Jacson and Joseph, 1973; Kullmann, 1969, 1972a; Kullmann et al., 1972; Kullmann and Zimmermann, 1971, 1974; Wickler, 1973), and *Mallos gregalis* (Burgess, 1978, 1979b; Jackson, 1977a, 1979b,c) are the best-known nonterritorial social species. These spiders live in vast nests, without distinction as to sex or age, and cooperate in the making of the nest, the capture of prey, the care of cocoons, and the care of the young, without any detectable division of labor comparable to that among social insects. *Achaearanea disparata*, of Gabon, has a special social organization. These spiders spin colonies of webs fastened one to another. Each web contains several spiders, of all ages and both sexes. In this organization they resemble at the same time nonterritorial and territorial social species (Darchen, 1968).

The social spiders are characterized by mutual tolerance, interattraction, and cooperation. Tolerance is undoubtedly a kind of interaction necessary for the appearance of social life among the spiders. The grouping drive differs somewhat from the interattractions seen in the social insects, in that it depends in part on the presence of the silky structures. Cooperation is more a consequence of group life, which is also explained by the important role that the silk plays in the transmission of various kinds of vibratory information. It is safe to say that the silk plays a determining role in the evolutionary origin of spider societies, since there is no social species among the webless wandering spiders (Krafft, 1979; Shear, 1970). However, the spiders also make use of tactile and chemical stimuli, as do the social insects.

Mutual tolerance. The predatory behavior of *Agelena consociata* is extremely rapid, because their webs, being nonsticky, do not hold prey. Therefore an efficient mutual tolerance mechanism that instantly inhibits bites between conspecifics is necessary. In *Agelena consociata* this tolerance depends on at least four types of stimuli: vibratory, behavioral, tactile, and chemical.

A spider moving around on the web originates vibrations that do not trigger any predatory behavior. This recognition of a conspecific from a distance avoids incessant confrontations and waste of energy. The vibratory stimulus is not of a stereotype unique to the species, since it depends on the type of activity of the conspecific. Undeniably, then, this signal is recognized in spite of its variability. When a frightened spider moves rapidly on the web, however, it can trigger the predatory behavior of a conspecific—that is, pursuit. When the two spiders then come into contact, other factors will inhibit them from biting each other. Spiders of related species emit the same signals: *Agelena labyrinthica,* a solitary spider, can move freely on the web of *Agelena consociata,* as can certain *Tegenaria,* as long as there is no direct contact. The same phenomenon is seen in *Anelosimus eximius* (Theridiidae), which ignores solitary theridiids (Brach, 1975). The vibratory information that allows a conspecific to be recognized must therefore consist of disturbances provoked by a way of moving that is common to several species within a family. Spiders of other families can be detected from a distance, pursued, and captured.

A different mechanism can be invoked to explain the absence of reciprocal pursuit in *Mallos gregalis.* Resonance in the web of this species amplifies vibrations between 30 and 700 Hz, whereas the web seems to attenuate signals above and below these limits. Vibrations produced by spiders (lower than 30 Hz) therefore seem not to be transmitted, while those produced by prey are broadcast throughout the colony, attracting a great many individuals (Burgess, 1979b). However, the same explanation may not hold for other social and subsocial spiders, nor for the solitary species in which the male and female live together. The webs of *Tegenaria* (Agelenidae) transmit perfectly the vibrations produced by the movements of spiders and the movements of prey (Krafft, 1978; Krafft et al., 1978; Leborgne and Krafft, 1979). On the other hand, this mechanism cannot explain why, in *Mallos gregalis,* a female reacts in one way to a male and in another way to prey; in fact, the males, in their courtship display, vibrate the abdomen at a frequency of 10 Hz (Jackson, 1979c).

It does occasionally happen that a spider pursues a conspecific. At the moment of contact, it strikes the conspecific with its first pair of legs, immobilizing it instantly. This behavioral response reduces the risk of reciprocal bites by eliminating any vibratory stimulus that might maintain the predatory behavior. This is the second mechanism that contributes to mutual tolerance between the individuals of this species. When two individuals make contact, tactile and chemical information come into play. The involvement of these two types of stimuli was demonstrated by the technique of artificial vibrations (Krafft, 1970b), which consists of triggering the predatory behavior of the spiders by making a lure vibrate on the web at a frequency of about 100 Hz: crickets or balls of elder pith are attacked and bitten without hesitation, whereas live or CO_2-killed spiders are tolerated. This demonstrates the involvement of a tactile or chemical stimulus. However, the vibrations thus animating the tested spiders cause them to be minutely inspected, as their conspecifics are simultaneously receiving a stimulus that normally triggers biting (vibrations) and one or more stimuli that inhibit biting. On the other hand, spiders washed in a mixture of alcohol and ether are bitten; this result is in favor of the involvement of a chemical stimulus corresponding to a pheromone that inhibits biting. The existence of this pheromone was confirmed when balls of elder pith were made to vibrate after having been covered with ground-up spiders. A count of the number of spiders that bit the lures after a contact shows that the lures dabbed with ground-up abdomen were statistically less often bitten than the control lures ($P < 0.001$) (Table 2.3). This bite-inhibiting pheromone is perceived by contact chemoreception.

Several convergent observations show that a tactile stimulus also enters in. Spiders washed in alcohol and in ether are bitten less often than similarly washed crickets, possibly because of a differ-

TABLE 2.3. Inhibition of reciprocal biting among *Agelena consociata*, suggesting the action of a pheromone (from Krafft, 1975b).

	Lure	
Spiders	*Untreated elder-pith*	*Elder-pith + ground-up abdomen*
Number touching lure	181	173
Number biting lure	146	85
Number not biting lure	35	88
(Percentage biting lure)	(80.7%)	(49.1%)

ence in the structure of the integument between these two arthropods. Solitary Agelenidae such as *Agelena labyrinthica* can move around on the web and make brief contact with the social species without being attacked, whereas the Araneidae, whose integument is very different, are bitten at the first encounter. If the same solitary Agelenidae are made to vibrate on the web, they are immediately bitten, although less often than crickets and about as often as *Agelena consociata* washed in an alcohol-ether mixture. The partial tolerance enjoyed by *Agelena labyrinthica* therefore seems to be related to the resemblance of the structure of its integument to that of *Agelena consociata*. But it lacks the pheromone that complements the effect of the tactile stimulus in the social species.

On the basis of these observations, the various stimuli that enter into the mutual tolerance mechanism can be ranked in an order of increasingly accurate inspection of the conspecific. Vibrations transmitted by the web are usually enough to prevent reciprocal biting. If that fails, there is the behavioral reaction of immobilization and the tactile stimulus. And when the conspecific thrashes about in agitation, like prey, it is inspected in detail with the legs and the pedipalps. It is in this last stage that the chemical stimulus enters in, as is shown by the fact that *Agelena labyrinthica* is inspected and finally bitten only if it is made to vibrate on the web.

The same experimental technique was used to show the involvement of a tactile and a chemical stimulus in *Stegodyphus sarasinorum* (Kullmann, 1972a; Kullmann and Zimmermann, 1971). However, *Stegodyphus mimosarum*, *Stegodyphus lineatus*, and *Eresus niger* are also partially tolerated by *Stegodyphus sarasinorum*. These results could be explained in terms of the involvement of a tactile stimulus common to all the Eresidae, corresponding to the structure of their integument. It is possible that a chemical stimulus also enters into the mutual tolerance mechanism in *Anelosimus eximius*, which bites solitary Theridiidae as soon as contact is made (Brach, 1975).

Interattraction. The social life of spiders implies that there must be one or several mechanisms that assure group cohesion. The grouping drive was demonstrated in *Agelena consociata* by comparing the spiders' distribution in an open field with a chance distribution (Krafft, 1970b). This grouping tendency is independent of thigmotaxis and of sex. Adult spiders from different webs group together, showing that the attraction is not familial, but is really a social phenomenon. The silk plays an important role, but it is not indispensable, since the grouping starts well before the spiders weave their webs and persists when the base of the open field is covered with a uniform layer of silk. We can conclude that the grouping drive in *Agelena consociata* is a genuine *interattraction*, which is dependent on reciprocal and symmetrical signaling between the individuals, as defined by Grassé (1952) and Gervet (1968).

The exact nature of the stimulus responsible for the interattraction is not yet known. For *Agelena consociata*, it is not the odor of a group of spiders that is attractive (Krafft and Roland, 1979). The same is true for *Stegodyphus sarasinorum*. Neither ground-up spiders nor a group of two dead spiders is attractive, whereas a group of two live spiders together attracts conspecifics (Krafft, 1970c). Thus the stimulus for the interattraction is related to the spiders' being alive. It could be a tactile or chemical stimulus, or vibrations transmitted by the silk. Indeed, the disturbance of the web provoked by an individual could be a signal that helps to call spiders together. A tactochemical stimulus may enter in when there is direct contact.

If each spider in a group is individually marked, it can be seen that certain individuals are isolated more often than others. Spiders can thus be ranked according to their grouping drive. The rank is stable from one experiment to another, and is the opposite of the hierarchical order found by observing the agonistic relations that arise when prey are being transported. The interindividual variability as a function of the grouping drive depends on differences in reactivity to the social stimuli of interattraction. This reactivity also increases after a previous absence of social contacts, showing that the interattraction depends on a genuine social drive. Spiders that have been isolated for 85 days and then reassembled form tighter groups than spiders kept in groups during the same period.

Other mechanisms besides interattraction assure the cohesion of the group when the animals are active. Here silk plays an essential role. A uniform carpet of silk attracts spiders. Silky structures, such as the galleries of the colonial web, are even more attractive. The cocoons also contribute to group cohesion; they attract the spiders even more than silk does. Silk, galleries, and cocoons, then, are elements that reinforce group cohesion in *Agelena consociata*.

This is an example of communication in the broad sense of the term, since these structures are established by the spiders, and each individual adjusts its behavior to them. Thus they have an effect analogous to that of the pheromones of certain insects. The worker bee that marks the entry to the hive or a source of nourishment with its Nasanoff gland and the ant that lays a trail of odor are the sources of signals that can modify the behavior of their conspecifics by orienting them. Spiders spin threads and construct various silky structures that also allow the orientation of their conspecifics in relation to the social group. The use of a T-maze (Krafft and Roland, 1979) shows clearly the importance of the silk in the gathering together of individuals in *Stegodyphus sarasinorum*. A female can follow the trail left in the maze by a conspecific of the same sex. This ability to follow the trail of a conspecific depends on information related to the dragline, and is only partially species-specific. Indeed, females of *Stegodyphus sarasinorum* follow a thread of *Araneus cornutus* placed experimentally in the maze. However, if the spider is given a choice between a thread of *Araneus cornutus* and one of *Stegodyphus sarasinorum*, both placed in the maze at once, she discriminates and prefers to follow that of her conspecific. Such discrimination is still possible between the threads of *Stegodyphus sarasinorum*, *Amaurobius*, and *Eresus*. It is not yet known whether the stimulus responsible for this social orientation is tactile or chemical.

In practically all insect societies, recognition of the colony's members is linked to a particular group odor. Any foreigner trying to enter the colony triggers aggressive reactions (Wilson, 1971). Among spiders, the phenomenon of interattraction and above all the involvement of a chemical stimulus in mutual tolerance could suggest the existence of the same mechanism in spiders. However, all the experiments made thus far show that spiders do not discriminate between individuals of the same species coming from different colonial webs: neither cannibalism nor segregation appears. The social signals underlying the interattraction and mutual tolerance must therefore be common to all the members of a species and independent of the biotope in which each colony develops.

Cooperation. One of the main functions of communication in the social insects is to ensure the coordination of the activities of individuals, so that cooperation to carry out common tasks is possible. "There is coordinated activity whenever individual activities, whether the same or dissimilar, reinforce one another or combine, producing a common piece of work of any sort" (Le Masne, 1952). Cooperation allows a group of individuals to carry out a task that could not be accomplished by an isolated individual (Chauvin, 1961), or it is the acts of an individual that are beneficial to other members of the group (Wilson, 1975). The term "active cooperation" is usually reserved for labors resulting from coordination by direct interactions. The dance of the bees, the antennal stimulation of wasps and ants, the alarm and recruitment pheromones (Wilson, 1971), and the stridulation of ants (Markl and Hölldobler, 1978) are classic examples. Among spiders, each individual through its activities creates a structure or a situation that has a particular stimulus value and orients the behaviors of other members of the group (Krafft, 1970a, 1979). This mechanism resembles the indirect interactions governing the transport of twigs by ants (Chauvin, 1971). The construction of the web by the colony as a whole amounts to the establishment of an original system of transmission of information, created entirely by the spiders. The silk is not an inert element, but is an element transmitting a whole gamut of vibratory and tactochemical information. It is used not only to capture prey, but also to establish a connection between the spider, its environment, its prey, and its conspecifics. It is the main medium for cooperation among individuals.

Cooperation in web-building. During the twilight weaving activity of *Agelena consociata,* each spider adapts its behavior to the structures established by its conspecifics (Krafft, 1970c). Such cooperation in the social spiders, which is mainly the consequence of mutual tolerance and interattraction, results in the construction of a large, structured web that assures the protection of individuals and facilitates the capture of prey. The same phenomenon is seen in the social Dictynidae, Eresidae, and Theridiidae.

Cooperation during hunting. The major role of silk among the social spiders is demonstrated by coordination during hunting. The predatory behavior of *Agelena consociata* can be divided into three phases: pursuit of the prey, short bites that exhaust it, and long bites that paralyze it. These various activities are the source of vibratory signals that allow each individual to adjust its behavior to the situation at hand. The prey struggling in the web is the source of vibration that attracts the spiders, the number attracted depending on the prey's size and strength. It is located immediately. The intensity and the form of the vibration allow each new arrival to adapt its behavior to the level of exhaustion of the prey. A spider attracted by an already exhausted insect immediately inflicts long bites upon it. There is indeed cooperation, since the individual activities reinforce each other to allow the capture of large insects that a single spider could not hope to subdue.

Other signals could also play a part among *Agelena consociata.* When a lure is made to vibrate on a web, the first spiders that touch it hesitate to bite it, whereas the following ones plant their chelicerare without delay. This facilitation of predatory behavior could be explained in two ways. It is possible that the venom impregnating the lure after the first bites lowers the bite-release threshold in the spiders that arrive later. But the possibility must also be considered that predatory behavior is facilitated by signaling among the spiders surrounding the prey (Krafft, 1975a, b). Furthermore, when an *Agelena consociata* is disturbed by an intense impact, it sometimes oscillates vertically, hitting the web with its abdomen. This behavior is close to that of the orb-spiders, which shake their webs after a loud sound, but its function is not known; it could be an alarm signal, in view of Lubin's (1974) observations on *Cyrtophora moluccensis*.

Cooperation and competition during the transport of prey. Occasionally, paralyzed prey may be carried into the galleries of *Mallos gregalis* (Burgess, 1979b). Such carrying, which is the rule in *Achaearanea disparata*, is performed by a group of several individuals (Darchen, 1968); however, this cooperation, as in hunting, is the sum of individual activities directed toward a common goal.

Among *Agelena consociata*, medium-sized prey are often carried, but always by a single spider. Repeated observations made on 53 groups, each of 5 spiders, showed that the individual spiders could be ranked according to their success at carrying such prey away from the group (Krafft, 1971a). When the prey had been paralyzed, certain individuals slid their legs between the insect and the other spiders and then, raising their legs, tried to break the contact between their conspecifics and the prey. The spiders separated from the prey sometimes seized debris from desiccated victims lying around on the web, but as these are not edible they quickly dropped them. Furthermore, a spider that never or only rarely carried prey in the presence of its conspecifics always carried it when alone on the web. These observations show clearly that there is competition for the possession of prey with a view to carrying it, since the motivation is found in all the individuals. Consequently, those animals that take over prey more often than the others can be considered dominant individuals (Bouissou and Signoret, 1970).

This competition between individuals depends on direct interactions, and is genuine communication. The dominant spiders, by their presence and their behavior, are the source of information that modifies the probability that carrying behavior will appear in the subordinate animals. Some of this information is tactile in nature, but the mutual-tolerance pheromone could also play a role. A spider that is simultaneously touching the prey and a conspecific perceives at the same time stimuli that favor its biting the prey and stimuli that inhibit biting. The dominant spiders are presumably less responsive to the social stimuli than are the subordinate spiders. This hypothesis is strengthened by the fact that subordinate spiders are also more responsive to interattraction stimuli. There does not seem to be any correlation between rank and size in *Agelena consociata*.

Group feeding. Once the prey that was carried has been deposited at a new location on the web, the spider that carried it will again tolerate the presence of its conspecifics. Several tens of individuals can participate in the meal. This behavior permits certain spiders to feed without ever taking part in the capture of prey. But alone on individual webs, such spiders are all capable of capturing insects. The group meal, too, may be a form of cooperation, in that spiders' digestion is external. The injection of digestive juices into the prey by several individuals may facilitate its digestion and, therefore, its ingestion (Jackson, 1979b; Krafft, 1965, 1971a,b).

Cooperation in the care of cocoons and of the young. The care of the cocoon and of the young is no more highly developed in the social species of spiders than in the subsocial species, and probably depends on the same mechanisms. The adult females all care for the cocoons and all feed the young, from whatever source (Krafft, 1970a; Kullmann et al., 1972). The resulting cooperation is simply the consequence of the mother's inability to distinguish between her own young and those of another female. However, the social life of the nonterritorial spiders has the advantage that the nourishment of the young, and thus their survival, no longer depends on the survival of their mother, unlike the case with the subsocial spiders.

Group effect. Group effect is a modification of the physiology or the behavior of animals that live in a group (Grassé and Chauvin, 1944; Grassé, 1946). It appears in these animals in response to mutual stimulations all of the same type, specifically exercised by their fellow creatures (Gervet, 1965). However, the expression *group effect* "can, with little effort, be stretched to become synonymous with communication in the broadest sense. In fact, the label covers a particular set of communicative phenomena that are of considerable importance in insect societies" (Wilson, 1971). These phenomena depend on signals with an effect less strong than a "releaser effect." The "primer pheromones" are signals of this type, but there is a wide variety of other categories of signals, very poorly understood, that also contribute to the group effect in social insects. A classic group effect in insects is the better survival of grouped individuals than of isolated ones. This modification of the physiology of the individuals is a more interesting criterion for the evaluation of the species' level of socialization than is the complexity of the system of communication. It shows us the strength of the bond that ties the individual to the group.

This last-described level of socialization has been attained by *Agelena consociata* and perhaps by other social spiders. Young spiders that are isolated as soon as they emerge from the cocoon have a higher mortality, slower growth, and longer first instars than young raised in groups of five individuals in the same conditions including high humidity (Table 2.4). Similar results have been obtained with adult and subadult spiders: isolation increases mortality, reduces the digestive metabolism, and increases the number of eggs laid (Table 2.5). This group effect is linked to specific, reciprocal stimulations (Krafft, 1971a), whose nature is not yet known. These signals are not directed signals that evoke a specific response from the conspecific, but are signals that modulate their physiology. Such modulatory signals, which are still very poorly understood, may play an important role in all animal societies by allowing the behaviors of different members of the group to be harmonized (Markl and Hölldobler, 1978).

TABLE 2.4. Group effect in spiderlings of *Agelena consociata*.

	Isolated Spiders	*Grouped Spiders*	*Significance*
Number of spiders tested	98	105	
Mortality (%)	35.7	22.9	$P < .05$
Length of tibia (first pair of legs) (mm)	0.71	0.81	$P < .01$
Duration of first instar (days)	27.8	22.4	$P < .001$

Development of spiderlings, isolated or grouped, immediately after hatching and during 30 days. From each cocoon, some spiderlings are isolated and others grouped (groups of 5 individuals). All are fed with killed crickets (from Krafft, 1971a).

TERRITORIAL SOCIAL SPIDERS The territorial behavior, the exploitation of individual webs, the frequency of agonistic interactions, and the spectacular absence of cooperation during the capture of prey in the territorial social spiders contrast strongly with the behavior characteristic of societies of nonterritorial spiders. However, the complexity of the interactions necessary for the organization of these societies must not be underestimated. The support threads and the barrier webs provide for relations between individuals and allow an exchange of vibratory information (Burgess and Witt, 1976). This peculiarity of the social structure probably results from the persistence of a territorial behavior (Jackson, 1979c). Such per-

TABLE 2.5. Group effect in adult and subadult *Agelena consociata*.

	Series A	*Series B*	*Series C*	*Series D*	*Differences*
Number of spiders	56	56	56	56	
Situation	7 per group	7 per group	isolated	isolated	
Prey	large	small	large	small	
Mortality	5	16	18	31	$(A+B) \neq (C+D)$, $P < .01$ $B \neq D$, $P < .01$
Mean number of excrements	36.4	28	15.8	15.6	$A \neq C$, $P < .05$ $B \neq D$, $P < .05$ $C = D$
Mean weight of excrements (mg)	0.111	0.091	0.069	0.070	$A \neq C$, $P < .05$ $B \neq C$, $P < .05$ $C = D$
Number of cocoons built	14	23	41	22	$(A+C) \neq (B+D)$, $P < .02$
Number of eggs	274	150	352	191	$A \neq B$, $P < .01$ $C \neq D$, $P < .01$ $B \neq D$, $P < .05$ $A \neq C$, $P < .05$
Number of eggs per cocoon laid	6.7	6.5	8.5	8.6	
Number of young hatched	40	7	61	36	$(A+B) \neq (C+D)$, $P < .01$ % of hatches

The spiders are isolated or grouped (groups of 7 individuals) during 85 days (from Krafft, 1971a). Series A: Grouped spiders, fed with large, killed prey. Communal feeding possible. Series B: Grouped spiders, fed with newly hatched crickets. Communal feeding hardly possible. Series C: Isolated spiders, fed with large, killed prey. Series D: Isolated spiders, fed with newly hatched crickets. Difference between A and C is the consequence of isolation and the possibility of communal feeding. Difference between B and D is the consequence of isolation only.

sistence, in the territorial social Dictynidae that make sheet webs (*Mallos trivittatus*, *Dictyna calcarata*, and *Dictyna albopilosa*), could originate partially in the necessity for preserving a large prey-catching area of web per individual. In fact, the area of the colonial webs of nonterritorial social species is always less than the sum of the areas of individual webs; and excessive extension of the trap is probably incompatible with the detection of the prey. On each web of *Mallos trivittatus* is found a single adult female, plus, perhaps, a male and several juvenile spiders (Burgess, 1978; Jackson, 1978d, 1979c; Jackson and Smith, 1978). Thus these spiders show at the same time a territorial social and a nonterritorial subsocial structure.

The mechanisms of territoriality in the orb-spiders are quite different. Each orb-web can only be efficiently exploited by a single individual; it is temporary and has as its primary purpose the capture of prey. Irregular or sheet webs, in contrast, are permanent and have several functions, including prey-catching, the protection of the adults, and the raising of young (Burgess and Witt, 1976). Interactions are frequent in colonies of the orb-weavers *Cyrtophora citricola, Cyrtophora moluccensis,* and *Metabus gravidus.* The spiders react to the approach of an intruder by giving tugs at the web—a vibratory signal that generally has the effect of making the intruder flee (Kullmann, 1959; Lubin, 1974; Rypstra, 1979). These agonistic interactions differ from predatory behavior, showing that each spider can distinguish web disturbances produced by the movements of a conspecific from those produced by prey. During the encounters genuine communication can be seen, with the exchange of stronger and stronger vibratory signals (Buskirk, 1975b). These aggressive reactions are related to the defense of a prey-catching area, since in *Metabus gravidus* they disappear when the individuals are on the support lines common to the whole colony. The threat signals of *Metabus gravidus* have a frequency of 5 to 10 Hz. They are emitted by means of movements identical to those used to locate immobile prey in the web. These movements seem to have acquired a function of communication in this species (Buskirk, 1975b). Furthermore, in *Cyrtophora citricola* aggressive relations are particularly frequent in the zone of the colony that is best for the capture of prey (Rypstra, 1979). The fact that the spiders tolerate each other outside their individual webs, even during contact with each other, suggests that there may be a second tolerance mechanism, probably dependent on tactochemical stimuli.

Group cohesion probably depends on the spiders' attraction to the silky structures. In *Cyrtophora citricola,* the young manifest a grouping drive that favors the construction of a colony. Later, the colonial web seems to be the element of attraction and appears to maintain group cohesion among subadults and adults. In fact, these show no grouping tendency if the colonial web is destroyed and the spiders are scattered on a bush; they do not then build a new colony, unlike the behavior of nonterritorial social spiders under similar conditions (Blanke, 1972). Cooperation does not seem to be evident in the territorial social spiders. The webs, however, are attached to one another or to foundation lines spun by several spiders working together. Insofar as the structure of the colony offers some advantages to the individuals, increasing the efficiency of predation, it may be considered a manifestation of indirect cooperation (Buskirk, 1975a; Lubin, 1974; Rypstra, 1979). But vibratory signals might also be involved. When prey arrives in a web, the spider gives an energetic jerk with its legs. This vibratory signal excites a similar response from more than half of the spiders of the colony, showing that they have been warned of the arrival of an insect (Rypstra, 1979). In addition, when a parasitic hymenopteran approaches a female of *Cyrtophora moluccensis*, the spider grips her cocoon and shakes it vigorously. Other females of the colony respond to the signal with a similar movement. This might be an alarm signal or a mechanism for protecting cocoons (Lubin, 1974).

The social organization of the territorial spiders is thus less well developed than that of the nonterritorial spiders. It nevertheless implies cooperation during the construction of the foundation lines and frequent exchanges of information between individuals.

Mechanisms of Communication

Mechanisms of communication are subject to constraints that are related partly to the morphology and physiology of the animals and partly to the environment. The environment may possibly be modified by the animal, for example, if it builds a structure that has a communication function, as in the case of the web-spinning spiders (Witt, 1975).

The signals used by spiders are varied. If one takes account, as we have just seen, of all the situations that bring about an exchange of information, one finds that the spiders use up to five channels of communication: visual, acoustic, vibratory (transmitted by the silk or the substratum), tactile, and chemical (airborne or contact pheromones). Depending on the species and its way of life, one or another of these channels may be favored.

Emission of Signals

The Salticidae, Lycosidae, and certain other wandering spiders emit visual signals in the form of movements of the legs, postures, or movements of the abdomen (see the chapters by Forster and Jackson). Many spiders can produce sounds by striking the substratum with their palps or their abdomens but most sound-producing species use stridulating organs (see the chapter by Uetz and Stratton). These often are present only in the males and undoubtedly have a sexual communication function (Gwinner-Hanke, 1970; Rovner, 1967); however, their presence in both sexes in certain species and the fact that these spiders stridulate as soon as they are disturbed suggest that in this case these organs can also provide a means of defense, as in certain insects (Masters, 1979).

The vibratory signals used by web-spinning spiders pose, for some of them, the same problem of motor coordination as would visual signals. The frequency at which the spiders tug on the threads of the web is often specific, and the temporal organization of the complex vibratory messages is often characteristic of the species. Most web-spinning species also send signals by vibrating their abdomens. Such abdominal vibrations are observed in many wandering species, too, particularly Lycosidae. Perhaps the movements of the male have the additional function of causing air movements that are perceived by means of the female's trichobothria (Weygoldt, 1977).

In insects, the emission of a chemical signal is often associated with the presence of a particular secretory gland. Paradoxically, while we know the role played by a number of chemical signals among the spiders, in particular the pheromones involved in sexual behavior, we do not know their origin. The pheromone incorporated in the dragline of the females is perhaps a metabolite of the sericigenic (silk-producing) glands. If so, the metabolites would seem to have acquired a function of communication, as has happened with the venom gland in several species of ants (Blum, 1966). Such an evolutionary mechanism would make the lack of specificity of certain signals understandable. But particular glands near the spinnerets of certain species could be the source of this pheromone (Kovoor, personal communication). The mutual-tolerance pheromone of *Agelena consociata* seems to be produced by all the cells of the hypodermis (Krafft, 1975b).

In addition, we know of a certain number of glands that are capable of emitting pheromones, but whose actual function is totally unknown; these are the gnathocoxal, epigastric, and clypeal glands (Legendre and Lopez, 1974a,b, 1975; Lopez, 1972, 1974; Meijer, 1976). In fact, as with other animal species, the description of many glands that probably secrete pheromones has preceded behavioral studies demonstrating the action of these pheromones (Shorey, 1976).

Transmission of Signals

The diffusion of air-borne pheromones defines a three-dimensional space within which any receiving individual perceives the signal directly. Instead of depositing most of their contact pheromone on the substratum, as insects do, spiders incorporate it into their silk. The association between pheromone and silk provides precise directional information, through chemical and tactile stimuli.

Silk is a material particularly well adapted to the propagation of vibrations (Denny, 1976; Langer, 1969). It transmits equally well the disturbances caused by prey and the signals sent by conspecifics (Leborgne and Krafft, 1979). In certain cases it can modify the amplitude of signals according to their frequency, increasing it in the range of the spider's maximum sensitivity (Burgess, 1979b). It may also contribute to the transmission of the signals produced by the spiders that stridulate (Chrysanthus, 1953).

Perception of Signals

Knowledge of the sensory organization of spiders is relatively recent. Much of this information is reviewed in the chapters by Barth (mechanoreception), Forster (vision), and Tietjen and Rovner (chemoreception).

It is important to distinguish between the reception of a stimulus, which depends on the sensitivity of the sense organs, and the perception of a signal, which produces a change in the psychophysiological state of the animal. Electrophysiological techniques applied to the sense organs permit the measurement of a receptor's response to a stimulus. On the other hand, the perception of a signal can often be evaluated only from the animal's behavioral response. The results of this indirect method do not always agree with the electrophysiological data collected from the sense organs. For example, the vibration receptors are sensitive to frequencies much higher than 2000 Hz, but above 700 Hz most spiders no longer respond with predatory behavior.

Though for physiological reasons a chemical receptor can respond to a particular stimulus (Dumpert, 1978), the same may not be true of receptors for vibratory phenomena. Male spiders can emit vibratory signals at frequencies from 4 to 200 Hz, while the wing vibrations of insects are from 5 to 500 Hz (Greenwalt, 1962). Therefore, the distinction that the female makes between prey and a male spider depends less on a transducing mechanism than on a central mechanism of integration (Krafft and Leborgne, 1979). Because of the stereotyped appearance of the male's signals, one can assume that there is a preprogrammed central decoding mechanism; but this is certainly not the only mechanism involved, since social and subsocial spiders recognize a conspecific at a distance on the basis of not very stereotyped vibratory information. The perception of vibratory phenomena is certainly complex. The analysis of such information is not limited to the perception of a fundamental frequency, but requires an integration of the factors of frequency, intensity, and duration that define a signal. In addition, spiders may learn from experience to distinguish among various vibratory signals (Bays, 1962).

Not all signals necessarily provoke an immediate, specific response from the receiver. A stimulus may simply change the animal's responsiveness to another signal. In the Salticidae and the Lycosidae, chemical stimuli can play a role as secondary releasers. In the absence of chemical stimuli, the visual stimulus represented by the female may incite the male of Salticidae to make a threat display instead of a courtship display (Crane, 1949b). Males of *Lycosa rabida* sensitized by contact with a substratum on which a female had lived followed the trails of females in 84% of cases, whereas nonsensitized males followed the trails in only 65% of cases (Tietjen, 1977). Sometimes the response of the receiver may be subtle: the odor of the female of *Schizocosa saltatrix* merely reduces the male's rate of locomotion (Tietjen, 1979a). Our knowledge of the systems of communication among spiders will unquestionably be enriched once we are able to demonstrate other such unobtrusive responses of individuals to the signals of their conspecifics.

The Complexity and the Specificity of Communication

Little has been published about the whole chain of events leading to an exchange of information, i.e., the signal's emission, transmission, reception, and analysis by the receiver (Witt, 1975). There have been no studies of the way in which the various signals are integrated. Studies of communication are usually limited to a single sensory channel. However, in the context of a given behavior, spiders use several modalities to extract all the information they need. Only in that way will communication contribute optimally to survival.

As discussed in Barth's chapter in this volume, web mechanics can be invoked to explain the absence of reciprocal pursuit in *Mal-*

los gregalis. But the same explanation may not hold for other social and subsocial spiders, nor for the solitary species in which the male and female live together.

It does occasionally happen that a spider pursues a conspecific. At the moment of contact, it strikes the conspecific with its first pair of legs, immobilizing it instantly. This behavioral response reduces the risk of reciprocal bites by eliminating any vibratory stimulus that might maintain the predatory behavior. This is the second mechanism that contributes to mutual tolerance between the individuals of this species. When two individuals make contact, tactile and chemical courtship display provokes a sexual response in the female and a threat display in another male (Rovner, 1967). Conversely, two different signals can provoke the same response. A male's courtship display and his threat display both trigger a threat display from a male conspecific (Rovner, 1967).

The emission of certain signals such as pheromones, even though they may be mixtures of substances, is apparently simple; on the other hand, the patterning of tactile, visual, and vibratory signals is often complex. The courtship displays of the Lycosidae and the Salticidae usually include a succession of precise movements executed in a definite order. Likewise, the vibratory signals of the web-spinning spiders involve a succession of elements characterized by their frequency, their amplitude, their form, and their temporal distribution. The male of *Tegenaria parietina* makes frequent stops when he moves on the web of the female. At the end of each phase of immobility, with his feet spread wide, he exerts traction on the web and then lets go abruptly and moves. This abrupt release of the web gives rise to a signal of great amplitude. During his movements the male emits stereotyped sinusoidal vibratory signals (15 Hz) by vibrating his abdomen (Krafft, 1978; Mielle, 1978) (Figures 2.1 and 2.2). The male of *Amaurobius ferox* emits four types of vibratory signals. The two most characteristic ones come from a drumming of the pedipalps immediately followed by a vibration of the abdomen. The pedipalps produce a signal composed of four to five spike cycles (mean frequency = 4.5 Hz). The abdominal vibration has a mean frequency of 68 Hz, but this frequency is not stable, beginning high and then falling during the emission. Between the periods of emission of these two signals, which are always associated with each other, the male beats several times with his pedipalps and sometimes makes abdominal vibrations of weaker amplitude, at a frequency of 47 Hz (Krafft, 1978; Krafft et al., 1978; Leborgne and Krafft, 1979) (Figures 2.3 and 2.4).

A simple description of the signals is not enough. It is also necessary to determine the pertinence of their various elements and of

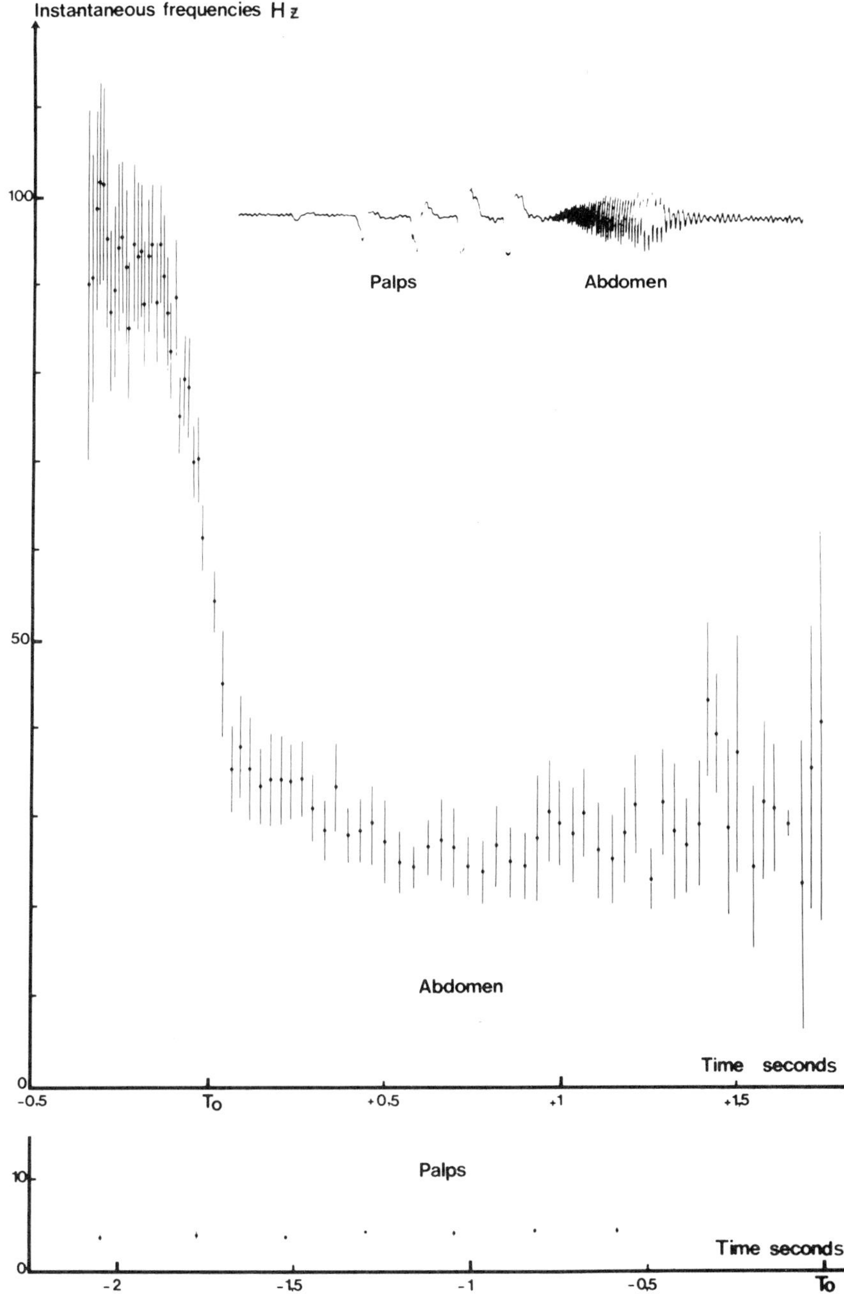

FIGURE 2.4. Changes in frequency of vibrations of the abdomen and drumming of the palps in *Amaurobius ferox*. T_0 = reference time, as the moment of maximum vibration amplitude. Each point is the mean of instantaneous frequencies. Each vertical line represents the 95% confidence limits of the mean (n = 33) (from Leborgne, unpublished data).

their chronological structure. One may ask, which elements are indispensable for the transmission of information? The problem is complicated by the redundancy within the overall display. Rovner (1968b) observed that in *Lycosa rabida* the male's drumming and rotation of the palps are not essential prerequisites for copulation. Experimental playback of the recorded signals of web-spinning spiders onto the female's web should make it possible to determine the significant elements of these messages (Leborgne and Krafft, 1979).

Chemical signals are often specific, as are vibratory signals, at least in their form. There are, however, exceptions (see the chapter by Tietjen and Rovner). It is probably an absence of specificity in their signals that allows *Agelena consociata* and *Agelena republicana* to cohabit when they are experimentally put together in one nest. In these conditions, the spiders cooperate in hunting, constructing the nest, and feeding together. They even manifest an interspecific interattraction (Krafft, 1970a).

But it would be wrong to judge whether interindividual interactions are specific or nonspecific on the basis of studying only one type of signal. It usually is the successive or simultaneous use of several signals that is responsible for specificity. We have seen how a spider of a different species is inspected and then finally bitten by *Agelena consociata*. The specificity of interactions may also depend on the active response of the partners. Indeed, we have seen that communication is established between a male and a female. Lastly, there also exist mechanical, seasonal, and ecological barriers between species (Hollander, 1971; Hollander and Lof, 1972; Hollander et al., 1973).

Uniqueness of Communication Among Spiders

The uniqueness of spiders lies in their various uses of silk for communication. Other arthropods, such as scorpions, can make use of vibratory information transmitted through the ground to detect prey (Brownell and Farley, 1979b). Many spiders have perfected the exploitation of these vibratory phenomena by weaving silky structures. They modify their environment in an adaptive manner, by surrounding themselves with a more or less complex network of threads making up a link with their conspecifics or with the environment (Burgess and Witt, 1976; Witt, 1975). A spider can send or receive information by means of its web. The silk may also be in itself an element of information when it contains a pheromone. The dragline of the lycosid female contains all the information necessary for the male: it informs him of the presence of a female and gives him directional indications that allow him to approach her. Insofar as morning dew destroys the pheromone, the thread contains temporal information, too.

The social organization of spiders depends mainly on indirect interactions (Jackson, 1979c; Krafft, 1970a). But these indirect interactions are linked to the presence of silky structures formed from a solid, elastic chemical compound, entirely secreted by the spiders. Such a system does not exist among the social insects. For example, ants indicate a source of food to their conspecifics by laying down a trail pheromone that must be labile so as to disappear when the food is exhausted. The trail must be relaid every time an ant finds a new source of food. But spiders can put up a network of threads that functions permanently and that conducts information to the conspecific only when prey is present. It is a permanent system of transmitting information, which comes into play as soon as prey touches the web, and from which all the members of the society benefit.

Furthermore, while insects are obligated to use different pheromones for different ends, spider silk can transmit equally well information about collecting food, and information—sexual, social, or parental—coming from conspecifics. Thus, spider silk is a carrier of signals; it provides a unique means of communication and plays a role as important as that of the pheromones of social insects.

Ontogeny and Regulation of Communication

The initial studies of mechanisms of communication in animals usually concentrate on the generalities observable in the majority of individuals of a given age. Thus a model is built up that is valid for most of the individuals of a species. But it is necessary then to study the differences among individuals and the plasticity and ontogeny of communication.

Intraspecific Variability

Among spiders there is an intraspecific variability in the sending and the interpretation of signals, depending on morphological and physiological differences. The courtship behavior of male Lycosidae varies quantitatively and qualitatively from one individual to another, and for any one individual (Dumais et al., 1973; Vlijm and Borsje, 1969). There is, then, variability in the emission of signals.

During agonistic interactions among *Agelenopsis aperta*, it is usually the larger spider that wins (Riechert, 1978a,b), which implies that either the sending or the interpretation of the signals depends on the size of the individuals. The same phenomenon is observed with *Metabus gravidus* (Buskirk, 1975a,b). The order of dominance observed in agonistic encounters between males of *Pardosa amentata* (Dijkstra, 1969, 1978) and *Linyphia triangularis* (Rovner, 1968a) is directly related to the size of the individuals. The same is true for *Nephila clavipes* (Christenson and Goist, 1979). In *Agelena consociata* we have seen that there are interindividual differences in grouping tendencies and a hierarchy during the agonistic relations that precede the transport of prey. Lastly, not all the females of *Agelena consociata* are equally attractive to males, though this difference might be due to the physiological states of the females.

Variability According to Sex and Physiological State

Female spiders do not react to the pheromone of other females. Perhaps they are incapable of detecting it; this nevertheless indicates a differential sensitivity according to sex. But certain signals perceived by both males and females provoke different responses. In *Meta segmentata*, for example, the approach of a male triggers sexual reactions in females and threat displays in other males (Bristowe, 1929).

The physiological state of the animal modifies the sending and the reception of signals. Hungry *Pirata piraticus* spiders respond with predatory behavior to a wider range of frequencies than do well-fed spiders (Berestynska-Wilczek, 1962). Crane (1949b) showed that in the Salticidae, age, hunger, thirst, fatigue, and overstimulation modified the responses of the male to the stimulus of the female. The males of *Lycosa rabida* first display courtship behavior on a substratum traversed by an adult female 8 days after the male's final molt. On the other hand, when directly stimulated by a female, they begin to react 5 days after the final molt (Rovner, 1968b).

The emission of sexual signals by a female also depends on her physiological state. *Cyrtophora cicatrosa* gives off the airborne pheromone only between the fourth and twentieth days following the final molt (Blanke, 1975a,b). Female *Lycosa rabida* first give off the sexual pheromone as late as 16 days after the final molt (Rovner, 1968b). The effectiveness of females of *Pardosa amentata* in triggering and maintaining the courtship behavior of males increases progressively from the final molt up to the production of the cocoon, after which it declines greatly (Vlijm et al., 1970).

A female's physiological state can modify her response to various signals of males. Two days after copulation, the female of *Cyrtophora cicatrosa* no longer drums in reaction to the courtship display of the male, but reacts aggressively (Blanke, 1975b). Nonreceptive females of *Lycosa rabida* flee from or chase males making a courtship display (Rovner, 1968b).

Any one signal can therefore produce different reactions, depending on the physiological state of the receiver. As we have also seen, the maternal behavior of a female spider depends strongly on her physiological state.

Variability with Age

The emission and perception of signals vary according to the state of the spider's development. Usually, sexual communication is observed only among adults, while agonistic interactions appear even among immature individuals (Aspey, 1974, 1976). Social interactions develop differently in the subsocial and the social species.

The emission of the sexual pheromone by female lycosids begins only in the adult stage (Hegdekar and Dondale, 1969; Rovner, 1968b), and the same is true of spiders of the genus *Araneus* (Meyer, 1928). On the other hand, even subadult females of the Theridiidae incorporate the sexual pheromone into their webs (Kullmann, 1964; Ross and Smith, 1979), but the sexual signals emitted by these females must be different or incomplete compared with those emitted by adult females: the male interrupts his courtship display, and the two sexes cohabit until the female's adult molt. This response of the male in adaptation to the female's state was well described for *Phidippus johnsoni* by Jackson (1977, 1978b).

The male's reactivity to female signals depends also on his age. Subadult males of *Lycosa rabida* do not show any courtship display toward adult females (Rovner, 1968b). This is explained partly by the lack of stridulating organs in subadults (Rovner, 1975). However, one must also consider the effects of maturation of the systems of reception, analysis, and programming of the signals.

The interactions responsible for the grouping tendency and for mutual tolerance are particularly interesting in the subsocial spiders, since these phenomena change during growth. Social relations are gradually replaced by agonistic relations and territorial behavior. The disappearance of the social life is seen at the end of the second molt of *Theridion impressum*, at the end of the fourth molt of *Theridion sisyphium*, and in the adult instar of *Anelosimus studiosus* (Brach, 1977). Dispersal takes place after the fifth molt of

Stegodyphus lineatus, the sixth molt of *Eresus niger*, and the seventh molt of *Eresus* sp. (Kullmann et al., 1972; Kullmann and Zimmermann, 1974). The frequency of cannibalism increases with age in *Sosippus floridanus* (Brach, 1976); and it has been shown experimentally that mutual tolerance disappears after the seventh molt in *Eresus* sp. (Kullmann and Zimmermann, 1974).

This modification of the way of life of the subsocial spiders may be related either to the disappearance of the stimuli responsible for interattraction and mutual tolerance or to the loss of the creatures' reactivity to these stimuli. In *Eresus* sp., mutual tolerance depends mainly on tactile information related to the integumentary structure of the individuals, which varies little from one instar to another. One may therefore suppose that it is these animals' reactivity to this stimulus that is lost. The fact that the communal life of *Coelotes terrestris* can be prolonged in the laboratory also favors this hypothesis. Indeed, if the loss of mutual tolerance were linked to the disappearance of the stimulus inhibiting reciprocal biting, these spiders would display an intense cannibalism. Raised in groups, these spiders display an interattraction at least until the sixth instar, while in normal conditions dispersal happens in the fourth instar. It is therefore likely that the dispersal of *Coelotes terrestris* is not due to the disappearance of the stimuli responsible for the grouping tendency, but rather to the appearance of agonistic relations, perhaps linked with the amount of prey available (Horel et al., 1979). An analogous development is observed among solitary spiders at the end of the communal phase of the young. However, when the dispersal of a solitary species such as *Tegenaria parietina* is prevented, cannibalism nevertheless develops between the second and the third molt outside the cocoon (Mielle, 1978).

A comparative study of these two mechanisms might be useful to achieve a better understanding of the origin of spider societies. In fact, the grouping tendency develops with age in *Agelena consociata* and decreases in *Agelena labyrinthica* (Krafft, 1970b).

Development with age of the emission and perception of signals must also take place in the myrmecophile (ant-associating) spiders. Young spiders of the genus *Myrmarachne* (Salticidae) usually associate with one species of ant and the adult spiders with another: young *Myrmarachne foenisex* live in the nests of *Crematogaster castanea*, whereas the adults live in the nests of *Oecophylla longinoda* (Edmunds, 1978). It is a pity that we do not know the interspecific signals that regulate these associations. In this case, presumably, the changes in the emission and the perception of the signals are qualitative in nature rather than quantitative.

Variability According to the Individual's Experience

Most of the examples showing the variability of the mechanisms of communication that we have seen so far can be explained in terms of phenomena of physiological maturation, or linked to age or to interindividual differences. It is, however, important to emphasize that mechanisms of communication depend not only on genetic programming, but also on the history and experience of the animal. If we consider the systems of communication of animal species, we find that some signals and the responses that they provoke are very stereotyped, whereas others may be modified or even acquired during the ontogeny of the individual. For the present, very little information is known about such phenomena in spiders.

In the context of predatory behavior, it has been demonstrated that experience can modify a spider's response to vibratory information (Bays, 1962; Meyer, 1928; Szlep, 1964; Walcott, 1969). When two adult males of *Schizocosa ocreata* that have been reared separately meet for the first time, both adopt a threat posture with the front legs held out obliquely. After that, the subordinate individuals display prolonged leg-waving and hold their legs vertically, while the dominant ones threaten by holding their legs horizontally and vibrating them (Aspey, 1974). Conflicts between males of *Pardosa amentata* modify their subsequent behavior: courtship displays toward females become less frequent in the losers than in the winners (Dijkstra, 1969). Previous contacts with a female also have an effect: adult males isolated after the subadult instar generally defeat males that have been in the presence of females; there is a negative correlation between dominance and the frequency of previous access to females (Dijkstra, 1969, 1978).

Thus the experiences of an individual can alter the emission, and perhaps also the perception, of signals. These changes can have major consequences. Cooke (1965) reared two species of *Dysdera* in the laboratory and obtained interspecific crosses; but if the two species were collected as adults in the wild, such crosses did not occur. The differences between the two results are attributable either to the lowering of thresholds of responsiveness in the lab-reared individuals or to the animals captured in the wild having had previous contact with individuals of both species.

Conditions of rearing have little influence on mutual-tolerance mechanisms in *Agelena consociata*. Young isolated for three months after birth display mutual tolerance when they are put in groups again. On the other hand, the grouping tendency is influenced by previous social contacts. As we have seen, adult and subadult individuals isolated for 85 days show an enhanced grouping tendency when they are put together again.

Variability of Communication According to the Situation

The emission and perception of signals can also depend on external contingencies, particularly on the situation in which the spider finds itself. The arrival of prey in the female's web is a determining element for the release of courtship behavior in male *Meta segmentata* (Bristowe, 1929). The presence of a conspecific of the same sex modifies the courtship display of *Pardosa amentata*; the encircling movements and the cyclic movements of the palps of a male in the presence of a female increase in frequency when a second male is present (Dijkstra, 1969). In *Lycosa rabida* the stimulative effect of the sounds of courtship or threat usually triggers a threat display in males, but the effect is lost when the males are placed on a substratum covered with female pheromone (Rovner, 1967). In *Agelena consociata*, the aggressiveness of the males increases in the presence of females (Krafft, 1970a).

The reactivity of a male to a signal may also depend on his location on the female's web. The approach of a conspecific of the same sex does not excite any aggressive reaction from males of *Nephila clavipes* grouped at the periphery of a web, but when one of them is near the hub he responds with aggressive signals to the approach of any conspecific of the same sex (Christenson and Goist, 1979).

External contingencies can determine the type of signals used by the male during courtship behavior. A male of *Phidippus johnsoni* that encounters a female in an open space in daylight uses a courtship display consisting mainly of visual signals. In darkness, on contact with the nest of a female, he sends vibratory signals (Jackson, 1977a,b, 1978a).

Among the social spiders, the reactivity of individuals to various signals seems to depend on their position in the web or the colony. We have seen that *Agelena consociata* compete to transport prey. But once the prey is in a gallery, the spider that transported it tolerates all the individuals that come to take part in the meal together. Thus a single stimulus, the presence of conspecifics, provokes two different types of response, depending on whether it is associated with the presence of paralyzed prey on the outer web, or within a gallery (Krafft, 1970a). The reactivity of individuals to the stimulus responsible for interattraction is also higher when spiders can satisfy their thigmotactism. These conditions occur in the galleries of the web (Krafft, 1970b).

There are no agonistic relations among the individuals of *Metabus gravidus* on the foundation lines, even when the spiders come in contact with one another. These individuals are also more clustered than those that have individual webs. The structure of the colony can also have an influence. In colonies where the webs are very close to one another, the spiders tolerate their conspecifics at shorter distances (Buskirk, 1975a,b). In *Cyrtophora citricola*, agonistic relations are especially frequent at the positions in the colony that are the most favorable for the capture of prey (Rypstra, 1979).

The space and the amount of available prey can also modify the agonistic relationships between solitary spiders. The development of cannibalism in *Tegenaria parietina* is a function of the space available (Mielle, 1978). *Stegodyphus lineatus* (Millot and Bourgin, 1942), *Amaurobius ferox*, and *A. similis* (Cloudsley-Thompson, 1955) display an interindividual tolerance, even if their webs touch one another, if enough space is available and if prey are abundant. The territorial behavior of *Agelenopsis aperta* is also a function of the density of prey available. But in agonistic encounters the possession of a web by one of the individuals gives it a psychological advantage, providing that the difference in size between the spiders is less than 10% (Riechert, 1978a,b). It may be the availability of prey that is responsible for the fact that *Achaearanea disparata* lives in a social state in Gabon and in a solitary state on the Ivory Coast (Darchen and Ledoux, 1978).

The Evolution of Communication

The evolution of systems of communication can be observed both in the context of sexual behavior, from the primitive to the highly evolved spiders, and in the context of social behavior, from the solitary to the social species.

Chemical communication by means of a pheromone is a very ancient mechanism which was already present in protozoa and was undoubtedly necessary for the appearance of the metazoa (Shorey, 1976). It is even reasonable to consider that pheromones may have been the precursors of hormones (Wilson, 1970). Chemical communication linked to the perception of the female sexual pheromone at the time of a direct contact must therefore be considered a primitive system of communication among the spiders (Platnick, 1971; Weygoldt, 1977). After that, there appeared a way of communicating at a distance by means of pheromones incorporated into the dragline of the female and airborne pheromones; it is not yet known whether these pheromones are identical to or different from the pheromone covering the integument of the female. Then, depending on their mode of life and their sensory organization, spiders perfected their means of communication, by adding either vibratory or, less often, visual signals.

The leg movements in the courtship displays of the Lycosidae and the Salticidae probably were derived from threat movements, which themselves originated in chemo-exploratory movements (Bristowe and Locket, 1926; Farley and Shear, 1973; Platnick, 1971). In fact, the less evolved members of the Salticidae and many of the Lycosidae display a courtship behavior that includes postures identical to threat postures, while in the more evolved species the courtship behavior is clearly different from the agonistic behavior (Crane, 1949b; Koomans et al., 1974). The emission of vibratory signals by the males of web-spinning species doubtless has the same origin. This exploration movement has thus acquired a communication value by ritualization. The fact that *Metabus gravidus* uses very similar signals in prey-location and in agonistic relationships reinforces this hypothesis (Buskirk, 1975b). However, the origin of the abdominal vibrations of the males of numerous web-spinning and non-web-spinning species is harder to explain.

The social spiders probably evolved from subsocial spiders (Krafft, 1979). In the species that weave irregular or sheet webs, one can imagine that maternal behavior opened the way to the subsocial structure. Afterwards, failure of the young to disperse would have led to the social species. Evolution was undoubtedly different in the orb-spiders, since these do not show any highly developed maternal behavior. Their colonies presumably arose simply from the failure of the young to disperse.

The appearance of mutual tolerance seems at first sight to be the essential prelude to socialization among the spiders. The attraction exerted by the silk could be responsible for group cohesion in the absence of agonistic behavior, even if it is complemented by interattraction among the nonterritorial social species. In fact, we have seen that interindividual cooperation is mainly a consequence of the use of silk. Mutual tolerance at a distance depends on vibratory information reflecting a mode of movement peculiar to the species but characteristic of the family. On contact among *Agelena consociata* and *Stegodyphus sarasinorum*, tactile information about the integument of the individuals comes into play; this stimulus seems family-specific for Agelenidae and for the Eresidae. Mutual tolerance in the social spiders is therefore not necessarily linked to the evolution of a new stimulus, but depends on the development in the social species. Secondarily, this tactile stimulus, which alone is responsible for tolerance in *Eresus* sp., has been reinforced by a chemical stimulus in *Agelena consociata* and *Stegodyphus sarasinorum*. Therefore there has been a progressive increase in the number of signals involved in mutual tolerance.

The process of socialization, however, is more complex. It requires that territorial behavior, too, should disappear. Spiders are exclusively predatory creatures whose nourishment depends on prey that fall into their webs. The intraspecific aggressivity that makes possible the defense of an area of web compatible with energy balance therefore favors the survival of the species. To lead to a society, the development of interindividual tolerance must be linked to the presence of a large number of prey, allowing territorial behavior to be reduced, and must be accompanied by an increase in foraging efficacy (Riechert, 1974b). It is possible that in subsocial species the web is no longer capable of providing for the needs of all the individuals after they reach a certain age, which could explain their dispersal. The use of individual webs by the territorial social species may be attributable either to the necessity of keeping a large area of hunting web per individual or to the geometric structure of the web. This hypothesis accords with the fact that all the nonterritorial social species known at present are found in regions where prey are always abundant.

Silk has certainly played a determining role in the appearance of societies of spiders (Shear, 1970). All non-web-spinning species are solitary. The silk is directly responsible for coordination and cooperation in social spiders. The only evolutionary change required consists in the development of the spiders' aptitude for exploiting the vibratory information transmitted by the silk. If one can consider that "the road to insect sociality is paved with pheromones" (Blum, 1974), one must grant that the society of the spiders hangs by a thread. But socialization has also been accompanied by the disappearance of at least one form of communication. The males of *Mallos gregalis* do not react to the silk of the females (Jackson, 1978a); therefore, socialization has produced the disappearance of either the females' addition of sex pheromone to the silk or the males' reactivity to it.

Conclusions

Spiders are of fundamental interest in the study of animal communication. They use visual and vibratory signals that are often spectacular and are relatively stereotyped and easy to quantify. The quantitative and sometimes qualitative appearance of these signals varies with the age, physiological state, and past experience of the sender. The receiver's response is very often unambiguous, either negative or positive. The perception of a signal by the receiver also varies with the receiver's age, its physiological state, and its life experience, and with the particular situation in which it finds itself. Consequently, spiders can provide interesting experimental models for the study of the plasticity of animal communication.

Techniques of quantifying the sending and the receiving of signals make it possible also to study the way in which signals of different kinds are integrated to ensure the transmission of a piece of information in a given behavior. This should allow us to understand better the specificity of the interactions.

Compared with insects, spiders have developed an original means of communication: the transmission of vibratory information through silk. Sexual and perhaps social pheromones are frequently incorporated into the silk. In addition, the silky structures can transmit equally well the signals involved in agonistic or sexual, predatory or social, behaviors. The discriminations that spiders make among these various vibratory signals depend on a central decoding mechanism. Vibratory signals are easy to manipulate; they can be recorded and then retransmitted onto the web. The playback of normal vs. modified signals should lead to a better understanding of how the information-decoding mechanism works. As this mechanism is genetically programmed, and yet plastic, spiders should give us an opportunity to study interactions between a genetic program and the influence of the environment.

Finally, spiders are a zoological group particularly favorable for the study of the evolution of communication in conjunction with socialization. Their silky structures, which are a permanent means of communication among the individuals of a society, carry vibratory information with the same functions as a considerable number of social pheromones of insects.

Acknowledgments

Many thanks to A. Horel, R. Leborgne, and C. Roland for the help they provided in the wording of this chapter, to E. Miller for the translation into English, and to the editors, who amended the English version and provided valuable advice.

Chapter 3

SPIDERS AND VIBRATORY SIGNALS: SENSORY RECEPTION AND BEHAVIORAL SIGNIFICANCE

Friedrich G. Barth
Zoologisches Institut
Johann Wolfgang Goethe-Universität
Frankfurt a.M.
Federal Republic of Germany

Introduction

The outstanding importance of vibratory signals in spider behavior has been known for a long time, and much energy has been devoted to observations confirming this again and again. The field is large. All sorts of vibrations are relevant. Most obvious are the ones generated in the spider web and transmitted by it. Since the web is one of the distinctive features of this animal group, it deserves particular attention (Witt, 1975). Vibrations in more solid substrates such as plant leaves and wooden stems have generated much less interest. For some spiders even vibrations of the water surface carry information, while airborne vibrations or sound guide the behavior of others.

A correspondingly rich variety of behavior is shaped by vibratory signals, and to consider the various ethological contexts involved is, in a way, to consider the diversity of spider ecology. Prey-catching is the most spectacular behavior and well known; but the list includes courtship, territorial behavior, and social interaction in species sharing a common web.

Spiders do not only perceive vibratory signals; they also emit them by drumming with the palps and the abdomen, by stridulating, or by plucking threads of their own or other spiders' webs.

Many reports, particularly in the older literature, describe such mechanically guided spider behavior, as reviewed in a paper by Krafft and Leborgne (1979) which appeared after completion of this chapter. Due to the lack of proper control and measurement of the signals involved, however, the sensory part of many of these studies is very difficult to evaluate. For a full understanding of spider communication we want to know the physical properties of the natural and biologically meaningful signals, the functional morphology and the physiology of the sense organs receiving them, and the brain's integrating capacities in handling the input from its sensory periphery. Detailed behavioral analysis, done with the neuroethological questions particularly in mind, will have to accompany all stages of such an effort.

The main purpose of this chapter is to introduce the sensory mechanisms available to spiders for the detection of vibratory signals and to characterize and illustrate the behavioral situations in which spiders actually or potentially make use of them. While following along that line, a number of large gaps in our knowledge will also become apparent. A rather long list of publications has been considered in order to provide a useful basis for future research.

A Variety of Behavioral Contexts

Vibrations in the Spider Web

PREY CAPTURE Fascinated by the geometry of the orb-web, many a naturalist has lured a spider from its retreat or the hub of its web to a vibrating blade of grass. In species such as the common cross spiders, *Araneus sclopetaria, A. diadematus* et al. (Araneidae) the female sits in the center of its web, the hub, for long periods of time. With the grass blade, it is easy to imitate entangled insect prey. The spider turns toward the vibration source and attacks it. If it was not already sitting in the hub, it first returns to it upon vibration and then orients toward the presumed or actual prey. After strong mechanical disturbances some spiders sit in a retreat in a niche or corner of the web instead of the hub. In this case the tarsus of one front leg is placed on a signal thread which transmits the vibrations from the hub (Peters, 1931, 1933a). A similar behavior is known for many other web spiders (Grünbaum, 1927; Liesenfeld, 1956; Robinson and Mirick, 1971; Wiehle, 1928; Witt, 1975; and others).

In araneid spiders, prey capture consists of a series of steps: localization and at least rough identification of the prey, attack of the victim, its immobilization, and finally its removal from the capture site and transport to the feeding place.

The overwhelming significance of vibratory signals for the first steps of such behavior has been known for a long time. Since vibrating tuning forks and doorbells were attacked even in the dark, the necessity of visual input was easily excluded (Barrows, 1915). Later workers stressed the mechanical heterogeneity of the web. Prey capture was found to depend largely on vibrations of the radii and, according to some reports, cannot be elicited even by high amplitude vibrations of the spiral thread (Grünbaum, 1927). Notwithstanding contradictions in the literature (Barrows, 1915; Meyer, 1928), the main point is a significant difference in the mechanics of radii and spiral threads.

Liesenfeld (1956) started to quantify such mechanical properties in the araneid *Zygiella x-notata*: the radii show a much larger range of elastic deformation, and a straightforward explanation of the difficulty of eliciting prey-catching behavior by vibrating the sticky spiral threads is offered by the better stimulus-conducting properties of the radii. The radii of the *Zygiella* web are still within the elastic (Hookean) range with loads bigger than 30 mg, whereas the sticky spiral thread is irreversibly deformed by loads as small as 3 mg, and ruptures easily.

Another example of such striking mechanical differences is known for viscid and framework silk in the orb-web of *Araneus sericatus* (Denny, 1976): breaking strain (strain = increase of length/original length) is consistently close to 3.0 in viscid silk but measures only 1.2 in framework silk. Breaking stress (stress = load or force/area) is close to 1 GN/m^2 in both cases, but viscid silk relaxes to a greater extent. Also, the stress-strain curve of viscid silk contrasts with that of framework silk in exhibiting a very low initial Young's modulus. This underlines its shock-absorbing function. Other impressive values in the literature give the extensibility of the dragline as 30% and that of the sticky spiral thread as 517% (*A. diadematus*). In still another case (*Meta reticulata*) the spiral thread could be reversibly extended even 14 times its initial length (de Wilde, 1943; Lucas, 1964; Work, 1976).

The particular importance of the hub in orientation behavior reflects the fact that all the radii converge onto it. It is both a geometrical and a mechanical center of the orb-web. There are other web types, however, and spiders use their silk in a wide range of cases and ways for signal transmission and as a means to enlarge their sensory range (Kaston, 1964; Kullmann, 1972a). Figure 3.1 illustrates a few representative examples.

1. *Zygiella x-notata*. Because of our awareness of the obvious mechanical heterogeneity of the orb-web and of various predominant functions of its parts, the web of the sector spider has appropriately received particular attention. In this web (Figure 3.1a) a single signal thread runs through a sector devoid of radii and spiral threads and mechanically links the hub to the spider's retreat in the web periphery. *Zygiella* sits in its retreat with one front leg tarsus on the signal thread and thus receives vibrations from all

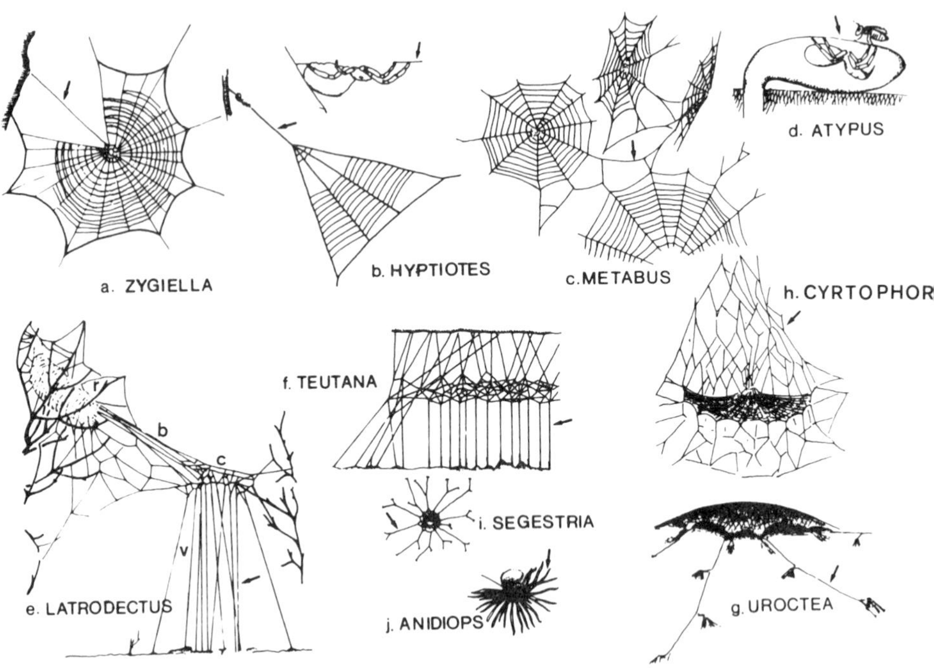

FIGURE 3.1a-j. Examples of spider webs chosen to illustrate the close relationship between web structure and vibration detection (note arrows). (a) *Zygiella x-notata*, sector web with signal thread (after Kullmann, 1972a). (b) *Hyptiotes* sp., triangle web, into which the spider incorporates itself mechanically by an in-series position with the signal thread (after Kullmann, 1972a and Wiehle, 1928). (c) *Metabus gravidus* is a colonial spider; the orb-webs of many individuals are connected by common support lines, which also carry vibratory information from web to web (after Buskirk, 1975b). (d) *Atypus* sp. uses a closed tube web to detect vibratory signals when catching prey (after Bristowe, 1958). (e) *Latrodectus pallidus* builds a web which consists of a retreat (r), a bridge (b), a catching platform (c), and vertical viscid threads (v). These are used to entangle insect prey and transmit the vibratory prey signals to the web above (after Szlep, 1965). (f) *Teutana castanea* also attaches viscid threads to a catching platform for the same purpose as *Latrodectus* (after Wiehle, 1931). (g) *Uroctea durandi* attaches signal threads, which are suspended on elaborate suspension poles, radially from its tent-roof web (after Kullmann et al., 1975). (h) *Cyrtophora citricola*, regularly woven sheet web is attached to three-dimensional snare web (after Wiehle, 1928). (i) *Segestria* sp. stretches snare and signal threads radially around the entrance to its burrow (after Kullmann, 1972a). (j) *Anidiops villosus*, a trapdoor spider, arranges twigs radially around the entrance to its burrow (after Main, 1957).

over the web. Mechanically the situation is less complex than in the web of *Araneus*, since the signal thread is not coupled to neighboring radii by the sticky spiral. Upon vibration, *Zygiella* darts to the hub along the signal thread, turns in the direction of the stimulus, and then runs toward the prey along the proper radius.

2. *Segestria* (Figure 3.1i). This spider lives in a tubular retreat and stretches silken threads radially on the ground from its opening (Kullmann, 1972b). Other spiders, including liphistiids and trapdoor spiders (e.g., *Nemesia dubia*: Buchli, 1969) do the same.

3. *Anidiops villosus* (Figure 3.1j). Some trapdoor spiders even collect small slender twigs which they arrange radially outward from the rim of their burrow. Their leg tips rest on the twigs, thus enlarging their detection and capture area (Main, 1957).

4. *Uroctea durandi* (Figure 3.1g). This spider produces a tent-roof web. Signal threads from the tent floor lead far beyond the limits of the roof. They form an alarm system, again for a considerably enlarged prey-catching area with diameters up to 50 cm (length of female spider: 1.6 cm). The signal threads do not touch the ground directly. Instead they are attached to silken "telephone poles" and thus transmit vibratory signals to the spider's retreat under the roof more effectively than they could when lying on the substrate (Kullmann et al., 1975).

5. *Latrodectus pallidus* (Figure 3.1e). As do other members of the Theridiidae, this widow stretches vertical viscid threads between the catching platform of its web and the ground. Moving prey stick to and snap these threads, which are then pulled in by the spider (Szlep, 1965).

6. *Atypus* and *Sphodros* (Figure 3.1d). These atypid spiders await their victims within a closed tube web extending from the silken wall of their burrow within the ground (Bristowe, 1958). Insects crawling on the web's surface and vibrating it are caught through the web and pulled inside through a slit cut (at the time of capture) for that particular purpose.

7. *Cyrtophora* (Figure 3.1h). Still other spiders attach a three-dimensional snare web to a more regularly woven sheet-web. This certainly affects the take-up and conduction of vibratory signals (Burgess and Witt, 1976; Wiehle, 1931).

For three more examples see Figure 3.1b, c, f. No quantitative data are available on the signal-transmitting properties of these webs. The only case studied is the web of *Zygiella* and details are given later (Graeser and Markl, in prep.).

The virtues of a more physically minded approach to the problems of vibration detection in spiders were first seen by Liesenfeld (1956, 1961). Using *Zygiella* and its web, he pointed to the physical and physiological measurements needed and the difficulties involved in gathering them. Apart from quantifying a few mechanical properties of the web, he made the first measurements of its vibration with improved techniques, using a function generator and a modified loudspeaker at the input side and a microscope for amplitude calibration. Liesenfeld (1956) found fast transients in the vibratory signal particularly effective in eliciting the spider's attack behavior, but otherwise noticed an independence of the reaction from the stimulus frequency as long as it was above 5 Hz (resonance frequency of web loaded with 20 to 50 mg). Behavioral threshold values were given for the first time as absolute measures (10^{-3} to 10^{-4} cm). In the later paper (Liesenfeld, 1961), threshold sensitivity of the vibration-sensitive metatarsal lyriform organ of the same species was determined electrophysiologically and found to vary greatly with stimulus frequency. Particular attention will be paid in the following pages to other papers taking this type of approach and tackling the problem from various angles.

ACTIVE PREY LOCALIZATION An interesting behavior found in many web spiders and already noted by Barrows (1915) in *Araneus sclopetaria* is the active use of vibrations to locate and possibly identify prey and other particles in the web. If the first vibratory signal is short and not followed by a second one then *Araneus* often places its front legs on neighboring radii, pulls them toward its body, and then releases them abruptly. The web is thereby set into vibration. The spider turns through a complete circle. If a motionless fly is in the web, the radii pointing in its direction will differ mechanically from the rest, and this obviously tells the spider the object's location in its trap (Peters, 1931, 1933a). The cross spider thereby finds a motionless insect as well as dirt particles, which it removes from its web. *Zygiella* does the same and can detect objects as light as 0.13 mg (cf. *Drosophila*: 0.78 mg; Liesenfeld, 1956).

Similar searching behavior is known from other orb weavers, e.g., *Nephila clavipes* (Robinson and Mirick, 1971), *Argiope argentata* (Robinson and Olazarri, 1971) and *Metabus gravidus* (Buskirk, 1975b). A closely related case is given as a last example to stress the intimate relationship between web mechanics and spider behavior. When stimulated by weak vibrations, *Zygiella* was observed often not to pluck as described above but instead just slightly to stretch the radii as if actively to improve their stimulus-conducting properties by taking up slack and increasing their tension.

COURTSHIP Vibrations actively produced by male spiders in the web of females are an important part of spider courtship behavior (e.g., Gerhardt, 1926; Grasshoff, 1964; Locket, 1926; Witt, 1975; see also the chapter by Krafft). The cross spiders are again the most familiar example: the male, sitting in the web periphery, plucks the thread which the female touches with its front legs. The main functions of such behavior are to suppress the female's prey-catching instincts, to announce the presence of a conspecific male, and to synchronize the mates (Figure 3.2).

A recent paper by Robinson and Robinson (1978) vividly illustrates the significance of vibratory signals in courtship. By comparison of more than 50 species of araneid spiders these authors came to distinguish three main categories of courtship. The suggested evolutionary sequence implies an increasing importance of vibratory signals (Figure 3.3).

1. In the group regarded as primitive the males cross the female's web and court in the hub, starting with tactile stimulation by rapping the epigyne or scraping the abdomen with their pedipalps. Copulation eventually takes place in the hub. Examples are various species of *Argiope, Nephila maculata,* and members of the genera *Nephilengys* and *Herennia.*

2. Spiders of the next category include other species of *Argiope.* They start courtship as just described. Then, however, these males cut a hole in the web close to the hub, insert a mating thread, and vibrate it by tarsal rubbing, line slapping, bouncing, and jerking. The female identifies these signals and finally joins the male to copulate on the mating thread.

3. In the last category the male stays in the web periphery all the time and stretches a mating thread between nearby vegetation and often the peripheral end of a radius. In some species the mating thread is inserted in the web's prey-catching area. In that case the male cuts away the viscid spiral from a radius which is used as mating thread (*Gasteracantha* sp.; *Micrathena sexspinosa*; *Eriophora fuliginea*). The male again vibrates the mating thread, and the female finally approaches him on the mating thread outside the web proper for copulation.

Various features of the behavior described are remarkable. To walk on the female's web is potentially deadly for the male because he might be mistaken as prey. Courting from a distance, decreasing the female's mobility, and actively attracting it with vibratory signals traveling along a mating thread seem to have evolved as effective methods to reduce the risk (Robinson and Robinson, 1978). On the other hand, the existence of Type 1 courtship shows that the female distinguishes between the moving male and prey; vibratory clues successfully used by the male to announce itself as a mate are certainly involved. (As to chemical signals, see the chapter by Tietjen and Rovner.) The male, in turn, has obviously evolved a sophisticated way of handling the female web mechanically. Attaching the mating thread to a radius and taking away the sticky

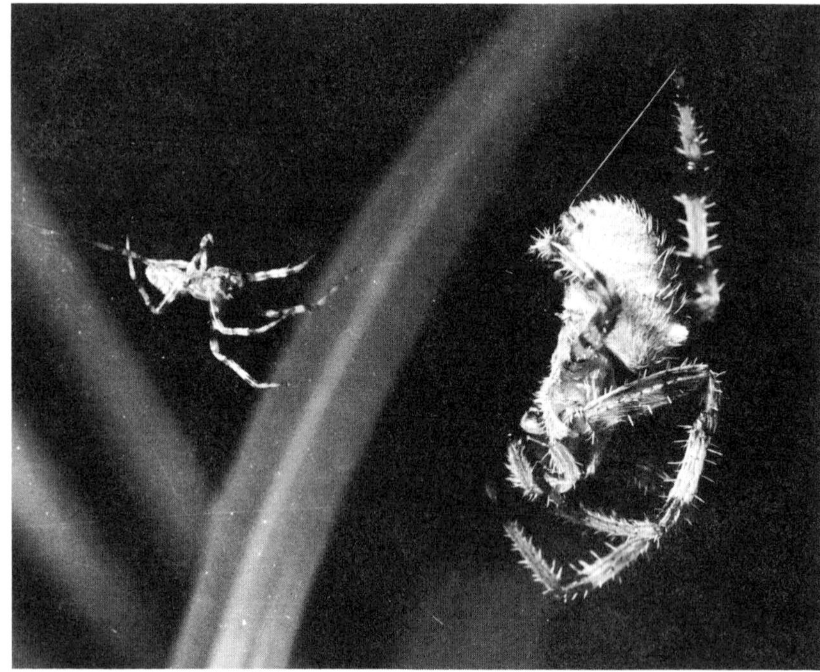

FIGURE 3.2. Male *Araneus pallidus* (left) courting the female in her web. (Courtesy of M. Grasshoff.)

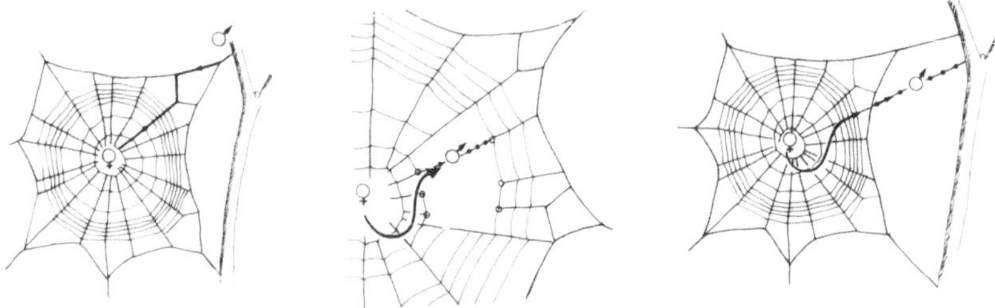

FIGURE 3.3. Male courtship behavior in araneid spiders is divided into three types, with increasing refinement (from left to right) in the use of vibratory signals; see text (after Robinson and Robinson, 1978).

spiral from the communication line are good evidence for this conclusion. The male *Zygiella* quite rightly cuts the signal thread of the female's web for courtship and replaces it with a communication thread from the upper part of the web to the female's retreat (Gerhardt, 1926).

Differences in male vibratory courtship signals may well be important for *intraspecific recognition.* Male spiders belonging to the genera *Amaurobius* (Amaurobiidae), *Tegenaria,* and *Coelotes* (Agelenidae) indeed differ with respect to the repetition rate of their drumming movements (Krafft, 1978). Male *Amaurobius similis* use palpal drumming at 4.1 Hz, whereas vibrations produced with the abdomen at a frequency of 150 Hz are presumed to be of particular importance in *A. fenestralis.* In *A. ferox,* on the other hand, frequencies of palpal drumming are very close to those of *A. fenestralis,* but its abdominal vibrations clearly differ, with a repeat frequency of only 47 to 68 Hz.

GREGARIOUS AND SOCIAL SPIDERS Most spiders are solitary, and their ferocious prey-capture behavior may even be cannibalistic. As we have seen, there are ways by which the male bypasses or eliminates the female's aggression during courtship. Also, when newly emerged from the cocoon, spiderlings of solitary species live together in harmony for a limited tolerant period before they start their aggressive individualistic life.

Some spiders, however, live together permanently and therefore are commonly referred to as social spiders (Brach, 1977; Burgess, 1976, 1979b; Krafft, 1970b; Kullmann, 1968b, 1972a; Lubin, 1974; Shear, 1970; Tretzel, 1961). Hundreds or even thousands of them may share one communal web; again, vibratory signals are involved in communication, tolerance, and even cooperation. Three examples will illustrate the scope of their significance.

1. In the Mexican spider *Mallos gregalis* (Dictynidae) the biggest adult female measures only a fraction of the size of the usual prey, flies. When a fly is entangled in the sticky surface sheet of the web and starts buzzing, up to 30 spiders come running from the web's interior, attack and overpower the prey in a joint effort, and finally feed on it together. The importance of web-borne vibrations is easily deduced from two simple observations: as long as the fly buzzes above the web without touching it, no reaction is observed; and a fly lying motionless on the web does not attract the spiders either. Although aggressive toward the fly, spiders were never seen to attack each other. Web mechanics is considered an important reason. Burgess (1976, 1979b), to whom we owe this story, found that the

web attenuates the vibrations caused by fellow spiders walking around, whereas prey vibrations contain frequencies which coincide with resonance peaks of the web between 30 and 700 Hz. When applied experimentally these frequencies (mainly 100 to 400 Hz) elicit the attack behavior most promptly. The web is to be considered a mechanical filter well-matched to the spider's demands. In another social spider, *Agelena consociata* (Agelenidae), mutual tolerance also depends, at least over a distance, on vibratory signals. The same may apply to all or most gregarious or social spiders; whereas at close range, tactile and chemical cues are more important (Krafft, 1970b, 1975b).

2. *Metabus gravidus* (Araneidae) is a gregarious spider which does not show any cooperative behavior. Nevertheless it lives communally in closely spaced orb-webs inhabited by up to a total of 70 individuals. These webs are linked together by common support lines (Figure 3.1c). *Metabus* uses active web shaking in a true communicative way to defend its individual feeding area (Buskirk, 1975b). When a conspecific intruder comes within the defended area *Metabus* bounces the hub, whereupon the intruder stops or changes direction. If the intruder does not respond to the warning and instead crosses a threshold distance of about 9 cm, then the web owner runs toward its rival and either chases it away or loses its web to the invader.

3. A last example brings us back to the tolerant phase in spiderlings. Females of *Theridion saxatile* share their ant prey with their young; the young remain in the maternal web for about a month after hatching. While the mother is capturing the entangled prey, the spiderlings remain in the retreat. Whenever they come too close to a struggling ant the adult spider turns and plucks the web with its forelegs, whereupon they return to their retreat. Later the situation changes. The youngsters then participate in the prey capture and are summoned by their mother with a new signal produced by different motions of the forelegs (Nørgaard, 1956).

COMMENSALS AND KLEPTOPARASITES There are several reports on insects having found a safe way into the spider web as commensals or kleptoparasites. Scorpion flies (Panorpidae) are among them (Thornhill, 1975); and, in particular, tipulids (Limoniinae) are known to hang from the tarsi of their two anterior pairs of legs from the non-sticky frame threads in the webs of several spider species (e.g., Theridiidae) (Alexander, 1972; Robinson and Robinson, 1976; Scott, 1910, 1958; Verdcourt, 1958). Their cryptic posture, with the legs perfectly aligned with the long axis of their body, resembles the resting posture of many stick insects. This is believed to be a visual defense against predators from outside the web. But how about protection against the spider itself? Several observations leave us with an attractive hypothesis.

Various observers have noted dancing and bobbing movements of tipulids in the spider web; Robinson and Robinson (1976) have pointed to a similarity of the tipulid gait in the web with the gait of emisinid bugs (Emisinae), which are also found in spider webs. Older observations on the kleptoparasitic spider genus *Argyrodes* (Theridiidae) (e.g., Kullmann, 1959; Legendre, 1960) and a recent experimental study on tropical representatives of these spiders in the web of *Nephila clavipes* and *Argiope argentata* (Exline and Levi, 1962; Vollrath, 1976, 1978, 1979a) all point in the same direction: the kleptoparasites were reported to move very slowly and "carefully," in particular during the time that the host spider is motionless (Figure 3.4). These intruders even seem to guard against a sudden release of tension when they cut out threads from the host web together with the stolen prey: "Then this radius is cut while being held by the first legs. These legs are then slowly stretched, gradually releasing the tension in the thread" (Vollrath, 1979a).

A remarkable ability to interpret vibrations produced by both the host spider and its prey (*Argyrodes* links itself mechanically to the hub of the host web and several radii by signal threads) and a perfection in adjusting the vibrations caused by its own locomotion are clearly part of the spider-kleptoparasite behavior. Apparently, non-prey vibrations are a safe route into the dangerous web not only for the courting male and the conspecific spiderlings but also for both insect and spider commensals and kleptoparasites. Low frequency vibrations and the lack of fast transients in the vibrations seem to be of particular importance in avoiding the host's attack. Essentially the same conditions may apply to spiderlings moving around in their mother's web (Tretzel, 1961). From the kleptoparasite's point of view it is also crucial to know the most effective timing of its raid. Vollrath (1979b) succeeded in measuring the vibrations produced by *Nephila* and *Argiope* during a prey-capture sequence. He found distinct patterns corresponding to the various actions like running toward the prey, wrapping, biting, taking it back to the hub, etc. The kleptoparasite's (*Argyrodes elevatus*) raid is clearly released by the wrapping vibrations and not by others. They signal the successful prey capture by the host and the presence of stored prey which can be stolen.

Clearly we need more such quantitative data. Other cases of sim-

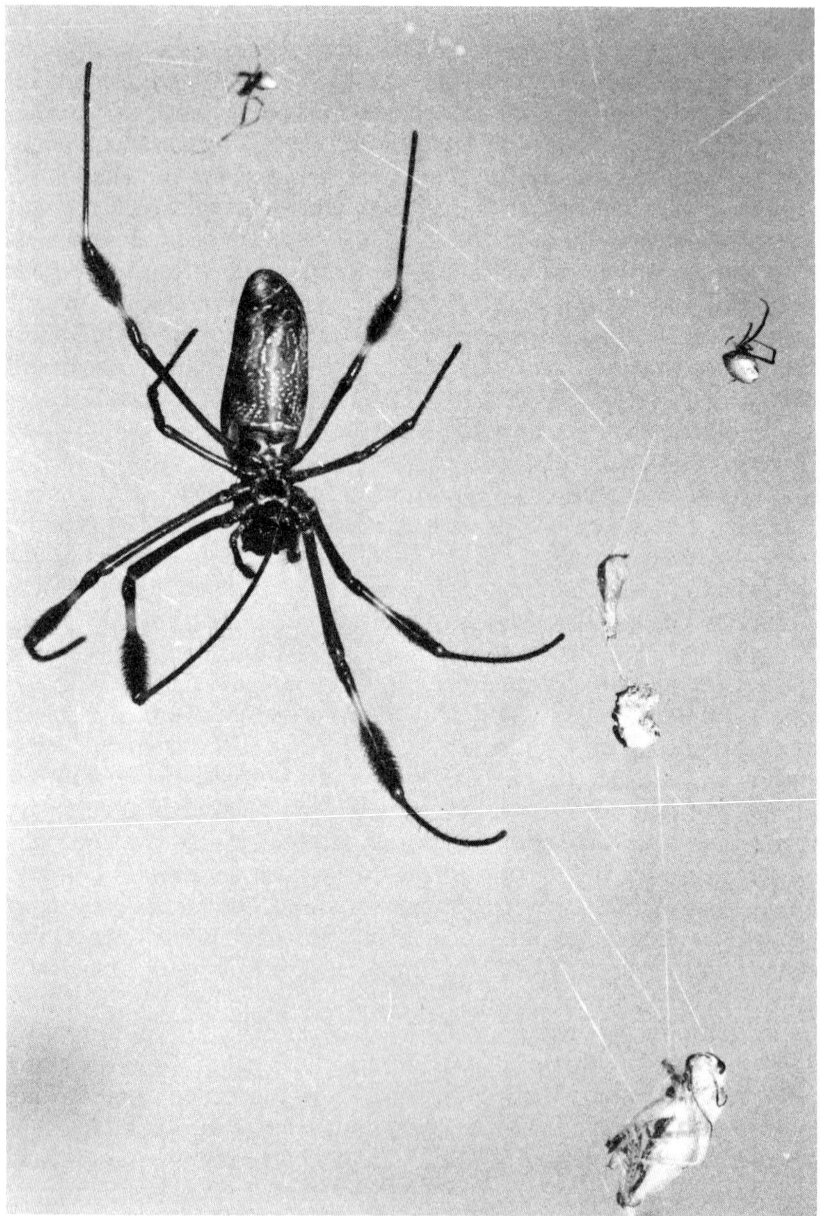

FIGURE 3.4. Two tiny kleptoparasitic spiders (*Argyrodes*) share the web of a female *Nephila clavipes*. (Courtesy of F. Vollrath.)

ilar behavior have been reported and await further analysis. Some of the most striking accounts describe various spiders (Mimetidae, Archaeidae) which emphasize fellow spiders in their diet (Bristowe, 1958; Gertsch, 1979; Legendre, 1961). Instead of using silk to trap prey, these spiders can "sneak" into the web of other spiders to catch and kill them. Again, a slow gait of the intruders is noted in the literature. *Ero furcata* even cuts an area of the web of *Theridion* free of threads without being identified. It then pulls the threads and mimics the vibrations of insect prey. The deceived fellow spider comes rushing toward the source of vibration, right into the reach of the predator's front legs, and is killed by the injection of venom. Obviously evolution has provided *Ero* with two sets of vibratory motor programs for just that purpose.

Vibrations of Solid Substrates and the Water Surface, and Airborne Stimuli

Reports on behavior guided by vibrations of media other than the web are scarce. Even among webs, however, very little can be said about a large number of types different from the orb-web. Often these webs are inhabited by small, inconspicuous, nocturnal spiders, which live their hidden lives in burrows or crevices without attracting much attention, even from arachnologists.

Many spider species either do not or only moderately rely on silk and webs; and some of them have given up the silken web or shelter completely. Instead they are vagabonds and are commonly called hunting, or wandering, spiders. Some of them, like the jumping spiders (Salticidae), rely to a large extent on their vision when catching prey and courting the mate. The majority, however, including the many nocturnal species, often depend on mechanical stimuli. (For a discussion of acoustic communication the reader is referred also to the chapter by Uetz and Stratton.)

DRUMMING AND STRIDULATION Male wolf spiders (Lycosidae) use their pedipalps to produce sounds during courtship (Allard, 1936; Davis, 1904; Lahee, 1904; Melchers, 1963); Melchers refers to *Cupiennius salei* as a ctenid spider. Some workers now place it among the lycosids (see Barth and Seyfarth, 1979; Homann, 1971). In descriptions of the partly percussive sound produced by *Lycosa gulosa*, this lycosid is often referred to as the "purring spider" (Kaston, 1936). Its sound, when generated on dry leaf litter, is audible to the human ear over distances of several meters (Harrison, 1969). Another wolf spider, *Lycosa rabida*, produces sounds with 0.5–3.5

kHz frequency pulse trains (produced on cardboard in the experiment and recorded with a contact microphone) lasting for about 2 sec and separated by intervals of roughly 14 sec (Rovner, 1967). With playback techniques Rovner was able to show that the female responds to the male sound with precopulatory display, even if it is emitted by a loudspeaker not touching the substratum. The same applies to male threat display, males responding to the airborne playback by threatening.

This is valuable evidence for the efficacy of airborne sound. Nevertheless, one should still beware of premature conclusions concerning the sensory pathway involved. One route of the playback sound to the receptors will have been through the air directly. The trichobothria in the near-field of the loudspeaker (suspended only 5 cm above the arena) are excellent candidates for such sound reception. Another possibility is an indirect route via the substratum which may well start vibrating when hit by airborne sound and in turn excite substrate vibration-sensitive receptors such as slit sensilla. There are, in fact, several observations in the account given by Rovner that point to the possible significance of substratum-conducted sound under natural conditions. Yet, these two possibilities in no way exclude each other.

For a long time sound production in lycosids has been referred to as drumming. This term still adequately applies to abdominal percussion as described for *Hygrolycosa rubrofasciata*. In this case a hard cuticular plate on the ventral surface of the abdomen is used for that particular drumming behavior (von Helversen, personal communication). A new aspect of so-called palpal drumming in lycosids, however, is the presence of a stridulatory file and scraper at the tibio-tarsal joint of the adult male (Rovner, 1975). The sounds produced by most lycosids are no longer considered as solely percussive. Instead, film analysis has shown for *Schizocosa* that the pedipalps are not used in a strict drumming fashion at all. The only movement seen is an anterior-posterior oscillation. In *Lycosa rabida* this is the predominant feature as well.

From a sensory point of view the two possibilities mentioned above still apply. On the one hand the moving pedipalp is a near-field sound source, as it is in other cases such as that of the male ctenizid spider *Cteniza moggridgei* which drums on the ground when approaching the female (Buchli, 1969). On the other hand, during stridulation the lycosid pedipalp typically is coupled to the substrate via stout spines. Its effectiveness as a source of airborne sound is thereby increased, and it simultaneously becomes a source of substratum vibrations to which female lycosids orient more promptly than to airborne sound (Rovner, 1975).

Stridulation is by no means restricted to the wolf spiders. Apart from their palpal type, there are at least seven other types involving various parts of the body in the formation of file and scraper (Legendre, 1963; see the chapter by Uetz and Stratton). In the number of species representatives, the Theraphosidae (Orthognatha) and the Theridiidae are the champions of the list of 22 families possessing stridulatory organs. Apart from Rovner's measurements, little is known about the behavioral and physiological implications of stridulation in these many cases. Exceptions are two theridiid spiders (Gwinner-Hanke, 1970): male *Steatoda bipunctata* and *Teutana grossa* both have a cephalothorax-abdomen-type stridulatory organ. They use it when stimulated by a chemical on the female web. In *Steatoda* the sound produced is a click with a frequency peak at 1 kHz and a duration of 23 msec. Its communicatory function is beyond doubt, since the female responds to it by searching movements of her front legs, body shaking, plucking the web, and cleaning. In addition to courtship, male *Steatoda* stridulate in agonistic behavior when another male is intruding into the web of the courting male and is chased away. In *Teutana* the behavioral significance of stridulation is still unclear. Female *Steatoda bipunctata* respond to stridulation even when the male is in a web 5 to 10 cm away. Again the answer to the possible routes taken by the signal is the one given before.

Summarizing our knowledge of spider stridulation, it would be inappropriate to connect it exclusively with either airborne or substrate-borne vibrations. Both the stimuli so far known and the sensory repertoire of the spiders do not permit an exclusion of either of these alternatives, and certainly more than one set of receptors may be involved simultaneously. A new type of sound production in spiders was first discovered in *Heteropoda venatoria* (Sparassidae) by Rovner (1980b). The courting male of this species produces humming noises, audible to the human ear up to at least 30 cm away, by vibrating its appendages (mainly leg 4) at frequencies often averaging 125 Hz. Coupling to the substrate by the tarsal claw tufts is thought to be an important feature. The female *Cupiennius salei* is courted similarly by abdominal and appendage vibrations, along with palpal drumming. Recent laboratory experiments have shown that a male courting on a banana plant (as in the wild) is identified by a hidden female at least 1 m away, even though the female is sitting on another leaf and the male's airborne sound is masked by the amplified output of a random-noise generator. Clearly substrate-borne signals are the important ones (Rovner and Barth, 1981).

PREY SIGNALS ON SOLID SUBSTRATES AND ON THE WATER SURFACE The diversity of the substrates carrying vibratory signals mirrors the diversity of spider ecology. Three examples are given to illustrate this.

1. *Cupiennius salei* (Lycosidae) is a hunting spider from Central America. At night it sits on the stem of banana plants and waits for a prey animal, which involuntarily signals its presence by vibrations (Barth and Seyfarth, 1979). Prey capture is elicited both by airborne stimuli (air currents) and by substrate vibrations. Although the corresponding sets of receptors are likely to be activated simultaneously in many a naturally occurring situation, each of them alone enables the spider to orient toward its victim (Barth and Rehner, unpubl.).

2. *Dolomedes* (Pisauridae). There are a number of spiders usually referred to as fishing spiders. They share the water surface with water striders and water bugs for prey and also prey upon aquatic animals such as tadpoles and small fish. Members of the genus *Dolomedes* are often seen resting on an object that is floating or projecting from the water; the spider's first and second pairs of legs contact the water surface. Alerted by insects that have fallen onto the water surface, they dart toward the source of surface ripples. No quantitative experiments are available, but observation of the behavior both in the field and in the laboratory strongly suggests an important role for vibration detection (Barth, unpubl.; Carico, 1973). Like *Dolomedes*, wolf spiders of the genus *Pirata* are often found with at least one tarsus of the first or second pair of legs resting on the water surface. They orient toward moving prey dummies even if separated from them by a glass jar, i.e., primarily visually. Vibrating prey, however, are bitten more frequently than non-vibrating prey (Gettmann, 1976).

3. *Argyroneta aquatica* (Argyronetidae). We know of only one spider that actually lives totally in the water. Being a land arthropod, *A. aquatica* accomplishes this remarkable behavior by taking a supply of air with it. Air is collected in a web, which like a tiny diving bell, keeps it from bubbling to the surface. From its underwater station the spider swims through the water to capture prey which hits the suspension threads of its bell. Basically this is a situation familiar from aerial-web spiders. An unusual aspect is the

effect of the surrounding water on the mechanics of signal transmission along the thread.

Defining Vibrations and Vibration Sense

The wealth of examples of spider behavior summarized in the previous pages implies a broad applicability of the term vibration and also points to the physical similarity of a number of mechanical events relevant to vibration detection. We speak of vibration detection if one of the following three types of stimuli is perceived by the animal (Markl, 1973).

1. The sensor is periodically moved by an object which touches it with varying pressure. Physically such vibrations, as all others, are described as sine waves, either pure (a laboratory stimulus) or as a mixture or spectrum of frequencies.

2. The second case is surface waves at the boundaries of media, occurring, for instance, between a solid substrate and air, or a body of water and air. Differing from the first category, the stimulus is here conducted and modified by a medium outside the animal before it reaches the sensory structures proper. Such surface waves are mainly known as Rayleigh waves and distinguished from the compressional waves found in the interior of a medium. Particle movement in a Rayleigh wave is in a plane perpendicular to the surface, whereas in a compressional wave it is in the direction of the vibrator motion. Also, in the same medium, propagation of a Rayleigh wave is much slower than of a compressional wave.

3. A third category is rhythmical vibrations of the medium in the near-field of a vibrating object. Of particular interest with respect to spiders and terrestrial arthropods in general are near-field vibrations in air. Obviously these three categories of vibrations come very close to other forms of stimulation not treated here. Category 1 is not far from touch, which is considered arhythmical, as opposed to the rhythmical vibrations. Category 3 brings us close to air current reception and hearing.

A vibrating object produces both pressure changes and particle displacements in the surrounding medium. Although these parameters are closely related to each other, there are biological (and technical) receptors clearly designed to measure essentially the pressure parameter and others essentially the displacement parameter. In the case of the latter receptor type, biologists have in recent years become increasingly aware of the necessity to apply the physicist's distinction between the near-field and the far-field of a sound source. (Sound is defined as compressional waves generated by a vibrating body and transmitted through an elastic medium. It can be airborne, fluid-borne, and solid-borne) (Markl, 1973). What is the difference?

In the far-field, particle displacement in air decreases as $1/r$ (r = distance from source), as does sound pressure. Particle oscillation becomes very small when $k{\cdot}r_0 < 1$ ($k = 2\pi/\lambda$; r_0 = radius of a spherical sound source). A sound source is very poorly emitting sound if its radius is smaller than about $\lambda/6$ (λ = wavelength). In the near-field, however, there is an additional component of rhythmical medium displacement which may reach high amplitudes well within the sensitivity range of displacement receivers such as spider trichobothria. The near-field, by definition, refers to conditions where $k{\cdot}r < 1$, or where distance under consideration is roughly $r < \lambda/6$. The far-field correspondingly applies to distances of $r > \lambda/6$. The additional displacement component found in the near-field lags behind the far-field displacement by $\pi/2$. The resulting overall displacement in the near-field is given as the vector sum $V_n = V_f/k{\cdot}r$. The important biological aspect is that particle displacement is not always linked to sound pressure; the amplitudes of near-field medium vibrations may be much higher than values calculated from the sound pressure under far-field conditions. The argument particularly applies to objects of small diameter and vibrating at low frequencies ($k{\cdot}r_0$ small; frequency $f < c/2\pi r_0$; c = propagation velocity). Near-field displacement decreases by $1/r^2$; therefore the distance of the receptor from the sound source is very critical (formula applies to monopole).

All sorts of vibrations can be described as a spectrum of sine waves. Since the parameters governing a sine wave are interrelated by a set of simple equations, one does not have to measure all of them to quantify the stimulus. Dealing with a mixture of frequencies, as is often the case, is much more complicated; and, in addition, in quite a few cases it is difficult merely to measure frequency and amplitude.

Depending on the receptor studied, different aspects of the stimulus may be of special interest: particle excursion amplitude (A) or its derivatives, velocity ($V = A{\cdot}\omega$; $\omega = 2\pi f$) and acceleration ($G = A{\cdot}\omega^2$); or sound pressure; or other features, such as temporal patterns of signal sequences, and amplitude and frequency modulation. In addition, the stimulus-conduction properties of the media carrying the stimuli to the receptors have to be taken into account: their damping effect, their resonance characteristics, etc. In a silken thread we expect longitudinal, transverse, and torsional vibrations simultaneously. A further complication, particularly pertinent to the water surface, is the frequency dependence on both the damping and the propagation velocity. Finally, the heterogeneity of natural material may introduce much complication.

The Receptors

Whenever we want to understand the mechanisms underlying the fascinating behavior of spiders, an examination of the sense organs involved must be an essential part of our effort. Sensory physiological experiments reduce or even end much confusion about the adequacy of a natural stimulus. Knowledge of the senses also enables us to observe animals from a specific sensory point of view, which is likely to reveal important aspects of their behavior otherwise not adequately appreciated or not even noticed.

We can expect *a priori* that the spiders' ability to localize and to identify vibrations is to a large extent reflected in the physiological properties of their sense organs. We are still a long way from a full neuroethological description even of bits of spider behavior. Above all, almost nothing is known about the physiology of the spider central nervous system. Our present knowledge of sense organs dealing with vibratory stimuli—although still incomplete as well—may illustrate the potential virtues of a more physically minded approach.

These sense organs are cuticular and are called trichobothria and slit sense organs. Chordotonal organs, such as those found in large numbers in both crustaceans and insects, are not known for spiders; internal joint receptors found in the spider leg are not very likely candidates (Foelix and Choms, 1979; Mill and Harris, 1977; Rathmayer, 1967).

Trichobothria

As the scientific name suggests, trichobothria are thin hairs (Gr. *trichos*, hair) articulated within and emerging from a conspicuous cup-like socket (Gr. *bothrium*, cup) (Figure 3.5). They are found only on the walking legs and pedipalps, and only on the tarsus, metatarsus, and tibia. It is easy to recognize trichobothria under the microscope or even with the naked eye: the slightest air current makes them vibrate. This has been known since the days of Dahl (1883). Typically trichobothria emerge from the dorsal aspect of the

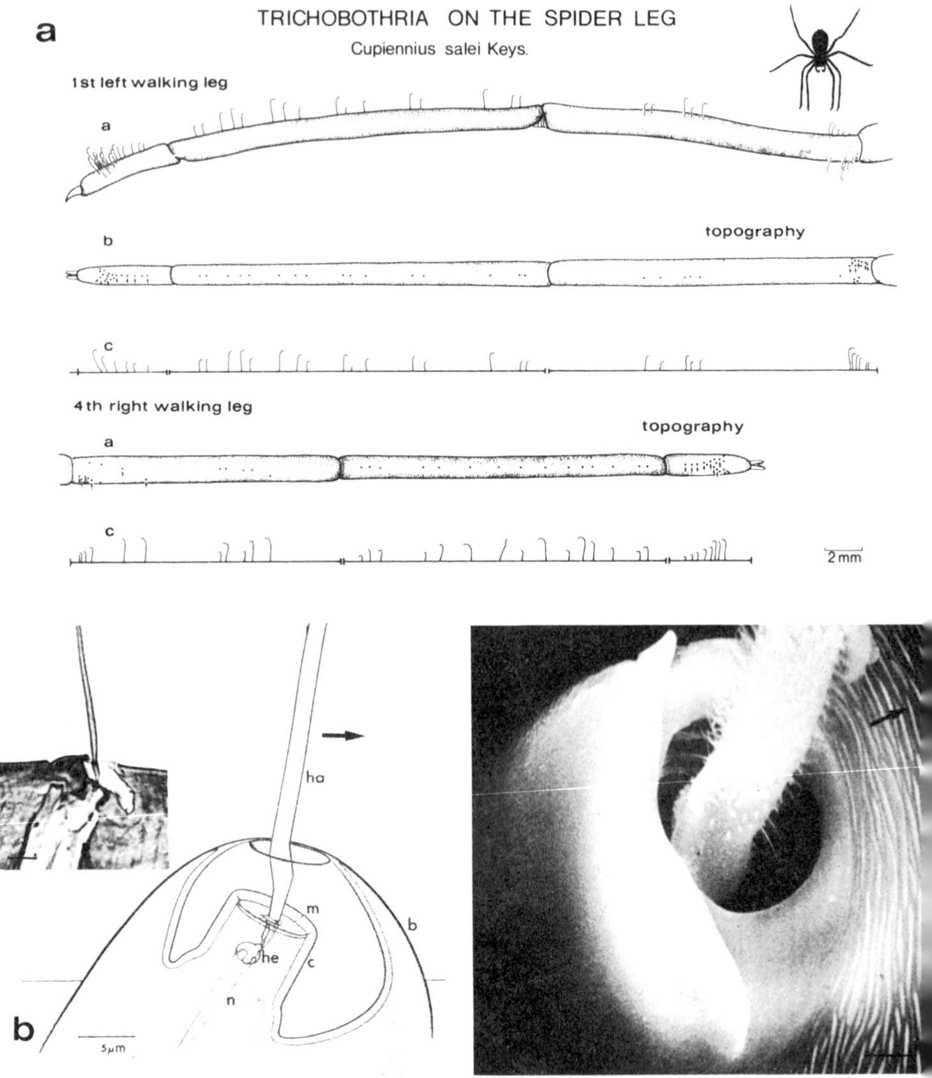

FIGURE 3.5a-c. Trichobothria on the spider leg. (a) Arrangement on the walking leg of *Cupiennius salei* (male, 8 months old) (Rehner and Barth, unpubl.). (b) Structure of hair base and dendritic attachment. Drawing: *Tegenaria*, ha, hair; b, bothrium; c, cuticular cylinder; m, membrane; he, helmet; n, dendrites (from Görner, 1965); inset: *Cupiennius salei*, scale 10 μm. (c) Base of trichobothrium, showing the cup and feathery structure of the hair (*Cupiennius salei* scale 2 μm). Arrow in (b) and (c) points distally.

extremity where they are most likely to be exposed to air currents (Figure 3.5). Only proximally on the tibia do they occur laterally as well. Often they are arranged in straight rows and form groups of up to 6 hairs with a conspicuous gradation in length.

The total number of trichobothria on an extremity varies considerably from species to species, as does their length. Thus *Agelena labyrinthica* has about 25 trichobothria per walking leg that vary in length from 0.1 to 0.7 mm (Görner and Andrews, 1969). The corresponding figures for *Cupiennius salei* (male, 8 months old) are 50 trichobothria ranging from 0.1 to 1.3 mm (Barth and Rehner, unpubl.). In a bird spider, or "tarantula" (*Sericopelma rubronitens*), they even reach 2.5 mm (Den Otter, 1974).

Due to its small mass, flexible articulation, and its exposure to the surrounding air, a trichobothrium is readily classified as a movement detector. It responds to airborne stimuli—not to sound pressure, however, but to particle displacement. Uniform airflow, air turbulences, rhythmically alternating air flow and the acoustic near-field, and particle movement in the acoustic far-field are all candidate stimuli.

The viscous forces of the moving air particles suffice to deflect the hair. It is taken along like a blade of grass in the wind. A rough surface improves its coupling to the air (Christian, 1971). Pressure can hardly develop, since the structure is so easily moved. Appreciable pressure gradients acting on the front and back surface of the hair cannot develop either, since its diameter is so small (in *Cupiennius salei* approximately 3 μm) as compared to the wavelength of the stimulus frequencies involved (e.g., $\lambda = 3.4$ m at 100 Hz). Hairs of this type are found in insects as well and are then called thread hairs or filiform hairs (Dumpert and Gnatzy, 1977; Gnatzy, 1976; Gnatzy and Schmidt, 1971; Markl and Tautz, 1975; Nicklaus, 1965, 1967; Tautz, 1977, 1978; etc.). They also occur in scorpions (Hoffmann, 1967; Linsenmair, 1968), myriapods (Haupt, 1970, 1978), and mites (Haupt and Coineau, 1975). All these hairs have basic features in common, although differences exist. The treatment of receptor properties given here largely refers to spiders and the work of Görner and his associates (Christian, 1971, 1972; Görner, 1965; Görner and Andrews, 1969; Reissland and Görner, 1978; Reissland, 1978), who have used *Agelena labyrinthica* and *Tegenaria* spp. as experimental animals.

DIRECTIONALITY The spider trichobothrium is mechanically nondirectional. It is bent into the various directions with equal ease (Reissland and Görner, 1978), in distinct contrast with the cockroach cercal hairs (Nicklaus, 1965, 1967), the curved thread hairs of the *Barathra* caterpillar (Tautz, 1977), and the scorpion trichobothrium (Hoffmann, 1967).

Mechanical isotropy coincides, however, with multiple innervation. The dendrites of at least two to four sensory cells insert on a helmet-like structure of the hair base (Christian, 1971, 1972; Görner, 1965; Harris and Mill, 1977a). They are stimulated effectively whenever the hair shaft moves from its resting position into the direction of the respective insertion area; then the dendrite is adequately deformed (Figures 3.5 and 3.6). The angular range of one sensory cell covers up to about 165°; all three ranges with some overlap cover the full circle of 360° except for small "dead" angles (Figure 3.6) (Görner, 1965). Within its range the response of a par-

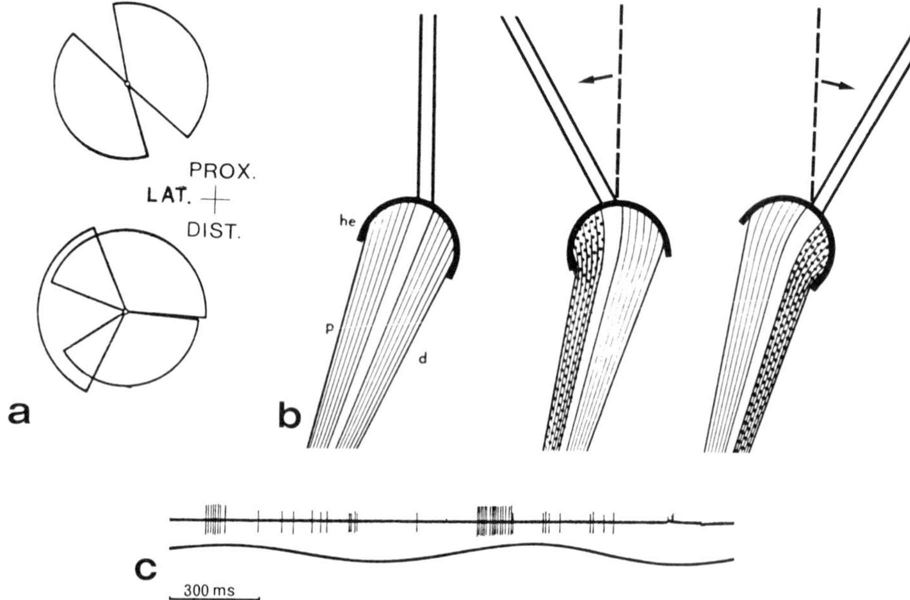

FIGURE 3.6a-c. Directional sensitivity of trichobothria. (a) Angular ranges of trichobothrium deflection effectively stimulating two (above) and three (below) sensory cells attached to two different trichobothria of *Tegenaria derhami* (from Görner, 1965). (b) The proposed mechanical basis for the differential directionality of the sensory cells of the same trichobothrium: dendrite deformation depends on direction of hair deflection due to the dendrite's asymmetrical attachment to the helmet (he helmet, p proximal dendrite, d distal dendrite; from Görner, 1965). (c) Impulse activity of two sensory cells of one trichobothrium (*Ciniflo* sp.) during sinusoidal deflection of hair; note difference in phase relationship which indicates difference in directional sensitivity (1 Hz; downward on stimulus trace is proximal) (from Harris and Mill, 1977a).

ticular receptor cell strongly depends on the azimuthal angle of hair deflection. This dependence is cosine-like, which implies that the response is much stronger with deflections in the middle part of its angular range than in its periphery (Reissland, 1978).

Trichobothria monitor change in hair position, i.e., the dynamic phase of a stimulus. There are several methods of extracting information on the direction of a stimulus from the nervous activity of displacement receivers like trichobothria. One extreme case would be to have a set of receptors, each with its own specific directional peak. Another extreme case would need only one sensillum, provided it were mechanically isotropic and supplied by several cells each responsive to hair deflection in a particular direction. Theoretically, then, the spider would need only one trichobothrium to tell at least roughly where the stimulus was coming from.

THRESHOLD SENSITIVITIES When a trichobothrium is deflected, the sensory cell covering that particular azimuthal angle responds with a series of nerve impulses. At angular deflection velocities of $10^{-2} \leqq \omega \leqq 1$ deg/msec the cell starts discharging at a hair deflection of roughly 1° to 3° from its resting position. With more rapid deflection ($\omega > 1$ deg/msec) this angle is larger (Reissland, 1978). When stimulating a trichobothrium with a small paper disk vibrating in air at 65 Hz close to the sensillum, the minimal value found is only 0.4° (Görner and Andrews, 1969). This last value closely approaches 0.1° found for the filiform hairs of the *Barathra* caterpillar (step stimuli; hair directly coupled to vibrator; Tautz, 1978). Clearly the length of a trichobothrium affects its absolute sensitivity, and the conspicuous length gradation found in a trichobothrium group reflects a corresponding physiological gradation of sensitivity (Figure 3.7).

FREQUENCY RANGE AND DISCRIMINATION It is tempting to assume that the striking differences in length among trichobothria of one group do not only result in different threshold sensitivities but also in different frequency tuning. Long hairs might be tuned to lower frequencies than short ones. This would indeed be an elegant way to provide a spider with frequency discrimination at the receptor level and on mechanical grounds.

The experiments in the literature are not in favor of this assumption as far as trichobothria are concerned. Between approximately 5 Hz and 2.5 kHz individual hairs do not show any obvious tuning in either the near- or far-field sound range. Taking sound particle velocity (mm/sec) as a measure, low frequencies (< 100 Hz) are

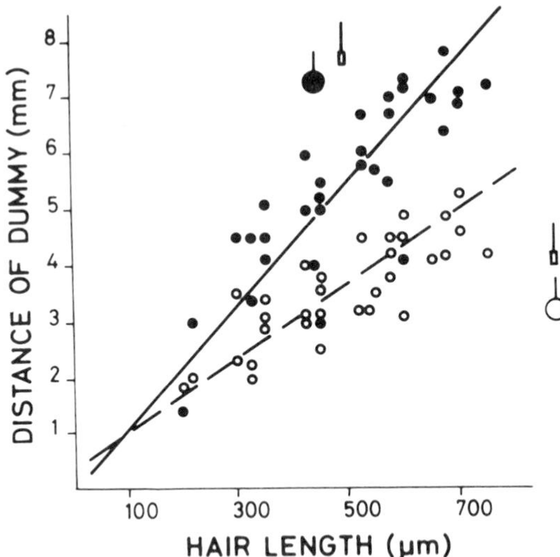

FIGURE 3.7. The effect of trichobothrium length on the hairs' mechanical sensitivity, measured as the distance of a small vibrating disk (65 Hz, ϕ 2.3 mm) at which hair starts being deflected. ○, disk above trichobothrium; ●, disk lateral to trichobothrium (from Görner and Andrews, 1969).

particularly effective. For a given hair deflection of, e.g., 1.64°, sound particle velocity must be four times as large at 100 Hz as at 5 Hz (Reissland and Görner, 1978). If one exposes a group of trichobothria to various sound frequencies, the potential of at least a rough frequency discrimination nevertheless emerges: the relative deflection amplitude distribution among the trichobothria varies with frequency. At 10 Hz the longest hair is farthest deflected; at higher frequencies the peak shifts to shorter hairs (Figure 3.8). Unfortunately, an electrophysiologically determined tuning curve is not available yet. The straight filiform hairs of the *Barathra* caterpillar behave physiologically as high pass filters with thresholds decreasing rapidly from 2.5° at 40 Hz to 0.5° at 150 Hz. Threshold stays constant at this low value up to 1 kHz, although at the same time a well-developed mechanical displacement resonance frequency is found at around 100 to 150 Hz (Markl and Tautz, 1975; Tautz, 1977, 1978).

HÖRHAARE? Ever since the first description of trichobothria by Dahl (1883, 1911) as "Hörhaare" and the report of Simon (1864) on the sensitivity of spiders to musical sounds, the question has been asked whether these sensilla provide a "real" sense of hearing (Chrysanthus, 1953; Legendre, 1963). The foregoing discussion on

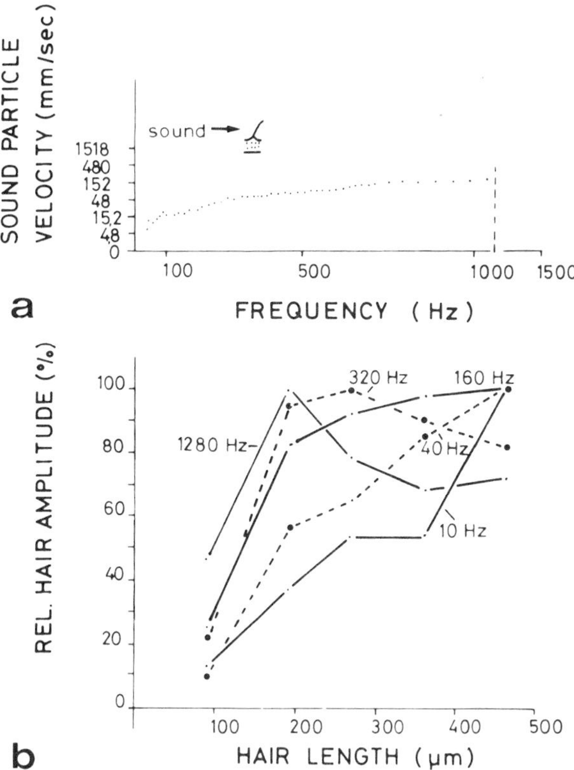

FIGURE 3.8a-b. Mechanical response of trichobothrium to far-field sound of varying frequency. (a) Loudspeaker was adjusted to give a constant hair deflection angle of 4.7° (pp). (b) Changes with various frequencies in the pattern of distribution of hair deflection amplitude (normalized to maximum amplitude found in the group) in a group of trichobothria differing in length (from Reissland and Görner, 1978).

the sound near- and far-fields and on the displacement receiver properties of the trichobothrium provides a rather straightforward answer.

Receptors sensitive to air currents will, if sensitive enough, also respond to rhythmical air particle movements in a sound field. It has been shown conclusively in behavioral studies that spider trichobothria are indeed sensitive to near-field medium vibrations (Görner and Andrews, 1969). Accordingly, the sensory capacity linked to the trichobothria was called "Ferntastsinn" or "touch at a distance" in analogy to comparable systems found in many aquatic animals (Dijkgraaf, 1934; Görner and Andrews, 1969; Harris and Bergeijk, 1962). Trichobothria will also respond to touch in the conventional sense of everyday language, but we have good reason to assume that they are not meant to deal with such signals biologically, at least not primarily.

On principle and due to the common physical basis, far-field particle displacement is a candidate stimulus for trichobothria as well, i.e., in terms of "Hörhaare." A threshold deflection of 1° (*Agelena, Tegenaria*), however, requires particle velocities in the sound far-field reached only with sound pressures as large as 90 dB SPL and more, even at low frequencies down to 50 Hz (Görner, personal communication; Reissland and Görner, 1978). (This value corresponds to the threshold pressure leading to a behavioral response such as front-leg extension in orb weavers (Araneidae) in response to airborne sound (Frings and Frings, 1966). The reaction of the spiders is attributed by these authors to "long, thin hairs" which cannot be the trichobothria, since they are said to occur on the body as well. The simultaneous stimulation of the trichobothria does not seem to have been excluded, however. Since the experiments were done in the acoustical near-field, and data on particle displacement are not given, the actual stimulus is hard to evaluate.) The experiments on the trichobothria of scorpions permit the same conclusion (Hoffmann, 1967). It is unlikely that a spider will encounter biologically meaningful sounds of that intensity other than in a laboratory. No case is known of a spider stridulating as loud as a cricket, which indeed emits sounds of approximately 90 dB SPL (peak at 4.5 kHz) measured at a distance of 5 cm (Nocke, 1971).

On the other hand, the term "touch at a distance" should not lead to the unjustified idea that the biological range of adequate input to the trichobothria is necessarily restricted to a stimulus source only a few millimeters away. There are observations indicating that the trichobothria may reach much farther than the maximum distance of 1 cm found with the dummy prey vibrating at 65 Hz above the leg of *Agelena* and *Tegenaria* (Görner and Andrews, 1969). We have watched *Cupiennius salei* jump toward flies passing by in flight and producing air currents at a distance of about 7 cm. We could also lure blinded spiders to flies not even flying or humming, but just crawling, over distances up to 20 cm (simultaneous substrate-borne vibrations excluded). The high sensitivity in *Cupiennius* is at least partly attributable to the large size of its trichobothria.

Slit Sensilla

Slit sensilla may be thought of as biological strain gauges. They belong to a class of receptors which measure strains (displacements) in the exoskeleton of arthropods. They are analogous to the campaniform sensilla of insects and crustaceans.

Basically all these sensilla form holes in the cuticle, each hole covered by a thin membrane. The dendrite of a sensory cell attaches to this membrane. In cross-section the hole varies from round to elliptical in campaniform sensilla and is markedly elongated in slit sensilla. The latter are from 5 to about 200 μm long and only about 2 μm wide (Figure 3.9). When the exoskeleton is mechanically loaded, the covering membrane is deformed due to the strains caused in the cuticle surrounding the receptors. As a consequence, the dendrite end-structure is deformed in turn. This deformation by compressional forces roughly perpendicular to the long axis of the slit finally leads to nervous activity (Figure 3.10) (Barth, 1971, 1972a,b, 1976).

Experiments with models as well as electrophysiological studies of the functional morphology of the slit sensillum have revealed an intimate relationship between structure and function (Barth, 1972a,b, 1976). In short, the stimulus-conducting structures are all specifically well adapted to transform their mechanical input into deformation, which is finally focused onto less than 1 μm^2 of dendrite.

The mechanical directionality of the slit and its orientation close to perpendicular to the compression lines in the loaded exoskeleton, the increased moment of bending of the covering membrane by its prebent shape, and the presence of a special coupling cylinder all add to the slit's easy deformability (Figure 3.10). As a consequence, slit sensilla are very sensitive: in a compound slit sense organ on the tibia, nervous impulses are elicited by sideways deflection of the metatarsus of only 0.01° (Barth and Bohnenberger, 1978); the metatarsal compound slit sense organ responds to vibrations of the tarsus as small as 10^{-7} cm at 1 kHz; finally, a single slit on the tarsus can be stimulated by an airborne sound having pressure as low as 40 dB SPL.

Slit sensilla are a peculiarity of arachnids, and spiders in particular are richly supplied with them. In *Cupiennius salei* we counted about 3,300 slits. They are all built into the exoskeleton at sites of particular mechanical "interest." The majority are found on the extremities (Barth, 1978; Barth and Libera, 1970; Barth and Stagl, 1976; Barth and Wadepuhl, 1975; Vogel, 1923). Here we also find the lyriform organs, which consist of a close parallel arrangement of up to 30 slits. From recent studies (Barth and Bohnenberger, 1978; Bohnenberger, 1978, and in prep.) we know that the main physiological features of such composite organs and their functional advantages over a single slit are a marked gradation of

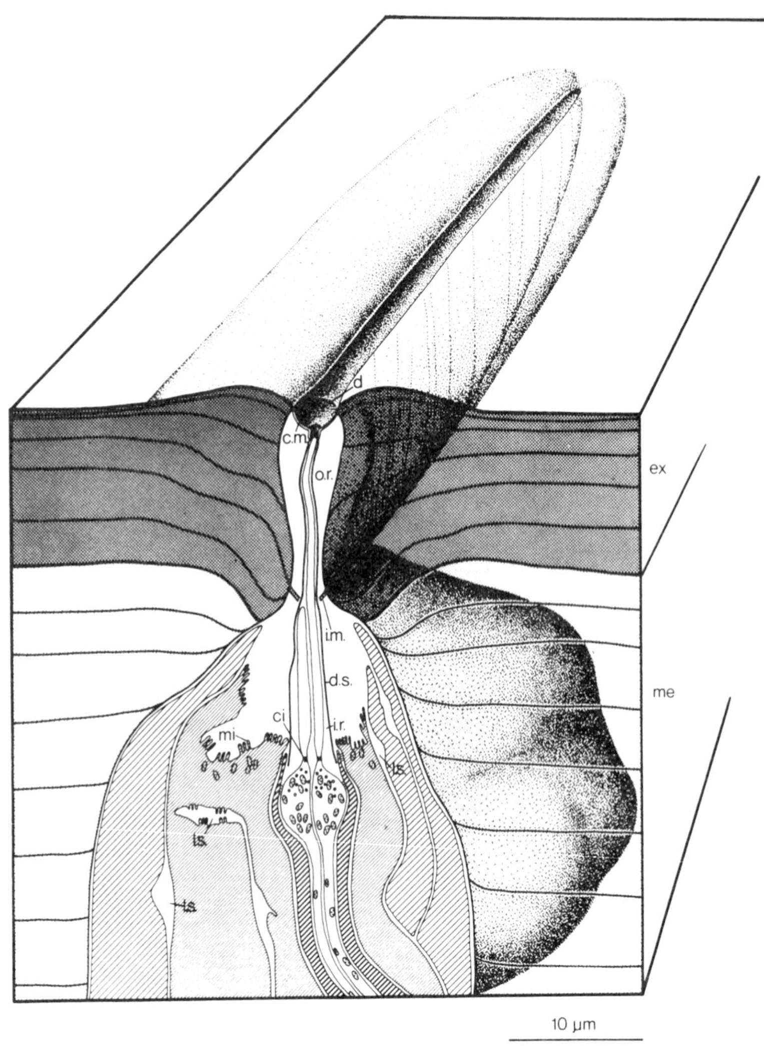

FIGURE 3.9. Fine structure of a spider slit sensillum as seen by a transverse cut through the cuticle. c.m., covering membrane; i.m., inner membrane; d, depression of c.m. around dendrite attachment; d.s., dendrite sheath surrounding two dendrites, one of which attaches to c.m., whereas the other one ends close to i.m.; ci, ciliary region of dendrites distal to mitochondrion-rich swelling; i.r. and o.r., inner and outer receptor lymph space; mi, microvilli of auxiliary cells; l.s., extracellular lacuna system between sheath cells; ex, exocuticle; me, mesocuticle (from Barth, 1976).

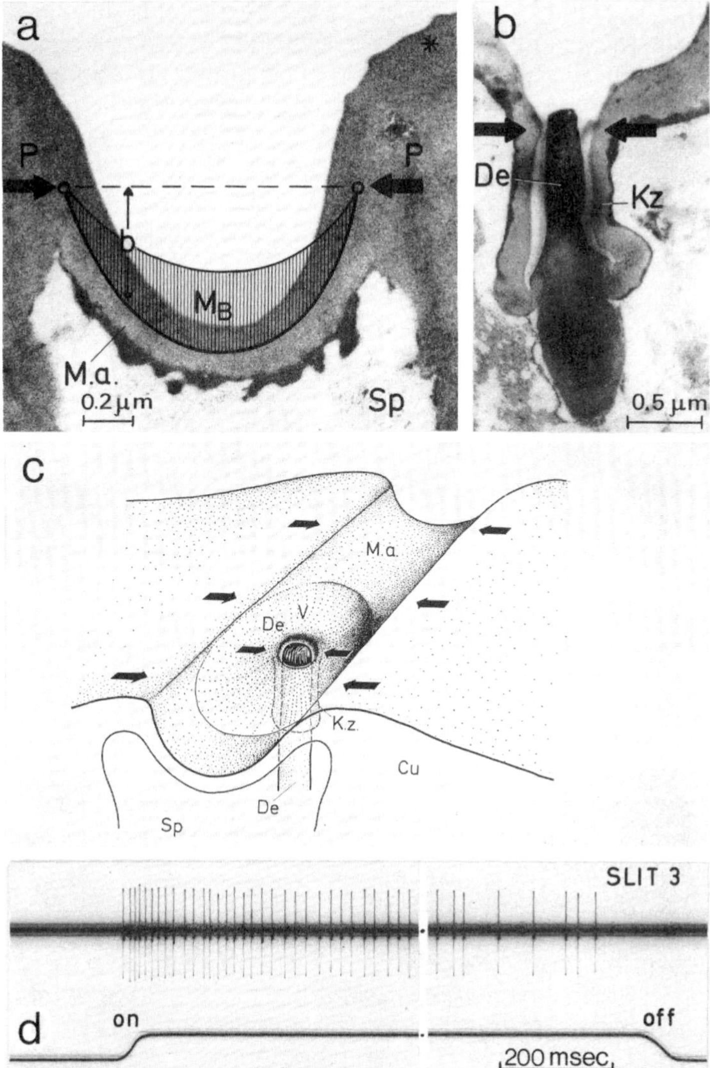

FIGURE 3.10a-d. Stimulus conduction and adequate deformation in the slit sensillum. (a) Covering membrane (M.a) of the slit with its pronounced curvature; M_B, moment of bending resulting from forces (P), which compress the slit. (b) Coupling cylinder (Kz) of the covering membrane into which the dendritic end (De) projects; arrows indicate deformation of dendrite by monoaxial compression forces. (c) Displacements taking place under adequate stimulation. (d) Spike response recorded from a slit of a tibial lyriform organ during compression (lower beam); interruption of recording 0.26 sec (from Barth, 1976).

threshold sensitivity, and an extension of the range of high-increment sensitivity and linearity of response.

It is among the lyriform organs that we find a vibration receptor, the metatarsal lyriform organ. A single slit behind the tarsal claw is vibration sensitive as well, and another single slit on the tarsus even responds to airborne sound (far-field). The overwhelming majority of slits, however, are believed to respond to cuticular strains (displacements) resulting from muscular activity, body weight, and hemolymph pressure. According to behavioral studies and electrophysiological data, we must nevertheless be prepared to find vibration sensitivity in still more cases.

THE METATARSAL LYRIFORM ORGAN *Topography and stimulation.* The metatarsal lyriform vibration receptor is found dorsally on the distal end of the metatarsus and with its slits oriented roughly at a right angle to the long leg axis (Figure 3.11). Both of these topographical features are exceptional among lyriform organs and of relevance with respect to the organ's sensitivity to substrate-borne vibrations.

In *Cupiennius*, movements of the tarsus both in the dorso-ventral and in the lateral direction result in a compression of slits as soon as the proximal part of the tarsus slightly lifts the distal part of the metatarsus (Figure 3.11). A discharge of nervous activity follows. The topography of the organ exposes the slits to compressional forces roughly perpendicular to their long axis. This coincides with the direction most effectively deforming them. A deep furrow on both its lateral sides also enhances the organ's deformability due to increased mechanical lability and a preferred transfer of compressional forces from the tarsus to the metatarsus at the site of the lyriform organ (Barth, 1972a,b). In *Cupiennius salei* the metatarsal lyriform organ consists of 21 slits (length 20 to 120 μm) (Barth, 1971). In *Achaearanea tepidariorum* it is made up of only 8 to 10 slits (Walcott and van der Kloot, 1959); in *Salticus scenicus* there are 11 slits; in *Zygiella x-notata* 20; in *Tegenaria larva* 16; and in *Nephila clavipes* 20 slits (van de Roemer and Barth, in prep.).

Tuning Curves. The metatarsal lyriform organ is very sensitive to solid-borne vibrations. Three electrophysiological studies are available to show this.

1. *Achaearanea tepidariorum.* Threshold vibration amplitudes rapidly decrease between 0.1 and 1 kHz and are minimal at around 2 kHz, where they measure only 2.5×10^{-7} cm. They increase again beyond 10 kHz (Walcott and van der Kloot, 1959).

2. *Zygiella x-notata.* A constant threshold displacement of the

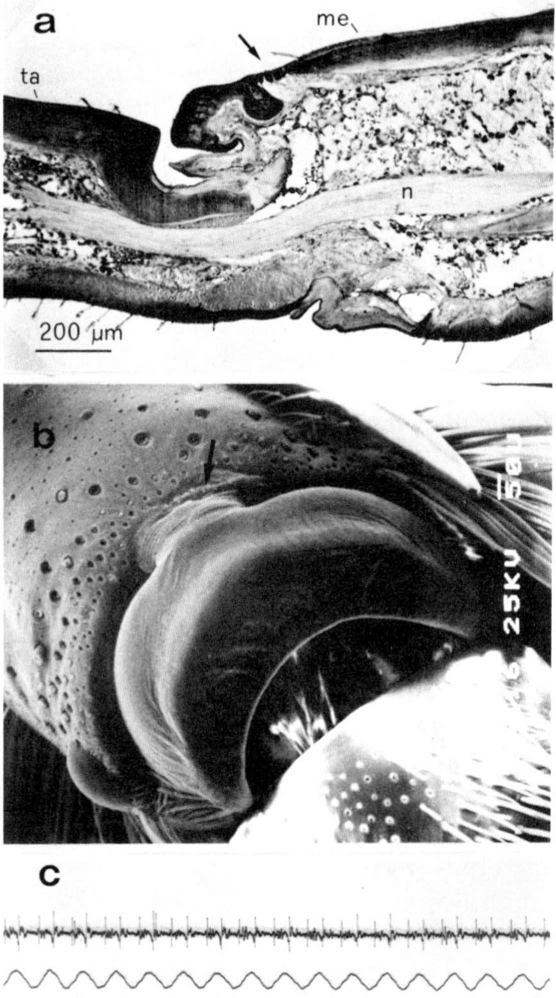

FIGURE 3.11a-c. The metatarsal lyriform organ (*Cupiennius salei*). (a) Longitudinal section through walking leg: ta, tarsus; me, metatarsus; n, nerve; arrow points to lyriform organ (from Barth, 1972b). (b) SEM showing location of the organ (arrow) dorsally on the metatarsus and furrow on both of its lateral sides. (c) Recording of impulse activity from one of the slits by vibration of the tarsus at 40 Hz and with an amplitude of 22 µm (pp, peak to peak dorso-ventral movement) (from Geetha Bali and Barth, in prep.).

tarsus of about 10^{-3} cm was found between 10 and 100 Hz. Thresholds decrease continuously at higher frequencies down to 10^{-7} cm at 5 kHz (*Tegenaria* spp.: between 0.6 and 4 kHz) (Liesenfeld, 1961).

3. *Cupiennius salei.* We have recently taken up the matter again and tested the sensitivity and tuning of the various slits of the metatarsal lyriform organ by recording from them for the first time individually (Geetha Bali and Barth, in prep.). Figure 3.12 presents the result. (a) With the tarsus tip firmly coupled to the vibrator and movement in the dorso-ventral plane, threshold deflection amplitudes amount to about 5×10^{-3} cm from 0.1 to 20 Hz; they decrease rapidly at higher frequencies down to 5×10^{-7} cm at 1 kHz. (b) If plotted as acceleration, the threshold stimuli instead indicate a high sensitivity at low frequencies. (c) There is no sharp tuning within the frequency ranges tested, and the tuning curves of different slits are remarkably similar. The relative thresholds in a group, however, vary slightly with frequency, which may be a basis for at least a crude frequency discrimination (cf. trichobothria, discussed earlier). (d) Essentially the same data apply to a tarsus standing loosely on the vibrating support, i.e., to the natural situation. (e) Sideways deflection of the tarsus is as effective as dorso-ventral deflection at least for some of the slits. As a consequence the organ detects both longitudinal and transverse waves in the vibrating medium. (f) Despite the differences in the techniques applied and the species used (hunting spider), a strong resemblance to the results obtained by Liesenfeld (1961) from nerve recordings with a web spider is evident. This also applies to the findings of Walcott and van der Kloot (1959) if restricted to the response to direct vibrations and frequencies up to 1 kHz.

Although both the electrophysiological findings and the functional morphology identify the metatarsal lyriform organ as a vibration receptor, it is clearly not the only vibration receptor available to a spider. In addition, the organ is good for more than vibration detection alone, since the majority of its slits are rather slow to adapt and may be important in monitoring tarsal movement during locomotion.

Very little is known about the vibration sensitivity of interneurons in the central nervous system of the spider. Summed responses (evoked potentials) in the ventral nerve mass of *Araneus diadematus* to tarsal vibrations were highest at around 100 Hz (Finck, 1972). The sensitivity of the metatarsal organ to airborne sound is a controversial matter. Particularly in the near-field the tarsus may be moved by airborne sound and thus excite the metatarsal organ. The most spectacular findings of Walcott and van der Kloot (1959),

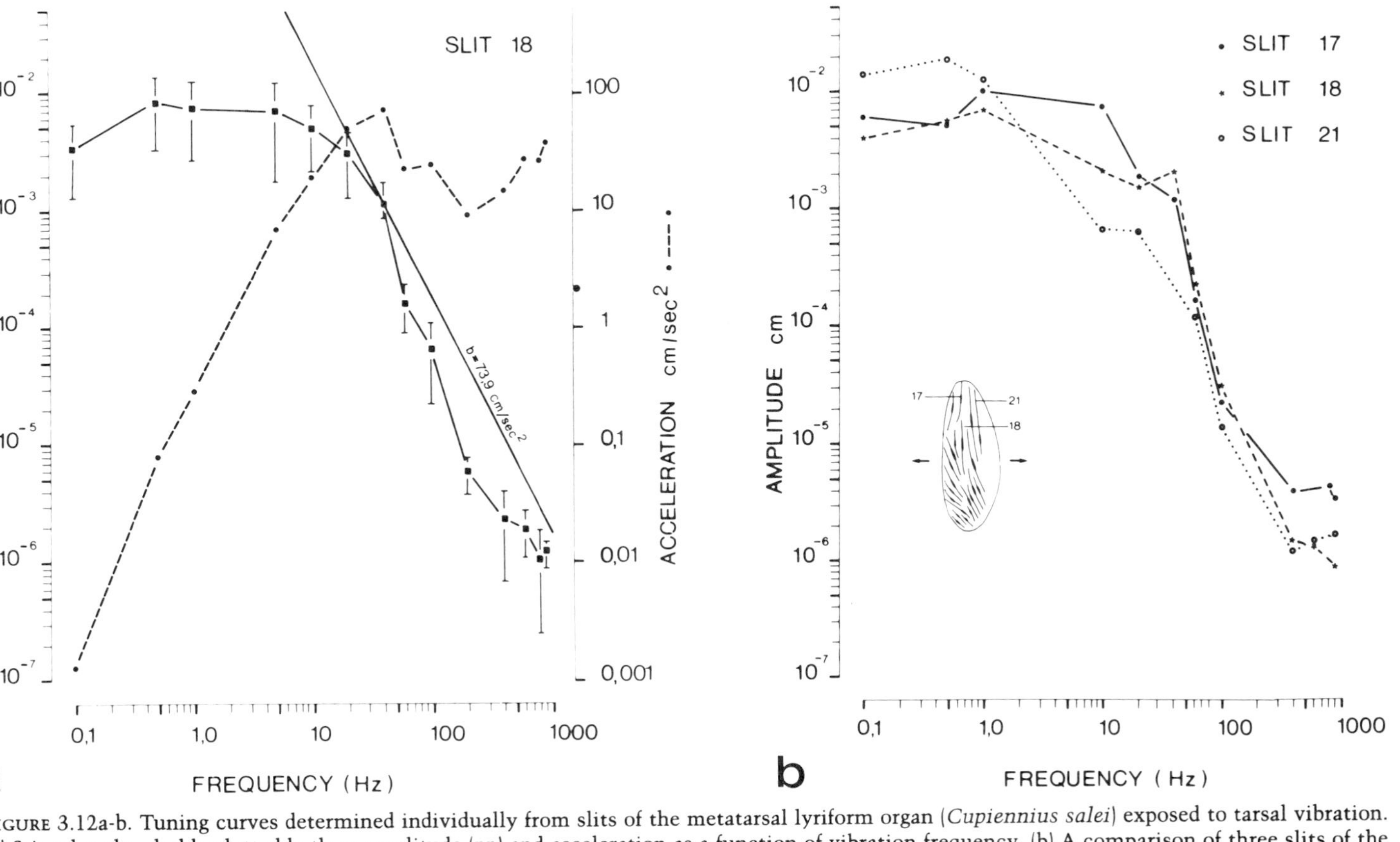

FIGURE 3.12a-b. Tuning curves determined individually from slits of the metatarsal lyriform organ (*Cupiennius salei*) exposed to tarsal vibration. (a) Stimulus thresholds plotted both as amplitude (pp) and acceleration as a function of vibration frequency. (b) A comparison of three slits of the same organ (from Geetha Bali and Barth, in prep.).

however, namely, an outstandingly sharp tuning of the units to various frequencies between 0.1 and 1.0 kHz, with an attentuation slope of up to 25 dB/100 Hz, are very doubtful in their significance. They were attributed by Liesenfeld (1961) to uncontrolled vibrations in the set-up. There are a number of puzzling contradictions in the paper of Walcott and van der Kloot, and it seems indeed likely that these remarkable peaks found with airborne sound are artifacts. They were not found when the tarsus rested on a web strand, which is the situation of interest in the present context.

THE CLAW SLIT AND A TARSAL SLIT SENSITIVE TO AIRBORNE SOUND One set of vibration-sensitive organs identified in addition to the metatarsal lyriform organ is the pair of single slits found in the cuticular fold behind the tarsal claws, one slit on each side of the tarsus (Figure 3.13). Although they are less sensitive to solid-borne vibrations than the metatarsal lyriform organ in *Cupiennius*, these slits are nevertheless well suited to measure vibrations (Speck and Barth, in prep.).

Another rather large single slit (length approximately 50 μm) on the anterior aspect of the tarsus is sensitive to airborne sound in the far-field; i.e., it is a "hearing organ" in the classical sense. This again was shown in *Cupiennius salei* (Figures 3.13 and 3.14). The organ could be effectively stimulated between 0.1 and 2.5 kHz with sound pressures ranging from about 50 dB to 80 dB SPL (0.063 to 2.00 μbar); its maximum responsiveness was between 0.3 and 0.7 kHz. The lowest threshold values found were about 40 dB SPL (Barth, 1967).

The high sensitivity of this tarsal slit can best be appreciated if we remember that similar values for threshold sound pressures are also known for tympanal organs, the classical organs of hearing in insects, in their optimal frequency range. According to the electrophysiological analysis, the response of the organ is a linear function of the logarithm of sound pressure over a range of about 15 dB above threshold. It saturates at about 20 dB above threshold. The impulse frequency recorded during continuous sound stimulation declines to zero about 10 to 20 sec after stimulus onset, the exact time depending on the frequency and sound pressure applied.

Natural Signals, Receptors, and Behavior

Three questions are the guidelines to the following pages. (1) What are the physical properties of naturally occurring vibratory

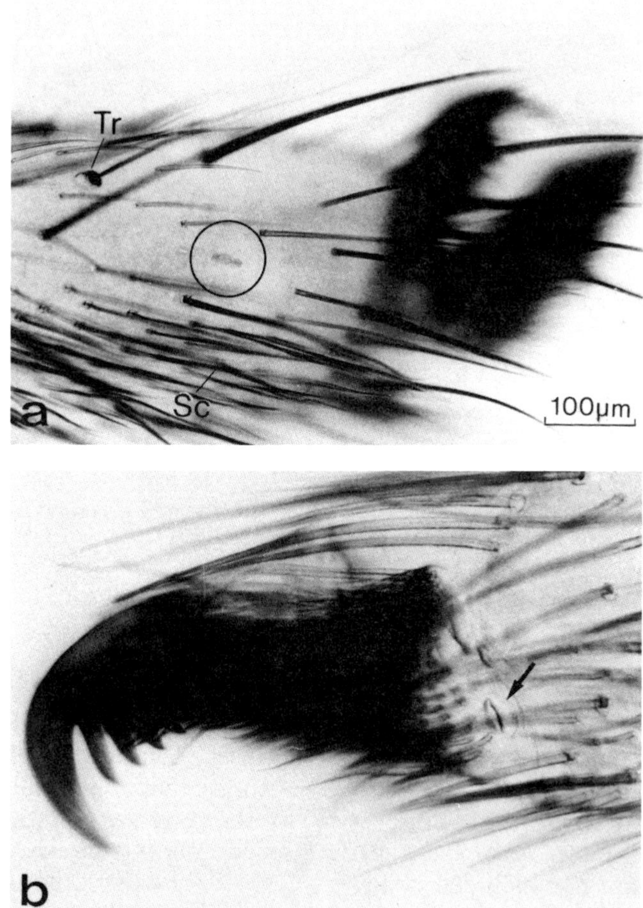

FIGURE 3.13a-b. Single slits involved in airborne sound and vibration detection. (a) Single tarsal slit (circle) sensitive to airborne sound (*Cupiennius salei*). Tr, trichobothrium; Sc, scopula hair (from Barth, 1971). (b) Single slit in cuticular fold behind claw, which is sensitive to substrate-borne vibrations (*Cupiennius salei*).

signals? (2) How do the signal properties relate to the receptor properties? (3) Can we relate specific sense organs to behavioral patterns directly?

We do not know the complete answers to these questions yet. The knowledge at hand, however, suffices to give a good idea of what we still need to know. It also justifies a first effort to characterize at least a few important interrelationships among the signals, the receptors, and behavior as parts of one system.

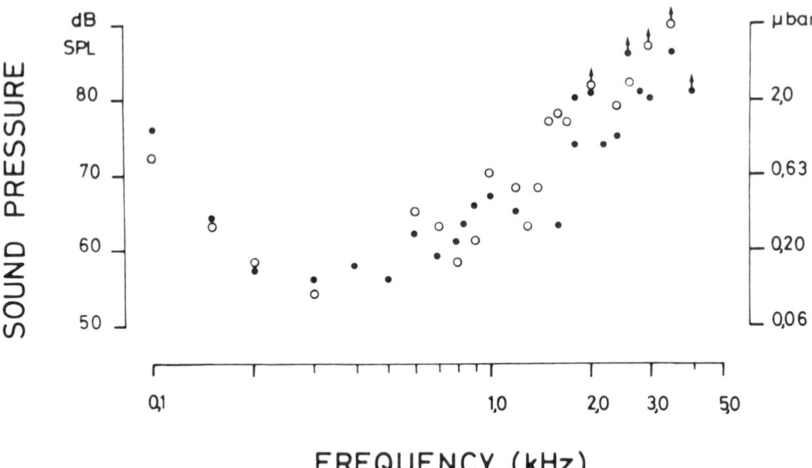

FIGURE 3.14. Tuning curve of the tarsal single slit (Figure 3.13a) when stimulated with airborne sound. Threshold stimuli given both as dB SPL and μbar (from Barth, 1967).

SIGNALS

FREQUENCY RANGES Table 3.1 summarizes the data available on the frequency ranges of naturally occurring, biologically relevant signals. Its main purpose is to give representative examples. In a number of cases it was not possible for the reviewer to extract unequivocal figures from the various experiments published. Such studies are not included in the list. (The data worked out by Walcott (1963) are not quoted since the author mentions that the vibration transducer was not calibrated and therefore the absolute form of the curves obtained for insect vibrations probably is not significant. Parry (1965) determined peaks of vibrations produced by cockroaches in the web of *Tegenaria* with components at 10 Hz and 100 Hz and fast transients. No further frequency analysis was done.)

In general there is a prevalence of low frequencies. With the exception of palpal stridulation of *Lycosa rabida* (Rovner, 1967, 1975), all signals are near or below 1 kHz. The airborne, web-borne, and solid substrate-borne vibrations of various sources were recently shown to have frequency peaks below 250 Hz and even down to a few Hz (Figure 3.15).

TABLE 3.1. Signals relevant to spider vibration detection and communication.

Type of signal	*Medium*	*Species emitting*	*Species receiving*	*Frequency (peaks, range)*	*Comment*	*Author*
A. *Courtship*						
1. palpal stridulation	cardboard	*Lycosa rabida* ♂	*Lycosa r.* ♀	0.5-3.5 kHz sonagram	A 2.2-sec train made up of pulses; 3-6 msec duration, 29/sec.	Rovner (1967, 1975)
2. abdominal stridulation	air	*Steatoda bipunctata* ♂	*Steatoda b.* ♀	0.7-1.4 kHz sonagram	Stridulation by movement of abdomen against file on cephalothorax.	Gwinner-Hanke (1970)
3. appendage vibration	air/substratum	*Heteropoda venatoria* ♂	*Heteropoda v.* ♀	80-150 Hz oscillogram	Humming sound mainly produced by vibrating posterior legs. As in lycosid stridulation, coupling to substratum is considered important.	Rovner (1980b)
4. web vibrations	orb web	*Zygiella x-notata* ♂	*Zygiella x-n.* ♀	10-14 Hz optical method, limited since thread was loaded	Values give transverse vibrations of courtship thread.	Liesenfeld (1956)
5. web vibrations	web	*Coelotes terrestris* ♂ *Tegenaria parietina* ♂ *Amaurobius* spp. ♂	*Coelotes t.* ♀ *Tegenaria p.* ♀ *Amaurobius* spp. ♀	8-150 Hz piezoelectric pick-up, oscillogram; amount of web loading not given	Frequencies are repetition frequencies of palpal, leg, and abdominal movements.	Krafft (1978)

TABLE 3.1 (*cont.*)

Type of signal	*Medium*	*Species emitting*	*Species receiving*	*Frequency (peaks, range)*	*Comment*	*Author*
B. *Agonistic*						
1. palpal stridulation	cardboard	*Lycosa rabida* ♂	*Lycosa r.* ♂	0.5-3.5 kHz	During vigorous threat display, long continuous series (up to 10 sec) of short bursts of pulses.	Rovner (1967)
C. *Prey*						
1. web vibrations	sheet web	flies	*Mallos gregalis*	30-700 Hz sonagram frequency spectrum	Web described as a mechanical filter which enhances prey signals but suppresses signals emitted by conspecifics; social spider.	Burgess (1979b)
2. web vibrations	orb web	*Ephestia kuehniella, Musca domestica*	*Zygiella x-notata*	3.5 and 30 Hz (Ephestia); 6, 20, and 230 Hz (*Musca*); optical measurement without loading web; line spectrum	Given are the frequency peaks of transverse vibrations ($\geqq 0.1$ μm) of the signal thread; they can be attributed to various types of movement of the entangled prey.	Graeser and Markl (in prep.)
3. web vibrations	orb web	*Agallia novella, Sitotraga cerealella, Halic-*	*Cyclosa turbinata*	5-250 Hz; optical measurement; load on web only	Measurement of hub vibration; the individual peaks found vary	Suter (1978)

4. plant material	agave leaf	cockroach	*Cupiennius salei*	5-30 Hz; power spectrum	Vibrations produced by crawling cockroach.	Barth, Seyfarth, Bohnenberger (unpubl.)
5. plastic	sheet of plastic	*Calliphora erythrocephala*	*Cupiennius salei*	2-50 Hz; power spectrum	Vibrations produced by crawling fly to which spider responds most readily with attack.	Barth and Bohnenberger (unpubl.)
6. water surface waves	water surface	*Apis mellifica*, *Chrysopa* sp., *Lucilia* sp.	*Dolomedes* spp., *Pirata* spp.	10 Hz (*Apis m.*) 28 Hz (*Chrysopa* sp.) 43 Hz (*Lucilia* sp.)	Surface waves produced by potential prey. Given are the marked peaks of the frequency spectra which contain frequencies from ca. 5 Hz to more than 150 Hz.	Lang (1979)
7. airborne	air	*Drosophila*, *Musca*	Agelenidae	160-500 Hz fourier spectrum	Sound particle velocity diminished by $\geqq 31$ dB ($\times 0.03$) from third harmonic on, with regard to fundamental oscillation (185 Hz in *Dros.*, 163 Hz in *Mus.*)	Reissland and Görner (1978)

SIGNAL SPREAD The spread of a signal much depends on the material conducting it. Signal attenuation and the sensitivity of the receptors determine the distance at which a spider can detect a vibratory stimulus. There is a large range of damping coefficients, and examples pertinent to the spiders are compiled in Table 3.2 (see also Markl, 1973). Natural materials like plant leaves and wood are generally complex structures due to their mechanical heterogeneity. Although rather incomplete in that respect, Table 3.2 gives at least a rough quantitative idea of the values to be expected under various conditions. When propagated through these media, the signal is attenuated due to geometric spread and a frequency-dependent absorption and scatter. In air these effects are comparatively small in the far-field, but they become more important in the near-field, where particle displacement decreases with $1/r^2$ (r = distance) instead of $1/r$ (far-field). High frequencies are much more attenuated than low ones; and, as seen from the given data, this is relevant within the frequency range of biological interest.

Again the silken web is of particular interest. Nevertheless, there exists only one study on the signal spread in a spider web done in some detail and with adequate methods (Graeser and Markl, in prep.). This study examines transverse waves conducted by the signal thread of the web of *Zygiella x-notata* (Figure 3.16). The best way of vibration measurement avoids any contact between the transducer and the web, and any other form of its mechanical loading. Graeser and Markl have applied such a method in the first and so far only case. Since webs are delicate structures with very little mass (web of *Zygiella x-notata*: 0.5 to 1.2 mg; same authors), all transducers applying a load to the web are of limited value only. The effect on the resonance frequency (-ies) is considerable (Figure 3.17) and such loads may even interfere severely with the measurement of vibrations in webs naturally loaded by entangled prey or a conspecific. The seriousness of the error depends on the question asked. The potential for error is greatest if one wants to measure the transfer characteristics of an unloaded web. It may well be serious if power spectra of natural signals are the aim. It may be tolerable if only the repetition frequencies of plucking and bouncing movements are of interest. In the case of the particularly heavy and tough communal web of *Mallos gregalis* (Burgess, 1979b) the error probably is rather small. The same is true for the experiments of Suter (1978): the additional mass (0.2 mg) applied to the web for measurement of prey signals (power spectra) was small both absolutely and—more importantly—relative to the mass of the prey insects studied (0.7 to 3.8 mg), and well within its range of variation.

Figure 3.18 shows what the signal thread detects when the web is vibrated at different places with frequencies varying between 20 and 160 Hz. A vibratory signal of 20 Hz attenuates on its way from the far end of the web to the signal thread at the spider's retreat to less than a 1/200th. The web is a low-pass filter for transverse waves but not tuned to a specific frequency within the range tested. The gain of the system only moderately decreases between 2 Hz and 100 Hz and even rises slightly between 100 Hz and 300 Hz; it then decreases rapidly at still higher frequencies. The transfer characteristics of the web are rather independent of the site of the vibration source.

In view of the high damping effect, the web presents itself not so much as a signal transmitter of great sensitivity (at least for transverse waves), but rather as a trap, which is soft and compliant enough not to be greatly damaged by high amplitude prey vibrations. The resonance of such a web varies greatly with the load it carries (Figure 3.17).

Signals and Receptors

Combining our knowledge of receptor properties and signal characteristics now permits some generalizations about the biological significance of various potential stimuli in behavior.

NEAR-FIELD VIBRATION Apart from being very sensitive to air currents in the ordinary sense, trichobothria are excellent near-field particle displacement receivers. Near-field vibration is significant as long as $r < \lambda/2\pi$ (r = distance from sound source; λ = wavelength) and as long as $r_0 < \lambda/2\pi$ (r_0 = radius of vibrating sphere). Natural signal sources relevant to trichobothria as near-field sources therefore have to be looked for primarily among objects (a) small relative to the wavelength emitted and too small to produce a relevant far-field component (i.e., vibrating at frequencies that are low relative to their size; $f < c/2\pi r_0$, where c is the propagation velocity) and (b) vibrating with relatively high amplitudes (Markl, 1973).

The actual range over which a spider can potentially detect near-field vibrations of the medium depends on the sensitivity of its detectors, in addition to details of the vibration source, such as its exact shape. Humming flies, prey struggling in the web, bouncing spider abdomens, and vibrating palps certainly are all among the near-field sources of interest to a spider. None of them has been properly studied in that role.

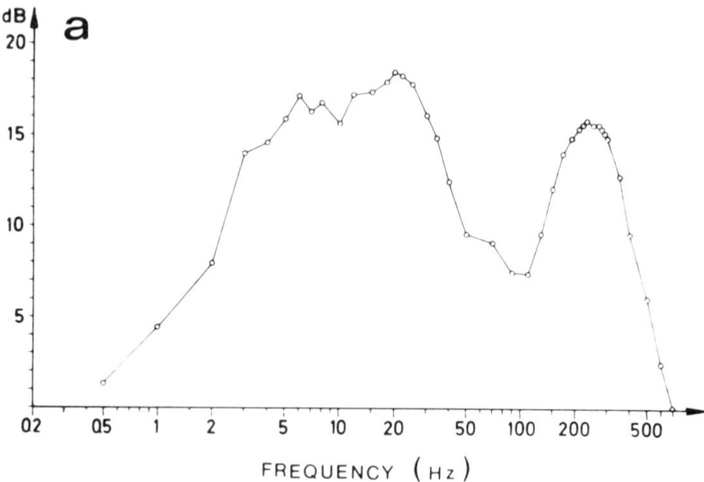

FIGURE 3.15a-c. Prey vibrations. (a) Transverse vibrations of the signal thread by a fly (*Musca domestica*) entangled in the web of *Zygiella x-notata*; line spectrum (from Graeser and Markl, in prep.). (b) and (b′) Prey vibrations in the web of *Cyclosa turbinata*. Averaged power spectra showing lower (b) and higher (b′) frequency contents. Numbers on right give sample size (from Suter, 1978). (c) Solid-borne vibrations which promptly elicit prey-catching behavior in *Cupiennius salei* and which were produced by a cockroach crawling on an agave. Left: oscillograms; right: power spectra (Barth, Seyfarth, and Bohnenberger, unpubl.).

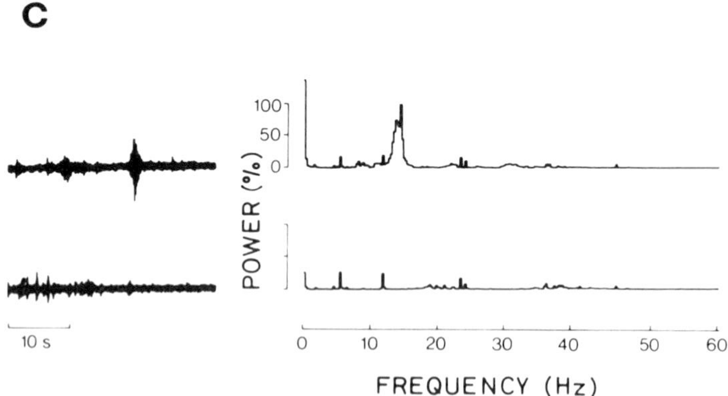

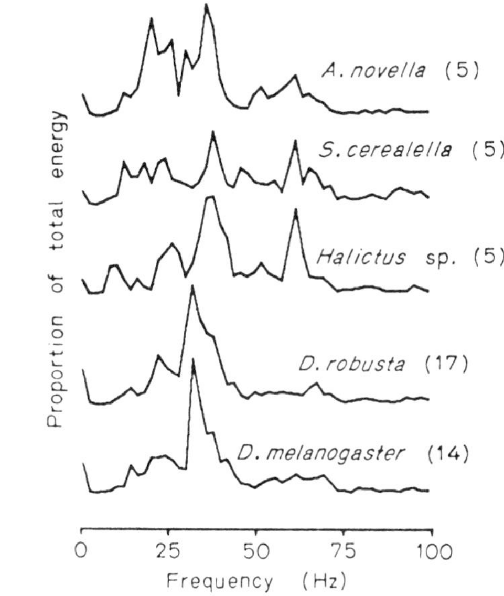
A. novella (5)
S. cerealella (5)
Halictus sp. (5)
D. robusta (17)
D. melanogaster (14)
Proportion of total energy
0
25
50
75
100
Frequency (Hz)
b

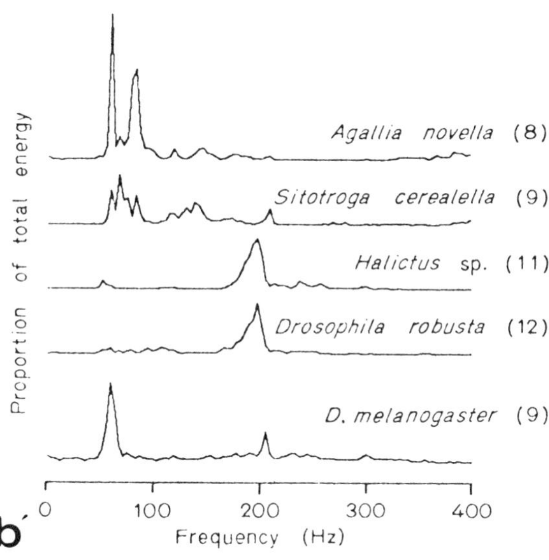
Agallia novella (8)
Sitotroga cerealella (9)
Halictus sp. (11)
Drosophila robusta (12)
D. melanogaster (9)
Proportion of total energy
0
100
200
300
400
Frequency (Hz)
b´

TABLE 3.2. Damping of signals in the various media relevant to spider vibration reception and communication.

Medium	Frequency	Damping	Comment	Author
A. *Air*	0.1 kHz 1.0 kHz	ca. 2×10^{-6} dB/cm ca. 6×10^{-5} dB/cm	These values give absorption coefficients and refer to dissipation of sound in air at 20°C and 70% rel. humidity. They add to the attenuation of sound pressure due to geometrical spreading which amounts to 6 dB per doubling the distance in the far-field, and may reach 12 dB/dd in the near-field of biologically significant types of sound sources.	Michelsen (1978)
B. *Water surface*	0.020 kHz 0.150 kHz	ca. 1.4 dB/cm ca. 2 dB/cm	Values refer to surface waves.	Markl and Wiese (1969)
C. *Spider web*				
1. single strand	0.050-2.0 kHz	ca. 1.5 dB/cm	Longitudinal vibrations measured on single strand of the web of *Achaearanea tepidariorum*; damping decreased to 1 dB/cm with tension experimentally increased close to breaking.	Walcott (1963)

TABLE 3.2 (*cont.*)

Medium	*Frequency*	*Damping*	*Comment*	*Author*
2. signal thread	0.020 kHz 0.160 kHz	ca. 2.0 dB/cm ca. 2.7 dB/cm	Transverse waves measured on the signal thread of the orb-web of *Zygiella x-notata*; damping of a signal on its way from the far end of the web to the spider leg on the signal thread is ca. 47 dB for 0.020 kHz and more than 55 dB for 0.150 kHz.	Graeser and Markl (in prep.)
D. *Wood*	4-5 kHz	1.8-2 dB/cm	Values refer to surface waves of oak and fir tree wood.	Markl and Fuchs (1972)
E. *Soil*	1-3 kHz	6 dB/cm	Measured in moist soil.	Markl (1968)

AIRBORNE SOUND In the work of both Rovner (1967, 1975) and Gwinner-Hanke (1970) the sound frequencies involved are within the sensitivity range of the tarsal slit sensillum (Barth, 1967). Compared to the long list of stridulatory organs given in the chapter by Uetz and Stratton, our knowledge about airborne sound production by spiders is rather scarce too.

SOLID SUBSTRATE VIBRATION The frequency range of solid substrate vibrations does not match the threshold curves of the metatarsal lyriform organ and those of the claw slits as long as these are plotted as excursion amplitudes. According to present knowledge natural signals are mainly in the low frequency range. It is there that these receptors are relatively insensitive. Possibly not all sig-

FIGURE 3.16. Web of *Zygiella x-notata*. The signal thread runs in the open sector, from the hub to the upper right corner (courtesy of K. Graeser).

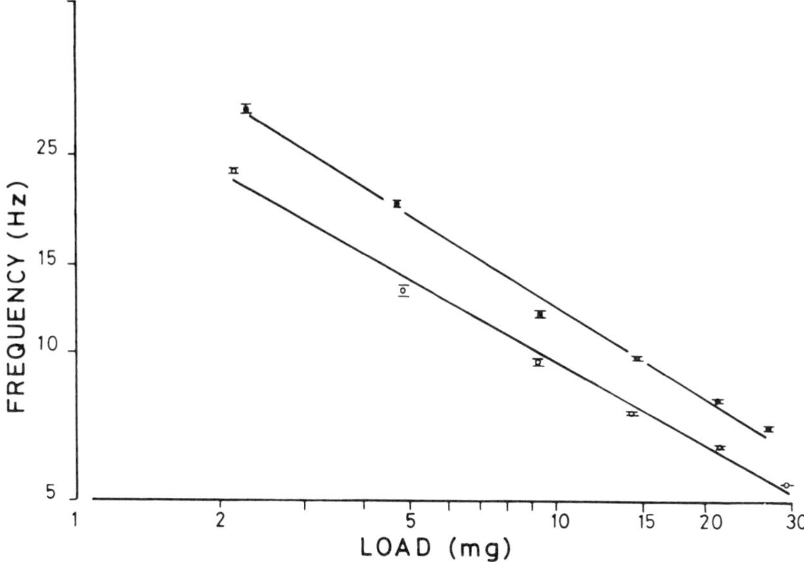

FIGURE 3.17. Change of resonance frequency of *Zygiella* web (two examples; measured as vibrations of signal thread) when loaded to various degrees in the hub (from Graeser and Markl, in prep.).

nificant components of natural vibrations have been adequately appreciated yet. The inconsistency of the picture disappears if we regard the receptors as acceleration detectors (Figure 3.12). Low frequencies are then correlated with the highest sensitivity. More experiments are needed on the relative importance of the various signal parameters. Special attention should be paid to fast transients such as impacts, ons, and offs, which have repeatedly been described as particularly effective in behavioral studies.

FREQUENCY DISCRIMINATION A sharp tuning of the receptors to different frequencies might be a basis for frequency discrimination. Neither the trichobothria nor the vibration-sensitive slit sensilla exhibit such sharp tuning within the frequency ranges tested. Instead, they offer at most the basis for a crude discrimination of larger frequency ranges, provided the spider brain understands the changing pattern of response (within a group of trichobothria and within the group of slits composing the metatarsal lyriform organ) going along with frequency changes. Resonance frequencies of delicate spider webs seem to be too liable to loading effects to be of

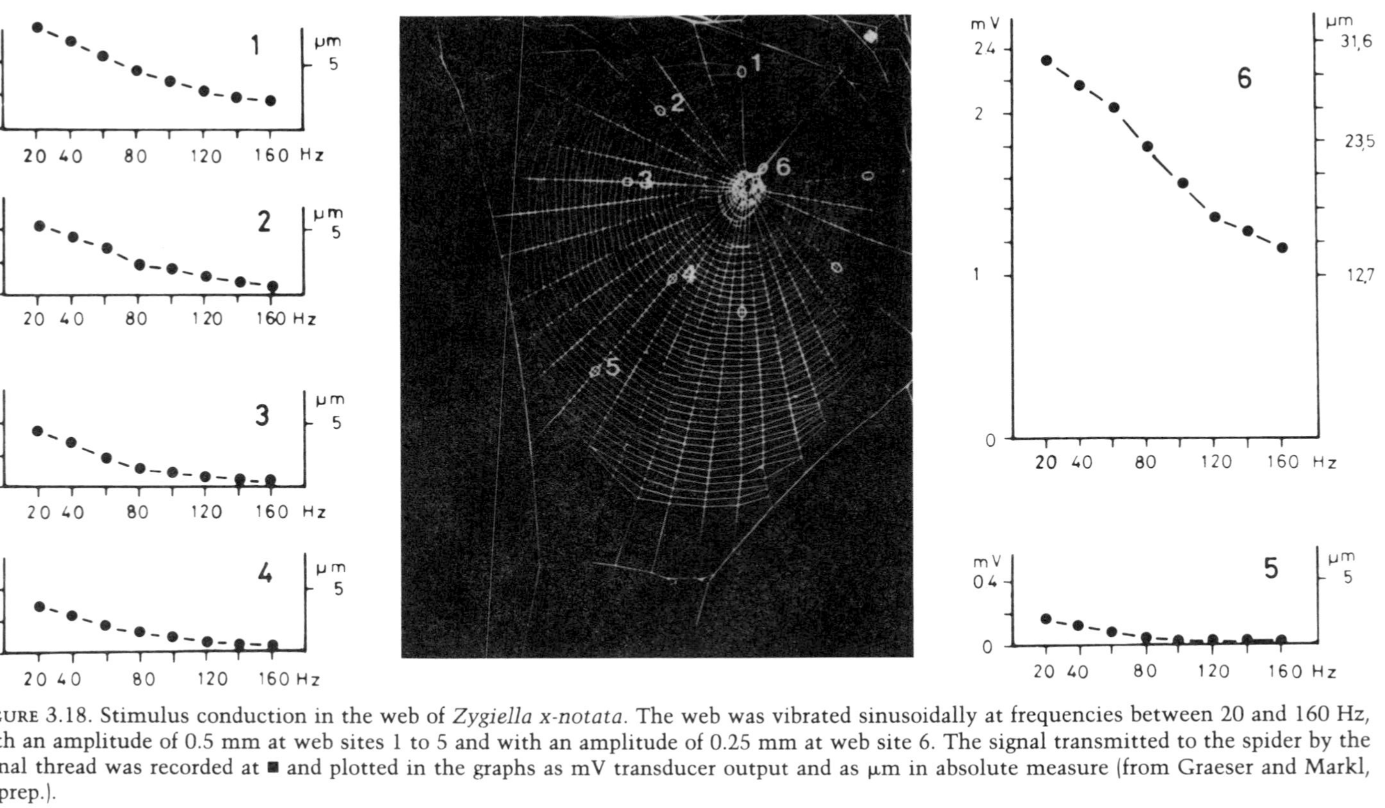

FIGURE 3.18. Stimulus conduction in the web of *Zygiella x-notata*. The web was vibrated sinusoidally at frequencies between 20 and 160 Hz, with an amplitude of 0.5 mm at web sites 1 to 5 and with an amplitude of 0.25 mm at web site 6. The signal transmitted to the spider by the signal thread was recorded at ■ and plotted in the graphs as mV transducer output and as μm in absolute measure (from Graeser and Markl, in prep.).

any use in frequency discrimination. The extremely sharp decline in gain beyond 300 Hz found in the web of *Zygiella*, as well as the filter properties attributed to the communal web of *Mallos*, however, again may be used to sort out roughly the biologically relevant frequency ranges even before the signal reaches the receptors.

Prey insects entangled in the web of *Cyclosa turbinata* all generate their own statistically distinctive vibrations. The approach rate of the spider changes significantly according to maximum signal amplitude at impact. Vibration frequency is an unlikely basis for prey species discrimination, however, because of considerable overlap and variability (Suter, 1978). In agreement with that conjecture, *Zygiella* attacks a vibration source over a wide range of frequencies, provided they are above 5 Hz (Liesenfeld, 1956).

RANGE Due to the attenuation of the signals on their way to the spider and despite the high sensitivity of the receptors, the vibration-detecting system of spiders will always be a close range system and used for relatively low frequencies. This is true for all vibration detectors. *Cupiennius* can be lured to a source of substrate vibration on a sheet of plastic over distances of up to 70 cm. The female responds to the vibrating signals of the male courting on a banana plant even when sitting on another leaf and 1 m away from him (Rovner and Barth, 1981). As to airborne stimuli, *Cupiennius* is attracted to a crawling fly over a distance of up to 20 cm. A desert scorpion locates vibrations in sand produced by a buried cockroach up to 20 cm away (Brownell, 1977). Finally, a record range is reported for a ghost crab, which perceives vibrations produced by conspecifics over distances up to 7.5 m and more (Horch, 1971).

Sense Organs and Behavior

Only a few studies have been performed which directly relate vibration-guided orientation behavior of spiders to the receptors involved. Receptor ablation experiments, some of them with surprisingly little effect, have added some complication.

TRICHOBOTHRIA *Agelena labyrinthica* accurately localizes a dummy prey vibrating in air about 1 cm away from a leg. It turns to the proper direction and grasps the dummy. Stimulus angle and turning angle are well-correlated over a range of at least 260° (Figure 3.19) (Görner and Andrews, 1969).

If trichobothria are removed on one side of the animal and the dummy held in front of it, then a clear deficiency in orientation results: the spider jumps toward its intact side. If, however, the stimulus is applied between two legs without trichobothria, a reaction is elicited in only 6% of the tests as opposed to 86% in intact animals.

Agelena also responds when outside its web; web vibrations are not important in this experiment. Clearly then, this is behavioral evidence for the role attributed to the trichobothria on receptor physiological grounds, as discussed earlier.

Curiously, dummy prey can be localized after removal of all trichobothria, even though the readiness to orient toward it is reduced (Reissland and Görner, 1978). Therefore another sensory system must be involved, and the slit sensilla are the prime candidates.

METATARSAL LYRIFORM ORGANS The same type of ablation experiment when done with the metatarsal lyriform organ and solid-borne vibrations again demonstrates that this organ cannot be the only receptor for substrate-borne vibrations either. When a hungry blinded *Cupiennius salei* sits on an experimental sheet of plastic as it would sit on plant leaves to catch prey in the field (Barth and

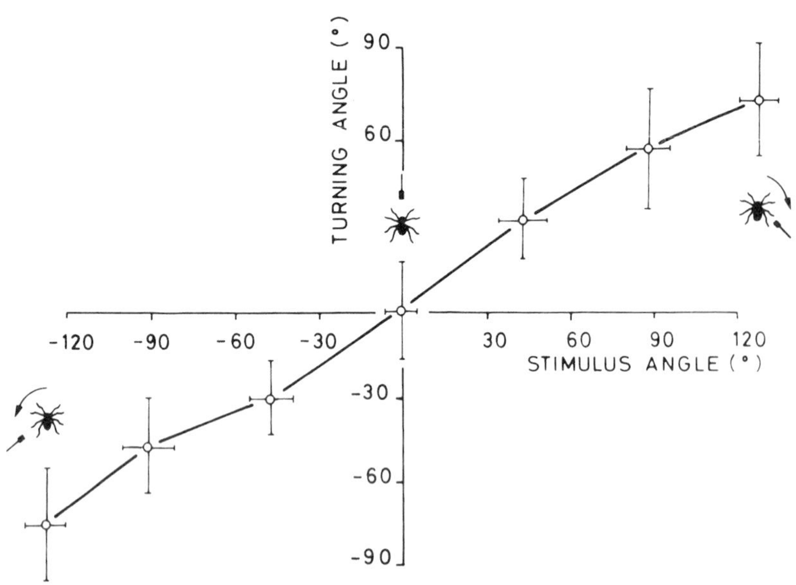

FIGURE 3.19. Turning reaction of *Agelena* when stimulated with a disk (—■, ϕ 2.3 mm) vibrating at 65 Hz close to the spider (from Görner and Andrews, 1969).

Seyfarth, 1979), it is attracted by vibrations of the substrate. These may be applied in several ways. Particularly effective is a fine brush, slightly touching the sheet from underneath to reduce airborne vibrations. A fly can be taken in the same way with its wings removed and just barely touching the sheet with its moving legs (Figure 3.20). The amplitude of such vibrations hardly exceeds 2μm (peak); and the corresponding power spectrum shows maxima below 20 Hz (Figure 3.20b). If stimulated at a distance (in our experiments up to 50 cm), the spider slowly walks to the vibration source and stops whenever the stimulus stops. If stimulated (through the plastic) underneath the tarsus of one of its legs or a few centimeters away from the tarsus, the spider immediately and precisely jumps toward the stimulus and its jump is completed in less than 50 msec (Figure 3.21) (Barth and Sperr, unpubl.).

Mechanical ablation of the metatarsal organ has no dramatic effect on this behavior. Only slight changes in the precision of the jump occur. When lacking the metatarsal organs on one side of its body and stimulated underneath one of its operated legs, the spider jumps too far with respect to both the rotational and translational component of its movement. Turning angles (β) and error angles (β′; Figure 3.21) change by 3° to 10°. The change in β′ is statistically significant. No changes are observed if the spider is stimulated underneath a leg of its intact side (Figures 3.22 and 3.23). Though the observed effects are small, not to have the metatarsal lyriform organ might be a real disadvantage to the spider, since even the smallest imprecision may provide the prey with the crucial milliseconds needed to escape.

CLAW SLITS Ablation of the claw slits changes both the turning angle β and the error angle β′ significantly and to an even greater extent than the ablation of the metatarsal organ (Figures 3.22 and 3.23) (Hener and Barth, unpubl.). In a more rigorous experiment the animal rested on an arena made up of quadrangular pieces of plastic sheet mechanically insulated from each other (attenuation better than 40 dB up to 500 Hz). Its legs could be stimulated individually, again from underneath. The spider localizes the vibration source under these circumstances as well. It also is successful with all trichobothria removed (Barth and Rehner, unpubl.).

In conclusion, the spider relies on a number of receptors when detecting solid-borne vibrations. Even more than the two receptors identified are to be expected. One of these may be the longest slit in the tibial lyriform organ (HS8) and equivalent slits in other lyriform organs. It is very sensitive and phasic in its response to loads

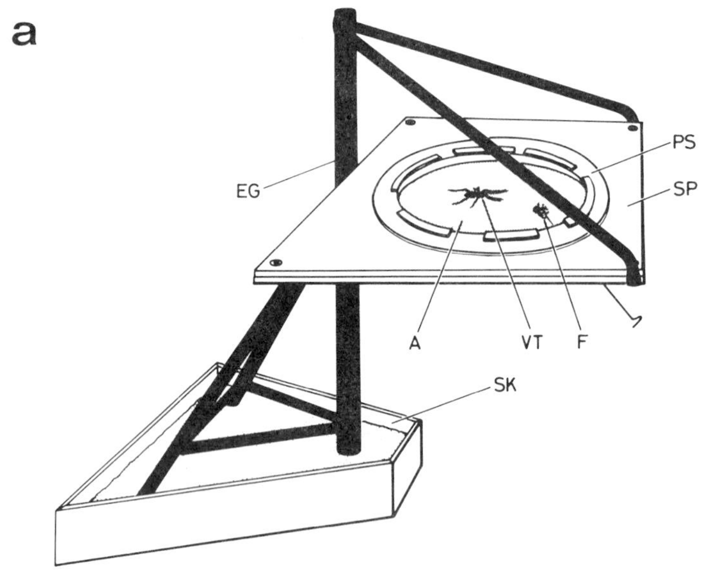

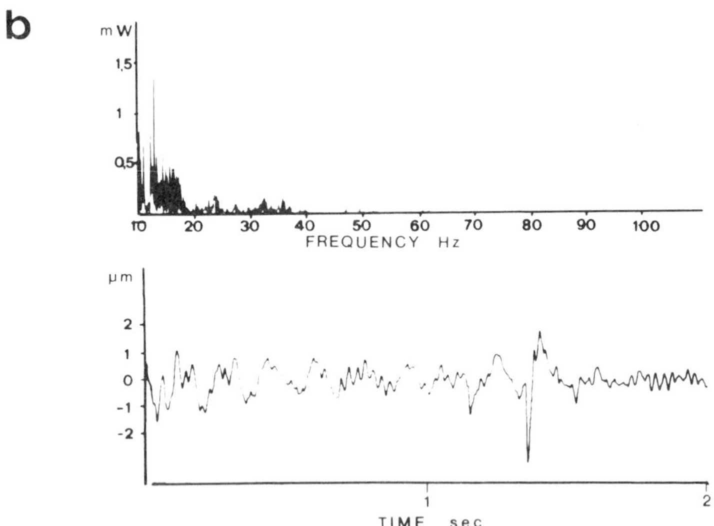

FIGURE 3.20a-b. Experimental set-up to test the response of *Cupiennius salei* to solid-borne vibrations. (a) Spider (VT) sits on sheet of plastic (A) and is stimulated from underneath. One of the most effective stimuli applied is a fly (F) crawling on the underside (to avoid airborne stimulation) of the arena (Hener and Barth, unpubl.). (b) Power spectrum and oscillogram of such a fly stimulus (Bohnenberger, unpubl.).

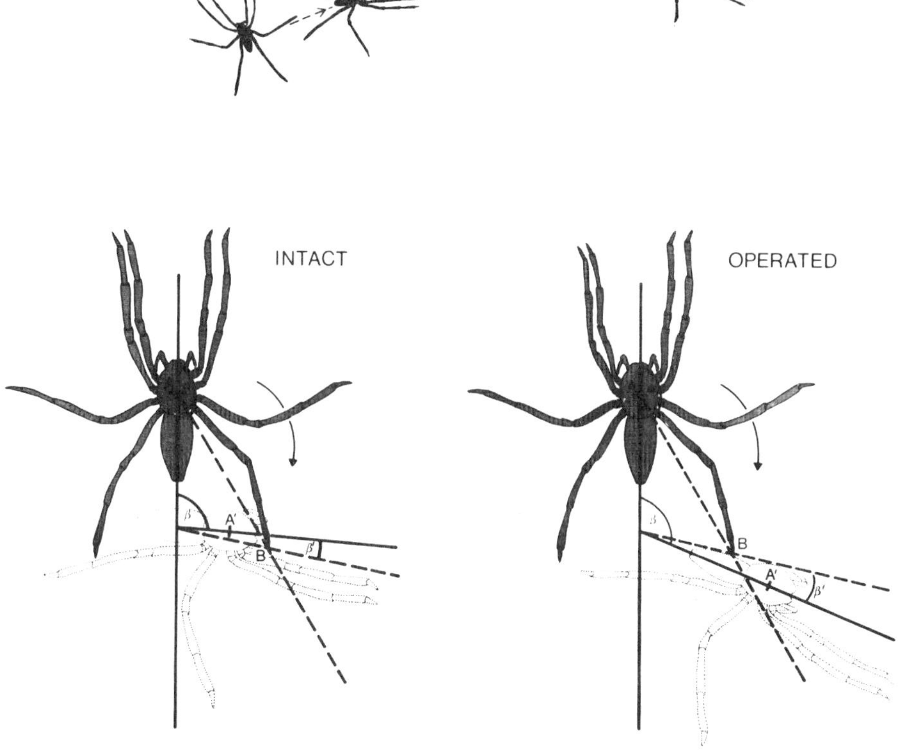

FIGURE 3.21a-b. Two different reactions are observed when *Cupiennius* is stimulated by substrate-borne vibrations (Fig. 3.20). (a) Reaction to distant stimulus (B). (b) Reaction when stimulated under tarsus. A,A′, center of prosoma before and after jump; β, turning angle; β′, error angle. In operated animals (right) substrate vibration receptors of stimulated leg are destroyed. As a consequence, the jump of the spider loses precision; it is slightly too far with respect to the site of stimulation (Hener and Barth, unpubl.).

applied to the metatarsus-tibia joint (Barth and Bohnenberger, 1978).

Summarizing the older literature on web spiders, Legendre (1963) concluded that stimulation of both the trichobothria and the lyriform organs signals prey to a spider and as a consequence releases attack behavior, whereas stimulation of the trichobothria alone would signal danger and release escape behavior. That conclusion cannot be maintained.

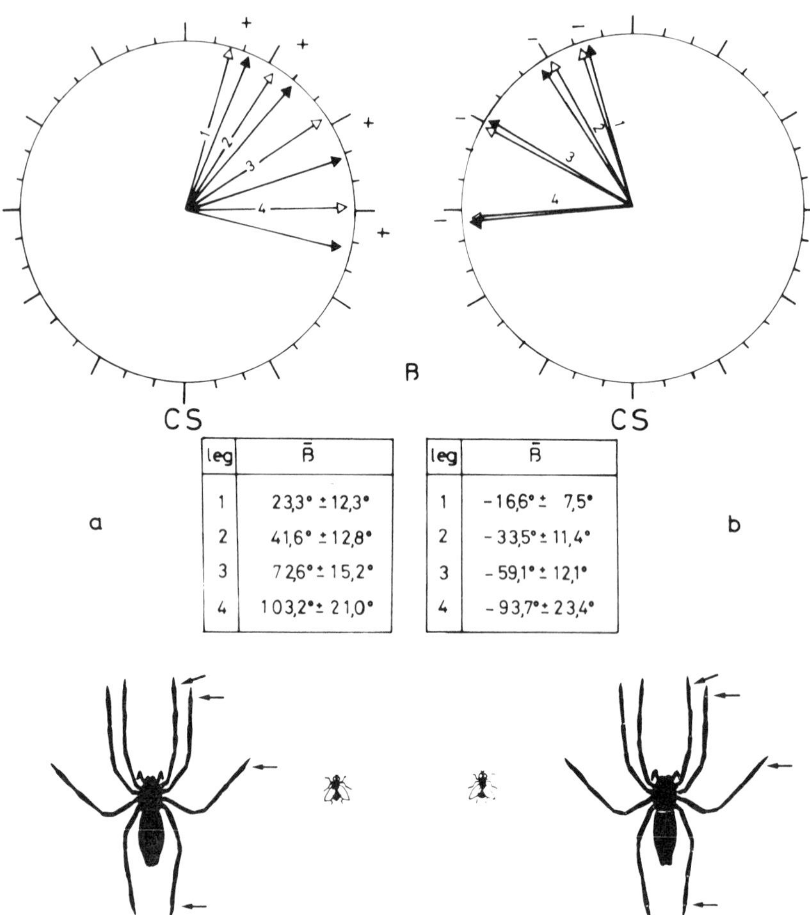

leg	$\bar{\beta}$
1	23,3° ± 12,3°
2	41,6° ± 12,8°
3	72,6° ± 15,2°
4	103,2° ± 21,0°

leg	$\bar{\beta}$
1	−16,6° ± 7,5°
2	−33,5° ± 11,4°
3	−59,1° ± 12,1°
4	−93,7° ± 23,4°

FIGURE 3.22a-b. Turning angle β of *Cupiennius salei* in response to nearby vibration (Fig. 3.21b) and with the claw slits (CS) destroyed on all right legs. (a) Stimulation under operated legs leads to turning angles (♦) significantly different (+) from those of intact animals (◊). (b) Stimulation under one of the intact legs is not followed by a significant change (−). Length of vectors (arrows) indicates small variation of response (r measured 0.91 to 0.99, n varied between 27 and 129). Numbers within circle refer to number of leg stimulated (Hener and Barth, unpubl.).

Conclusions

The spider may turn out to be a case particularly worthy of study as an example of an intimate relationship of vibratory signals and animal behavior in general. The reports summarized in the first part of this chapter deal with behavioral patterns which have at-

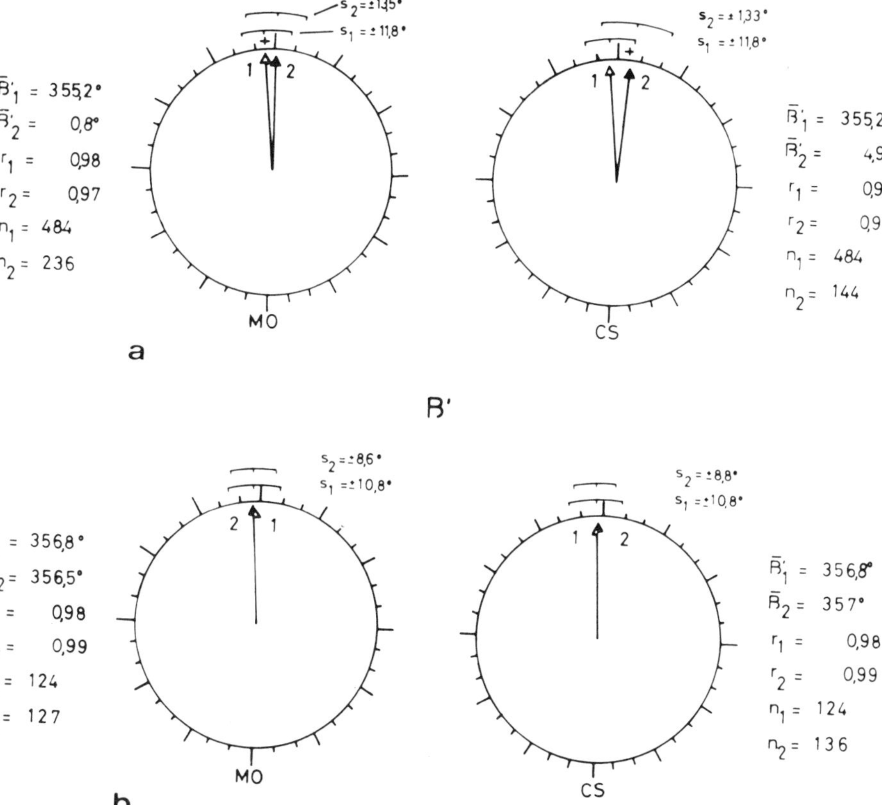

FIGURE 3.23a-b. Same experimental situation as shown in Fig. 3.22. Error angle β′ (Fig. 3.21b) after destroying the metatarsal lyriform organ (MO) and the claw slits (CS), respectively. (a) Stimulation under operated leg. (b) Stimulation under intact leg (Hener and Barth, unpubl.).

tracted naturalists for almost a century. If knowledge of the mechanisms underlying the physical, neurobiological, and ethological aspects of vibration detection is our goal, there is no way around a more quantitative approach. The experiments to be done are within sight. The bulk of the work lies ahead of us. Most of the fascinating observations in the literature, some of them several decades old, will have to be reexamined with the proper techniques.

Subjects which need further study include the following topics. We need more data on naturally occurring signals. No courtship signal has been fully analyzed; only little is known about the bioacoustics of stridulation. In no case has the effect of changing the signal parameters on behavior been studied in detail and with adequate methods. Not enough data are available on the near- and far-field components in the signals produced by drumming, bouncing, and jerking spiders. Another set of questions relates to the engineering of the web and in particular to its signal-conducting properties. The neurobiological part of the story is rather incomplete as well. Nothing is known about the central integration of the sensory input or the nervous organization of the motor output. Finally, we do not have enough quantitative behavioral studies with stimuli under control.

Why then a review at all, when it may seem too early with respect to the gaps in our knowledge still to be filled and too late with respect to the many bits and pieces found widely dispersed in the literature? Mainly, to show that a rich field of research lies ahead of us and to lay the ground for its harvest.

Acknowledgments

The author's laboratory was generously supported by the Deutsche Forschungsgemeinschaft. Thanks are due to H. Markl and K. Graeser for permission to quote unpublished material, to M. Grasshoff, F. Vollrath, and K. Graeser for photographs (Figures 2, 4, 16), to A. van de Roemer and H. Hahn for help in preparing the drawings, to E.-A. Seyfarth for valuable comments on the manuscript, and to P. Görner for discussing aspects of trichobothria physiology; and to Elsevier (Amsterdam), Springer Verlag (Heidelberg), and Cold Spring Harbor Laboratories for permission to use illustrations previously published.

The following publications appeared after completion of this chapter:

Barth, F. G. Strain detection in the arthropod exoskeleton. In *Sense Organs*, M. S. Laverack and D. J. Cosens (eds.), Blackie, Glasgow, 1981.

Bohnenberger, J. Matched transfer characteristics of single units in a compound slit sense organ. J. comp. Physiol., *142*: 391-402, 1981.

Masters, W. M. and H. Markl. Vibration signal transmission in spider orb webs. Science, *213*: 363-365, 1981. (In this study, laser Doppler vibrometry revealed particularly little attenuation of *longitudinal* waves, even at high frequencies up to several kHz.)

Chapter 4 ACOUSTIC COMMUNICATION AND REPRODUCTIVE ISOLATION IN SPIDERS

George W. Uetz
and Gail E. Stratton
Department of Biological Sciences
University of Cincinnati
Cincinnati, Ohio 45221

Introduction

Many arthropods communicate with each other using sounds. The chirping of crickets or the buzzing of cicadas readily come to mind as examples of arthropod sound production. Much interesting research has been done on the sounds produced by crickets, locusts, katydids, grasshoppers, and other members of the arthropod class Insecta. Since it is our intent to consider acoustic communication in spiders and not insects, we will refer all those interested to other sources (Busnel, 1963; Sebeok, 1977; Walker, 1964). It would appear that sound production is more or less common in all the major arthropod classes, including the Diplopoda and Chilopoda (Busnel, 1963) and the Crustacea (Salmon, 1965; Weygoldt, 1977). The Arachnida are not commonly thought of as producing sound, although it would appear that some orders are capable of making sounds of a variety of types (Berland, 1932; Busnel, 1963; Weygoldt, 1977), including scorpions (Dumortier in Busnel, 1963), and Solifugae as well as Araneae. Perhaps it is because the sounds made by spiders and their kin are barely audible to humans when compared to those produced by many insects (some of which can be almost deafening!) that they have not received as much attention. Because of the great diversity of the order Araneae, the study of acoustic communication in spiders should be interesting and might provide us with a few surprises.

In this chapter, we will look at the kinds of sounds produced by spiders and how these sounds are produced. To do this, we must examine the structure and function of sound-producing organs, and the nature of the sound produced—its amplitude, frequency, and pattern. (How spiders perceive sound is the subject of Chapter 3 in this volume by F. G. Barth, and will not be discussed here.) Later we will consider the biological significance of sound production in spiders, focusing on its significance for the behavior and ecology of the animal and its role in the evolution of species. We will report on some current research in our laboratory concerning sound production and courtship in wolf spiders (Lycosidae). Finally, we will speculate about future research on acoustic communication in spiders and point out questions that need answers.

Acoustic Communication in Spiders: A Review of the Literature

Types of Acoustic Communication in Spiders

There are 26 families of spiders that appear to be capable of sound production (Table 4.1). In our search of the literature, the references we found either describe a stridulatory organ, or the researcher mentions being able to hear sounds produced by the spider. Recordings have been made in only a few instances, and experiments demonstrating the purpose of sound production are also few. Within Araneae three sound production methods have been observed (Rovner, 1980b) and are here described in some detail. They are (1) stridulation, (2) percussion, and (3) vibration of structures. Twenty-two families of spiders have species that are capable of stridulating, six families have species that use percussion, and two families use leg or body vibration to produce sounds (Table 4.1). In several instances, a family has some members that use percussion and other members that use stridulation (Lycosidae, Salticidae, and Clubionidae). There are even a few examples in which a single species uses both methods (*Lycosa rabida* and others, Rovner, 1975; *Schizocosa rovneri*, Stratton and Uetz, in prep.). Clearly, sound production in spiders is much more common than is generally thought.

STRIDULATION "Stridulation" is a method of sound production involving the friction of rigid parts, which often are provided with a special surface (Busnel, 1963). It requires the movement of one piece (the scraper) across a second (the file). There are eight different types of stridulatory organs, based on the widely varying location of the file and scraper on the body of the spider (Figure 4.1; Table 4.1). Legendre (1963) made the original classification of then-known stridulatory organs into types a - g, while Rovner (1975) added type h. Rovner (1975) classified these organs in spiders into four basic forms: (1) the abdomen rubs against the prosoma (type a), or the pedicel (type b); (2) one appendage rubs against another, including chelicera-chelicera (type c), chelicera-pedipalp (type d, Figure 4.2A and B), pedipalp-leg I (type e), leg I–leg II (type f); (3)

TABLE 4.1. A list of the families known to produce sound or to possess organs that are presumably capable of making sounds. (Stridulatory organ types correspond to Legendre's [1963] classification; see text.)

Family	*How Sound is Produced*	*Reference*
1. Theraphosidae	stridulation (types c, d, e)	Wood-Mason 1876, Spencer 1896, Pocock 1895, 1897, 1898, 1899, Meyer 1928, Monath 1957, all cited in Legendre 1963. Chrysanthus 1953.
2. Barychelidae	stridulation (type d)	Pocock 1896 in Legendre 1963.
3. Dipluridae	stridulation (type d)	Pocock 1896 in Legendre 1963.
4. Uloboridae	stridulation (type d)	Opell 1979.
5. Sicariidae	stridulation (type d)	Chamberlin 1924, Simon 1893, both in Legendre 1963. Millot 1949 in Busnel 1963.
6. Scytodidae	stridulation (type d)	Pickard-Cambridge 1895 in Legendre 1963.
7. Diguetidae	stridulation (type d)	Chamberlin 1924 in Legendre 1963. Bentzien 1973.
8. Ochyroceratidae	stridulation	Legendre 1970.
9. Leptonetidae	stridulation	Bishop 1925 in Legendre 1963.
10. Segestriidae	stridulation	Gertsch 1979.
11. Pholcidae	stridulation	Gertsch 1979.
12. Theridiidae	stridulation (type a)	Westring 1843, 1858, Hasselt 1876, Campbell 1881, Pickard-Cambridge 1881, Kulczynski 1905, Crosby 1906, Bishop 1925, Meyer 1928, Scharrer 1932, Guibe 1943, Braun 1955, all in Legendre 1963. Gwinner-Hanke 1970.
13. Linyphiidae	stridulation (types d, f)	Bishop 1925, Campbell 1881, both in Legendre 1963.
14. Erigonidae	stridulation (type g)	Carpenter 1898, Falconer 1910, Bishop 1925, Guibe 1943, in Busnel 1963.
15. Araneidae	stridulation (type g)	Kolosvary 1929, cited in Legendre 1963. Hinton and Wilson 1970.
16. Agelenidae	stridulation (type b)	Pocock 1895 in Legendre 1963.
17. Hahniidae	stridulation (type a)	Opell and Beatty 1976.
18. Mimetidae	stridulation (type d)	Machado 1941 in Kaston 1948.
19. Lycosidae	stridulation (types g, h)	Rovner 1975, Kronestedt 1973, Hallander 1967, Stratton and Uetz in prep.

TABLE 4.1 (*cont.*)

Family	*How Sound is Produced*	*Reference*
19. *cont.*	percussion (tapping of abdomen, leg 1 and palps)	Burroughs 1881, Davis 1904, Lahee 1904, Bristowe and Locket 1926, Bristowe 1929, Chopard 1934 in Chrysanthus 1953, Allard 1936, Kaston 1936, P. Bonieau (18th century, in Bonnet 1945), Hallander 1967, Buckle 1972, Von Helversen in Weygoldt 1977, Rovner 1967, Harrison 1969.
20. Gnaphosidae	stridulation (type a)	Montgomery 1909.
21. Clubionidae	percussion	Bristowe 1929.
	stridulation (type a)	Montgomery 1909.
22. Anyphaenidae	percussion	Bristowe 1958.
23. Ctenidae	percussion	Melchers 1963.
	vibration	Rovner and Barth, 1981.
24. Sparassidae	vibration	Rovner 1980b.
25. Thomisidae	percussion	Kaston 1936.
26. Salticidae	percussion	Bristowe and Locket 1926, Bristowe 1929.
	stridulation (types d, h)	Petrunkevitch 1926 in Legendre 1963. G. B. Edwards, in press.

one appendage rubs against the abdomen (type g, Figure 4.2C); and (4) the file and scraper are on opposing segmental surfaces of a joint within a single appendage (type h, Figure 4.3).

Highspeed cinematography is necessary to describe a spider's movement during stridulation. In *Teutana grossa* (Theridiidae) the stridulating organ is on the posterior cephalothorax and the anterior of the abdomen. Gwinner-Hanke (1970) describes an "up-down" movement. This action occurs in 0.0079 to 0.0047 seconds. In the "up" movement the abdomen is raised up and rubbed against the ridges on the cephalothorax. The sound produced in another theridiid, *Steatoda bipunctata*, is described as "a rapid trilling series" which begins slowly and becomes more rapid (begins softly and becomes louder).

The stridulatory organs of *Micrathena gracilis* and *M. schreibersi* (Araneidae) are described by Hinton and Wilson (1970). The file is on the surface of the booklung cover and the scraper is at the base of the femur in the last pair of legs (type g, Figure 4.2C). Hinton and Wilson suggest that the booklungs could be acting as resonating structures. The sounds were produced on dead spiders simply

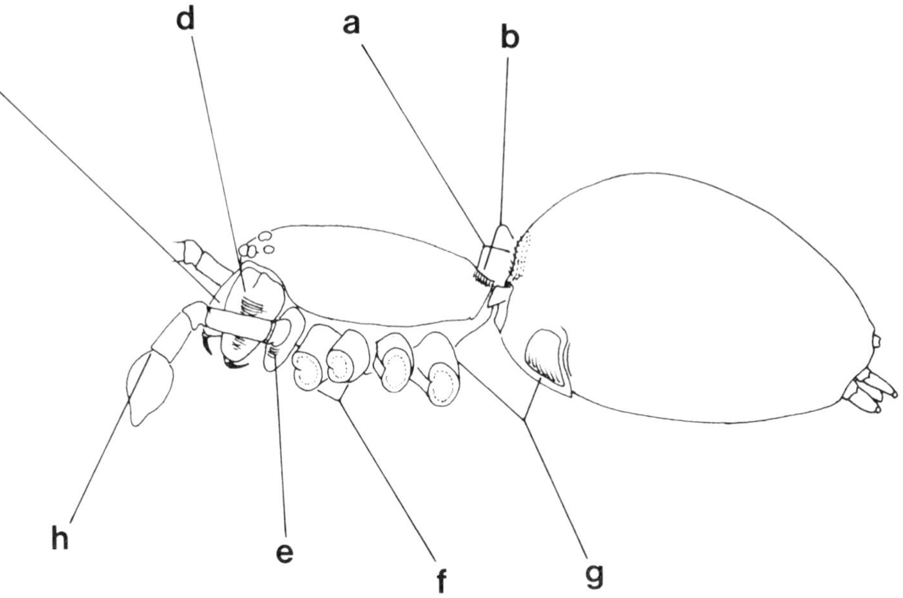

FIGURE 4.1. Composite diagram of a spider, showing location of stridulatory organs. Letters a-h refer to types of organs described by Legendre (1963) and Rovner (1975) (see text for explanation).

by moving the fourth leg across the abdomen. A defensive function is hypothesized, as the stridulatory organ was found only on the females; and the sound was produced when the spiders were touched or otherwise disturbed.

The movement during stridulation in many of the Lycosidae is described by Rovner (1975). The stridulatory organ is found on the pedipalps of the mature spider (type h, Figure 4.3). "The palpal movement is an anterior-posterior oscillation with maximum displacement at the tibio-tarsal joint that results in alternating, low-amplitude flexions and extensions." Palpal scrapes and taps were also seen in the stridulation sequence of *Lycosa rabida*. The macrosetae (stout spines) are in continual contact with the substrate during stridulation and probably serve as holdfasts to prevent or reduce slippage.

PERCUSSION "Percussion" can be defined as shocks to the substrate. In spiders this often involves tapping of the legs, the palps, and/or the abdomen. The substrate must be a hard surface for sound production; however, spiders appear to use a variety of other substrates for vibratory communication. For example, the silk of a

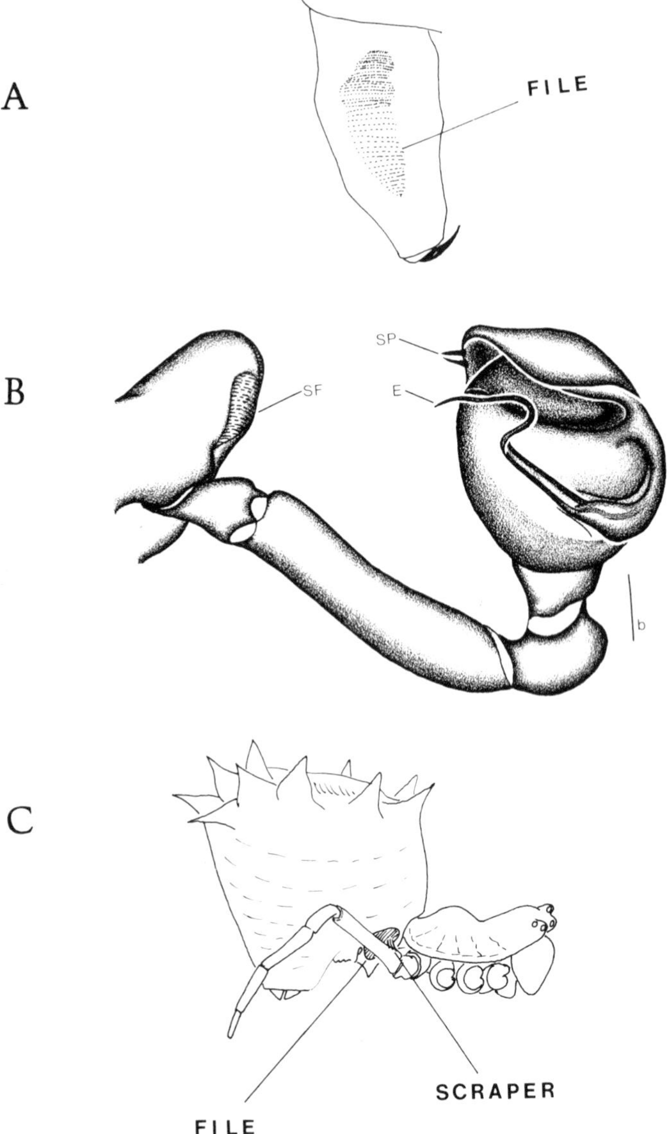

FIGURE 4.2. Examples of stridulatory organs in various spiders.

(A) Stridulatory file on the chelicera of a linyphiid spider, *Meioneta* sp. (redrawn from Kaston, 1948).

(B) Stridulatory organ located on the palpal coxa of *Tangaroa tahitiensis*, a uloborid spider (from Opell, 1979); SF = stridulatory file; SP = spines used to scrape file; E = embolus of palp.

(C) *Micrathena gracilis*, an araneid, with stridulatory organ on the booklung cover (redrawn from Hinton and Wilson, 1970).

web (Krafft, 1978) or a retreat (Jackson, 1976a) or even water (Bristowe, 1958) may be used to transmit vibrations.

Percussive sounds in spiders are produced in several ways. Male *Hygrolycosa rubrofasciata* (Lycosidae) (von Helversen in Weygoldt, 1977) have a sclerotized plate on the abdomen, which is covered with specialized knobbed hairs. Vibrations of the abdomen produce sounds when it hits the ground. Other spiders such as *Tarentula pulverulenta* (Lycosidae) (Bristowe and Locket, 1926), and *Lycosa gulosa* (Harrison, 1969) also produce sounds by abdomen-substratum contact, but do not appear to have specialized structures. *Schizocosa rovneri*, when courting vigorously, strikes the whole body against the substrate, in addition to stridulating. Other percussion methods involve striking and tapping the legs and/or palps on the substrate. This is seen in *Lycosa gulosa* (Harrison, 1969; Lahee, 1904), *Trochosa picta* (Bristowe and Locket, 1926), *Pardosa chelata* (produced a "distinct drumming or purring sound") (Hallander, 1967), and *Agroeca brunnea* (Clubionidae) (Bristowe, 1929), although some of these may be stridulating as well (Rovner, personal communication).

VIBRATION Production of sound by "vibration of appendages" has only been described in two families of spiders: Sparassidae and Ctenidae. Rovner (1980b) found that *Heteropoda venatoria* can produce sounds during bouts of leg oscillations (Figure 4.4A). The vibration movement occurring in *H. venatoria* was described with the help of highspeed cinematography. The filming revealed low-amplitude vertical oscillations of the posterior pair of legs and abdomen. Body jerks also occurred during courtship. The sounds are produced by a "tuning fork-like effect of appendage vibrations that set up regions of compression and rarefaction" (Figure 4.4B). *Cupiennius salei* produces sounds in much the same manner as *Heteropoda* (Rovner and Barth, 1981). Coupling to the substratum via tarsal adhesive hairs appears to be essential for sound production. *Heteropoda* may be using the substrate as a "sound board," amplifying the airborne component of the sounds produced. There are also examples of spiders "quivering" or "vibrating" that may possibly include sound production (*Aelurillus v-insignitus*, Bristowe, 1929; *Philodromus rufus vibrans*, Dondale, 1967; and *Philodromus pernex*, Kaston, 1936).

Nature of Spider Sound Emissions

There is no evidence that arthropods can perceive "tone" (Matthews and Matthews, 1978). Spiders as well as insects are probably

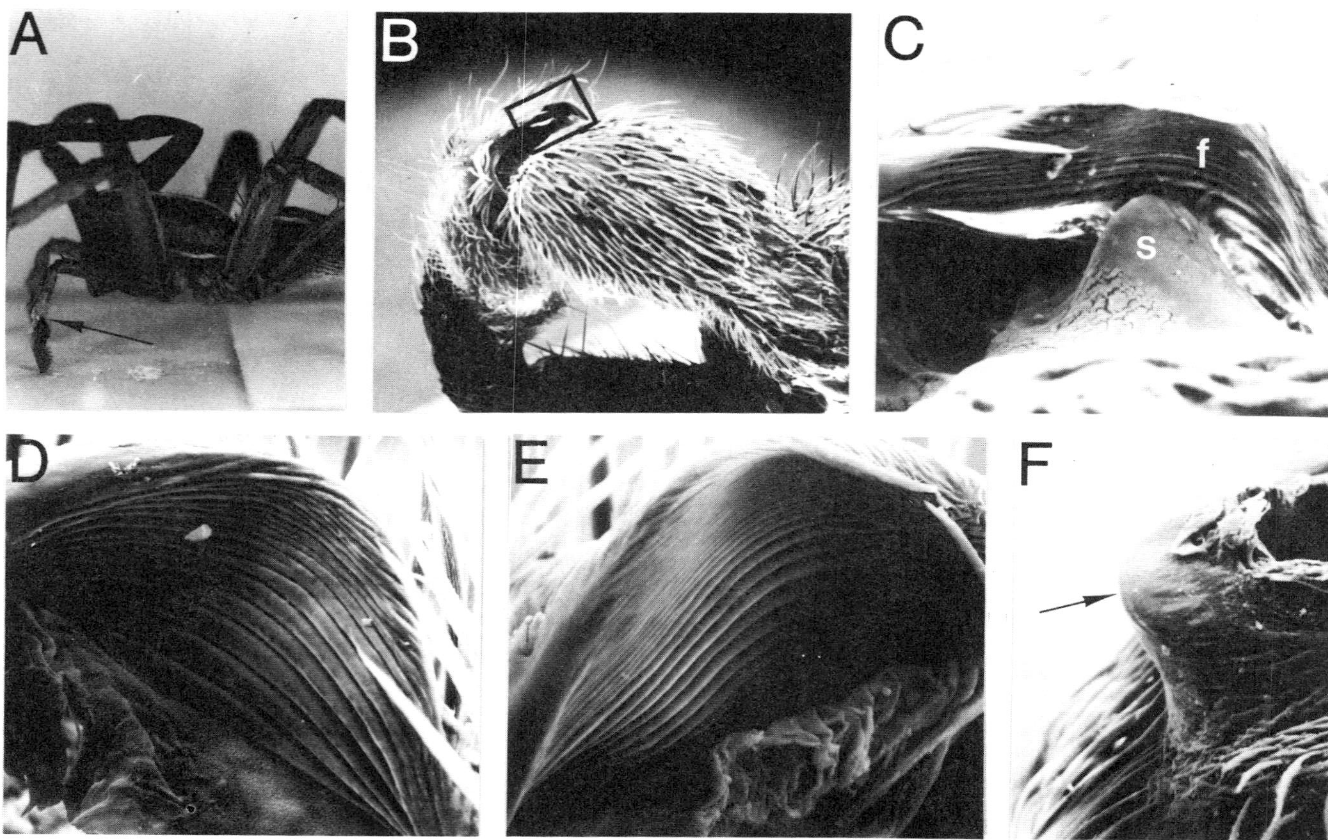
A
B
C
f
s
D
E
F

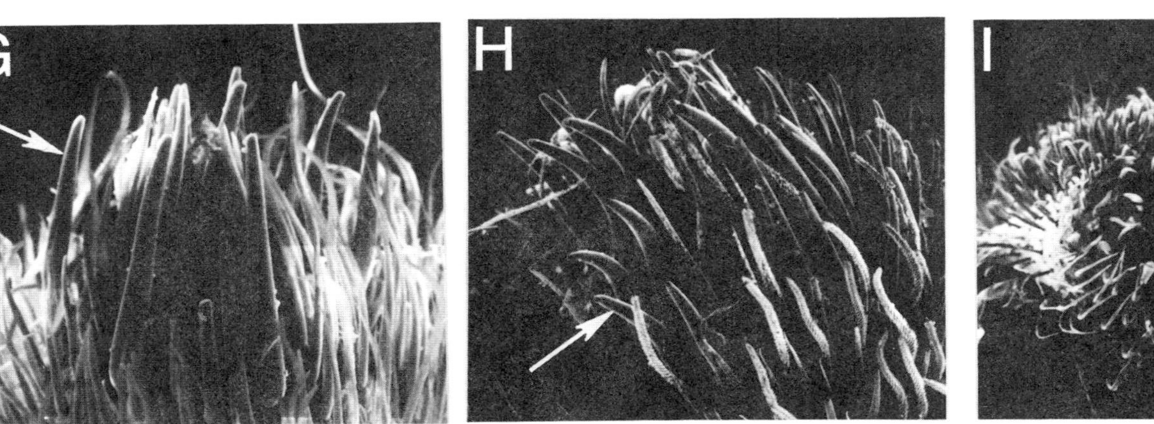

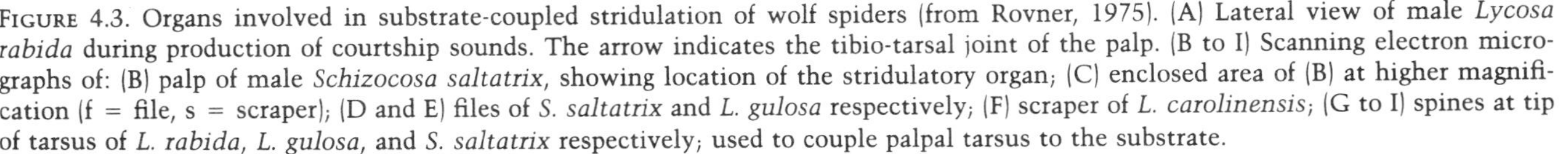

FIGURE 4.3. Organs involved in substrate-coupled stridulation of wolf spiders (from Rovner, 1975). (A) Lateral view of male *Lycosa rabida* during production of courtship sounds. The arrow indicates the tibio-tarsal joint of the palp. (B to I) Scanning electron micrographs of: (B) palp of male *Schizocosa saltatrix*, showing location of the stridulatory organ; (C) enclosed area of (B) at higher magnification (f = file, s = scraper); (D and E) files of *S. saltatrix* and *L. gulosa* respectively; (F) scraper of *L. carolinensis*; (G to I) spines at tip of tarsus of *L. rabida*, *L. gulosa*, and *S. saltatrix* respectively; used to couple palpal tarsus to the substrate.

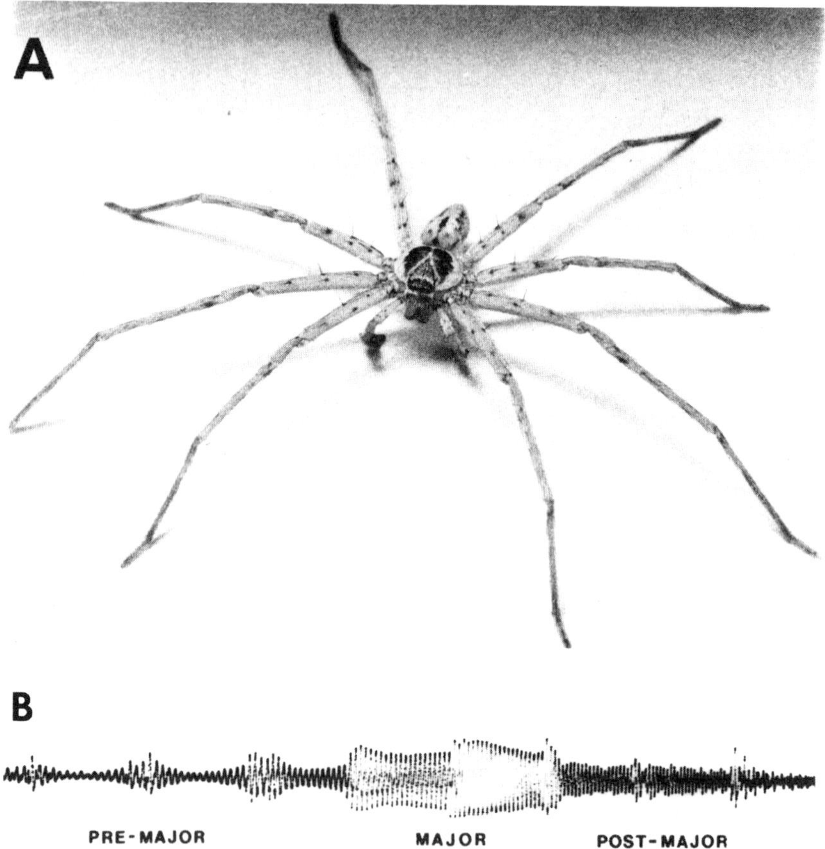

FIGURE 4.4. (A) *Heteropoda venatoria* in courtship position (vibrations are used to produce sound at this time). (B) Oscillogram of sound produced by leg vibrations (from Rovner, 1980b).

tone deaf. However, sound intensity and changes in the patterning and timing of the elements of sound production probably contribute to the informational content of the spider "song." Following are some definitions that will be important in the subsequent sound analysis.

Wave-train: one wave-train forms a pulse (Busnel, 1963).

Pulse (behavioral): the individual behavioral song component; a pulse is often all the sound produced in one cycle of stridulatory movement (Busnel, 1963).

(operational): the simplest element of amplitude modulation convenient to recognize; a wave-train isolated in time by a substantial reduction in amplitude or a silent interval (Matthews and Matthews, 1978).

Pulse-train: the first order grouping of two or more pulses, preceded and followed by a period of silence substantially greater than any of the time intervals between pulses.

There have only been a few instances of recording and analysis of spider sounds. These include examples in each type of sound production: percussion, stridulation, and vibration. Following are several examples of oscillograms and sonograms made from various groups of spiders. Sonograms show the carrier frequency on the vertical axis (expressed in kilohertz) and time on the horizontal axis. Amplitude is suggested by the darkness of the trace. Oscillograms also represent time on the abscissa; the ordinate expresses the pattern of sound intensity (loudness) fluctuations, without specifying the carrier frequency except for simple sound-waves such as those of various flying insects.

Rovner (1967) provided analysis of both courtship and threat display sounds in *Lycosa rabida*. (Sonograms of each are reproduced in Figure 4.5.) The courtship sound consisted of several sequences of the following: "first a brief burst of pulses followed by a relatively long continuous train of pulses which increased in pulse rate and intensity at the end of the train. Each display sound ended abruptly and was followed by a period of relative quiet." The threat displays differed: they consisted of short, frequent bursts of pulses which were highly variable in duration. The intervals between the bursts were very brief. The pulse rate, pulse and interpulse durations, and frequency spectra within each burst resembled those of courtship sounds. *L. rabida* is a stridulating wolf spider (type h). Each pulse possibly corresponds with a single scraping of the teeth of the spider's palpal file.

Lycosa gulosa produces sounds both by striking the tip of the abdomen against the substrate and by palpal stridulation (type h) (Harrison, 1969; Rovner, 1975). These sounds can be heard in a quiet forest at 6 meters. The palps oscillate to produce "strums" at intervals. The strums usually occur in pairs or triplets. Each strum lasts less than 0.25 seconds (a triplet would be slightly more than one second). Harrison provided both oscillograms and sonograms showing palpal strums and abdominal taps.

Schizocosa ocreata and *Schizocosa rovneri* are two closely related species of wolf spiders (they are discussed later). We have produced sonograms and oscillograms for both species (Figures 4.8, 4.9, and 4.10). The sounds they produce are predominantly stridulatory. *S. ocreata* also shows some leg tapping. The sound produced by *S.*

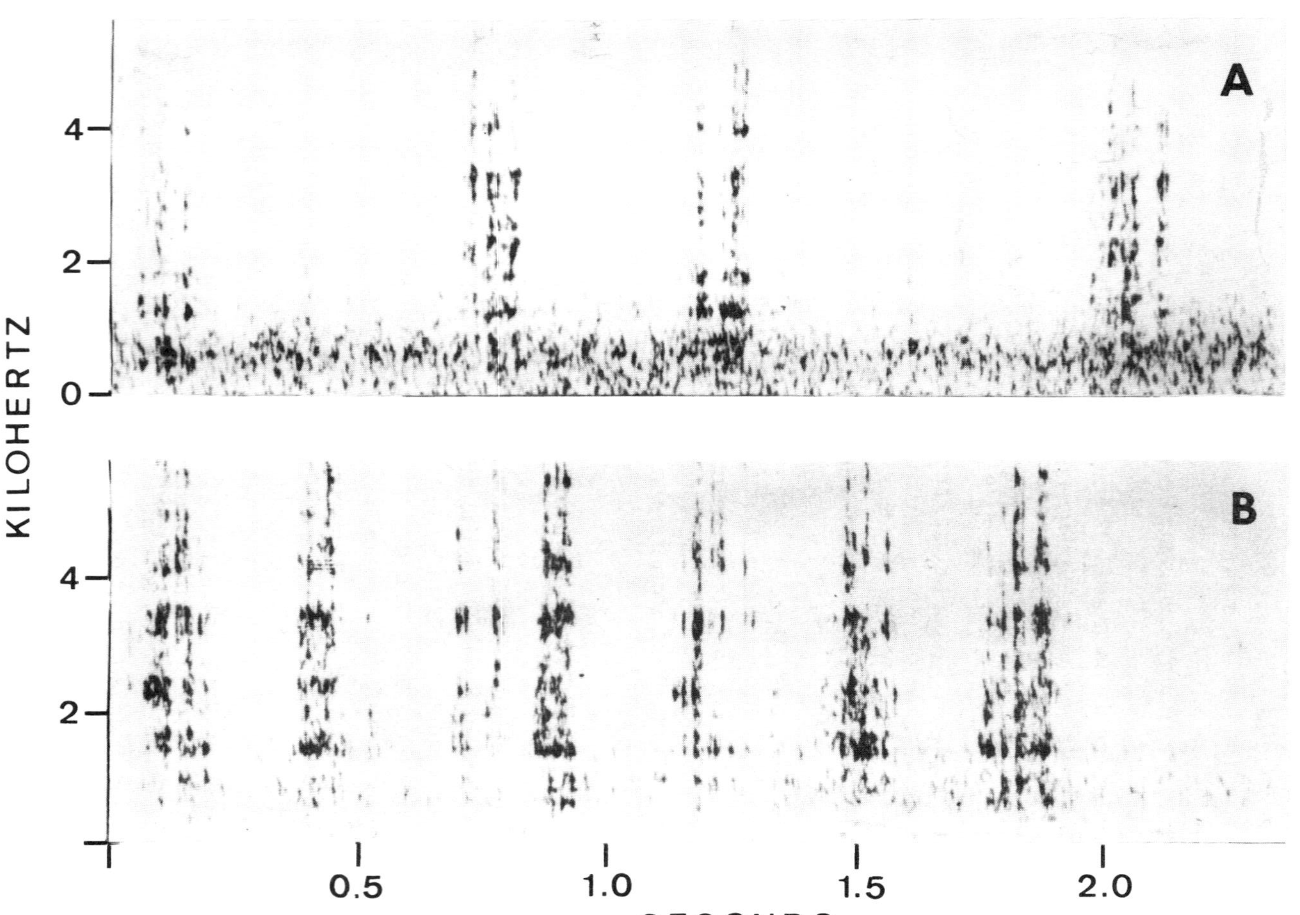
A
B
KILOHERTZ
4
2
0
4
2
0.5
1.0
1.5
2.0
SECONDS

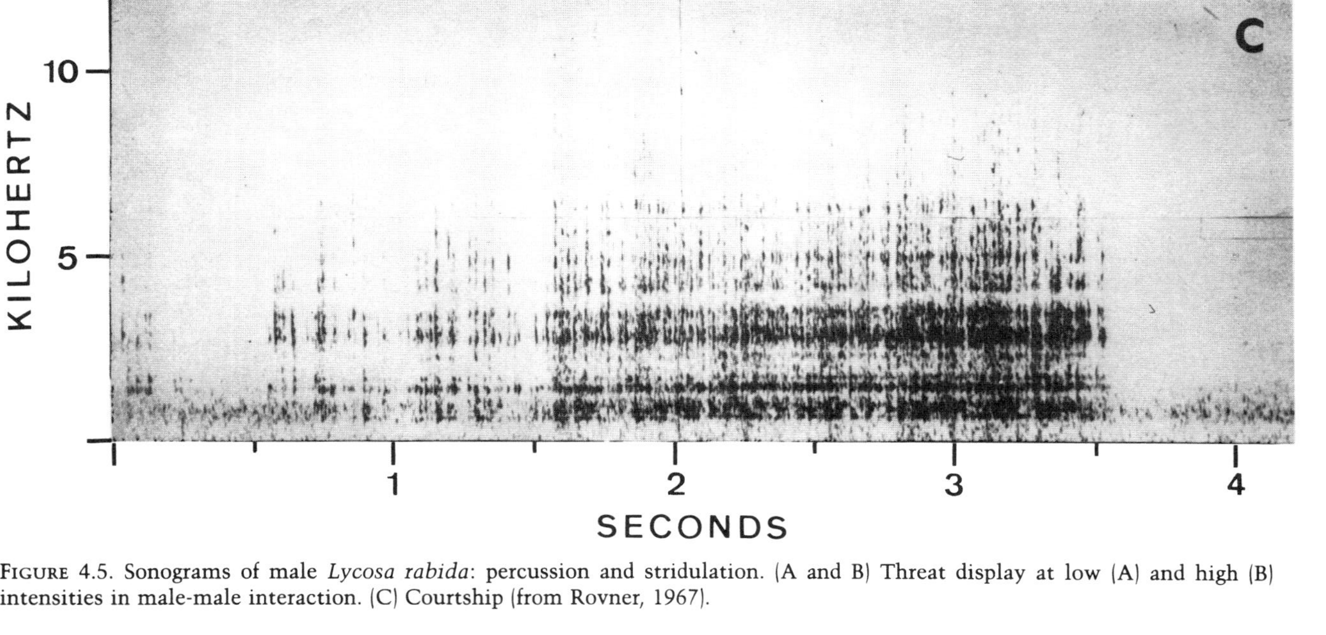

FIGURE 4.5. Sonograms of male *Lycosa rabida*: percussion and stridulation. (A and B) Threat display at low (A) and high (B) intensities in male-male interaction. (C) Courtship (from Rovner, 1967).

ocreata consists of several pulse-trains over several seconds. *S. rovneri* also produces several pulse-trains, but the sound is much more "concise"—the duration of a train of pulses is 0.125 seconds, whereas in *S. ocreata*, the train is more or less continuous for nearly one second. A more complete discussion of these sounds is provided in a later section.

Schizocosa mccooki uses percussion of palps to produce courtship sounds, along with its stridulatory organ. This species, misidentified as *S. avida*, was studied by Buckle (1972). We also have studied this species with specimens from New Mexico (Stratton and Uetz, unpublished). *Schizocosa mccooki* provides an example of drumming behavior to produce sounds during courtship. An episode of courtship by male *S. mccooki* lasts about 6-10 seconds. During this period the male shows a series of 6-9 bursts of percussion, each bout consisting of 2-5 individual taps of the pedipalps. The bouts are very regular in timing, with 2.2 bouts occurring every second (Figure 4.6A). The individual taps are also very regular with 4 taps in 100 msec (Figure 4.6B). Each tap shows a sharply defined transient peak at the beginning of the wave-train, with the amplitude decreasing over a period of about 15 msec (Figure 4.6C). Highspeed (60 frames/sec) cinematography has revealed that the bursts are produced by alternate drumming of palps in a right/left/right or left/right/left sequence.

There is a similarity in the frequencies of the sounds produced in

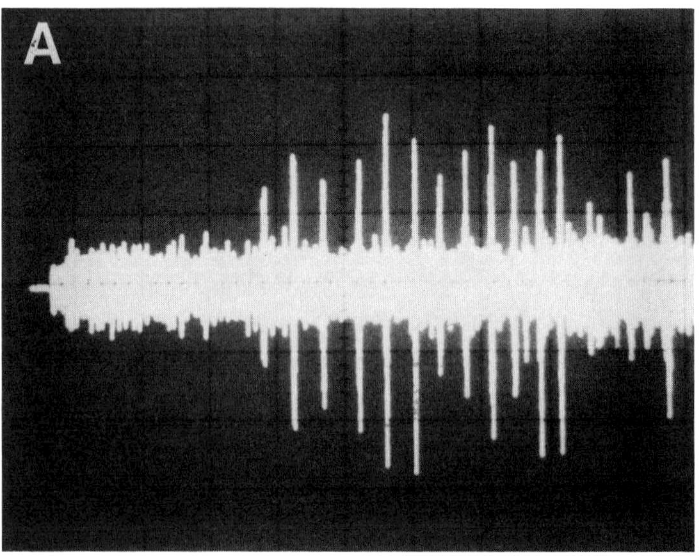

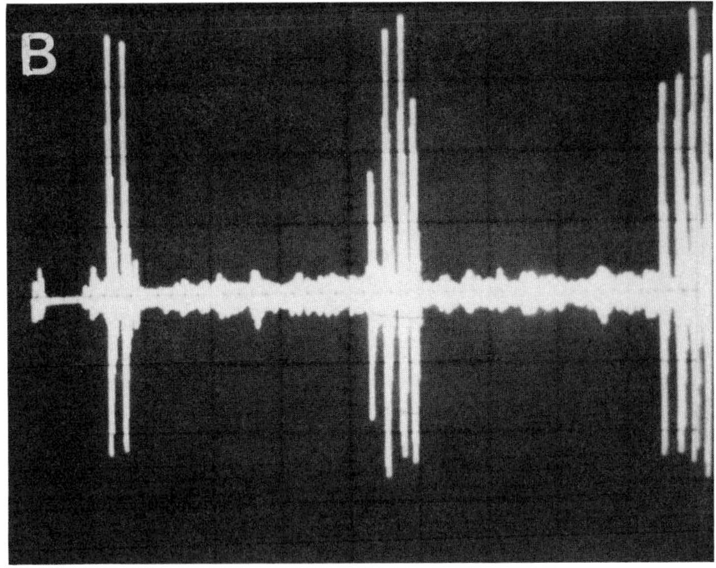

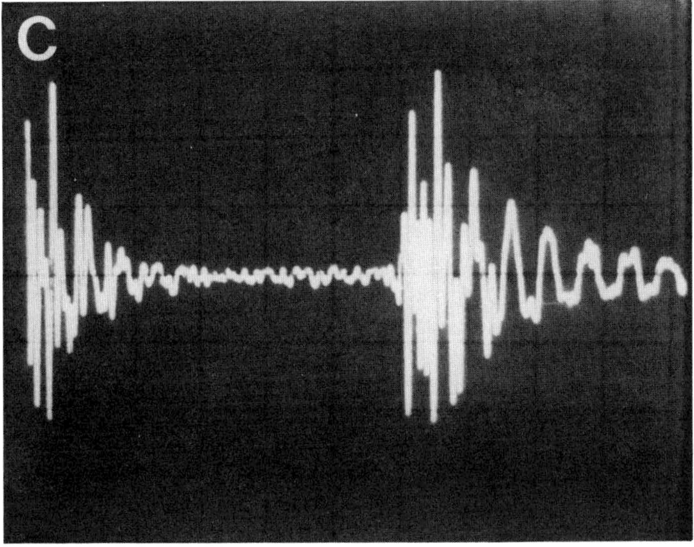

FIGURE 4.6. Oscillograms of the sounds produced by the percussing male *S. mccooki*. (A) 1 sec/div. (B) 0.125 sec/div. (C) 12.5 msec/div.

the Lycosidae. The range varies from 0.15 kHz to 3.5 kHz in *Lycosa rabida* to 2 kHz–6 kHz in *S. ocreata* and *S. rovneri*. The nature of the substrate is important in determining the sound spectra produced by their substrate-coupled stridulatory organs. The family Sparassidae produces lower frequencies: mostly 90 Hz to 150 Hz in *Heteropoda* (Rovner, 1980b); no stridulatory organ is involved in this case. Rovner describes in detail the nature of the sounds produced by *Heteropoda*:

> Sound production resulting from appendage oscillations involved a series of intermittent, very low amplitude "minor" wave-trains that culminated in a louder primary signal lasting approximately 2.5–4.0 sec (Figure 4.4B). The primary signal usually contained 4 distinct wave-trains. . . . The primary signal showed a trend of increasing frequency in successive wave-trains. After the primary signal ended, an inactive period or a bout of exploratory behavior often occurred during a relatively quiet period lasting 36.7 ± 19.60 S.D. sec. The pattern of courtship signaling is an alternation of a relatively quiet period that includes low-level sound-bursts with a period consisting of the louder primary signal.

Interaction of Sounds and Substrate

The question, "Can spiders hear?" was actively discussed until recently. Often airborne vibrations are closely linked with substratum vibrations in soil or leaf litter. It is often not clear whether the spider is perceiving (or responding to) airborne sounds, or substratum vibrations, or both. (There are also several recent papers discussing communication through vibration of silk (Burgess, 1979b; Krafft, 1978; Suter, 1978; Vollrath, 1979b). As these apparently do not deal with acoustic communication, they will not be discussed here.) In the case of wolf spiders using type h (palpal) organs, the vibratory energy is directed into the substrate (Rovner, 1975). In these and other instances, the substrate may act as a sound board, enhancing the loudness of the airborne component. Rovner (1980b) suggests as a generalization that sound production by wandering spiders always includes a substratum component, whether the mechanism be stridulation, percussion, or vibration.

There is both behavioral and electrophysiological evidence that spiders perceive airborne sounds. We shall consider the former type of evidence here. Through playback experiments, Rovner (1967) demonstrated that female *Lycosa* would show oriented responses to airborne stimuli, although these responses were not as strong as those elicited when the females were provided with substrate sound. Females of *Steatoda bipunctata*, a web-building spider, respond to stridulation from the male even when the male is not touching the female's web (Gwinner-Hanke, 1970). Spiders can perceive airborne sound via the single slit sensilla of the tarsi (Barth, 1967), and can also use their trichobothria to detect air movements generated by nearby vibrating objects (Görner and Andrews, 1969).

Biological Significance of Sound Production in Spiders

Importance of Sound Production

In considering the significance of sound production in arthropods, Busnel (1963) discusses the relative importance of acoustic communication in comparison to that of other communication modes. One important criterion mentioned by Busnel is distance—the physical limits to the perception of sound. In spiders, as in most arthropods, acoustic communication can be perceived at greater distances than can visual cues (which are limited by the range of spider eyesight), but the acoustic range is probably smaller than for chemical cues (airborne pheromones in other arthropods are detectable at distances up to several kilometers). Although acoustic communication falls into the mid-range of communication modes based on the distance over which it may be perceived, that range may be optimum when we consider the biological contexts of spider acoustic communication. Since acoustic communication is perceived as air movement or deformation of a substrate (e.g., a leaf surface), it does not require that the sender and receiver be in close proximity. This may be adaptive, given the cannibalistic nature of spiders. A spider capable of getting a message across at a distance (either a courtship song or a defensive sound) has a much greater chance of escaping being eaten!

Acoustic communication is also an efficient means of communication. It usually involves the use of morphological structures that are already present in simple body movements like leg-rubbing, palpal oscillations, etc. (although there may be developmental energy costs). This is in contrast to the energy required for the more vigorous body movements associated with visual displays or for the synthesis of chemical compounds (pheromones). Acoustic communication is also efficient with respect to the facility of diffusion through the environment. Most physical factors (light, temperature, humidity, etc.) do not affect its transmission or alter its parameters. Certain aspects of habitat structure may, however, modify sound transmission, i.e., some substrates may amplify or reduce sound (e.g., leaf litter). It nonetheless appears that acoustic signals are less affected by and less dependent on physical factors than chemical and visual signals. Another important aspect of the efficiency of acoustic communication is the temporary nature of sound emission—it is simultaneous with the behavioral phase of the animal producing it. This can be important, particularly with respect to reproduction. Pheromones may persist in the environment long after the sexually receptive phase of the female releasing them has passed. For these reasons, acoustic signals play a major role in the chain of communicative events preceding mating in a variety of spider species.

Biological Contexts of Spider Sound Production

Despite the numerous references to stridulatory organs and other mechanisms of sound production in spiders, only a few descriptions can be found of why acoustic signaling occurs. Spider sounds can be grouped into three broad categories, based on the apparent biological context of sound production: defensive sounds, aggressive sounds, and courtship sounds.

Among probable defensive sounds, Hinton and Wilson (1970) describe the stridulatory organ of *Micrathena gracilis*, and how it is used to produce a "low-pitched buzz" when the spider is disturbed. We are familiar with this species in Ohio, and have observed the same phenomenon when handling individuals of *M. gracilis*. The sound is audible from about 1 meter away if the spider is on its web, and can be felt as vibration if the spider is held in the fingers. Hinton and Wilson (1970) suggest that the sound is defensive in function, and we agree, although it is not clear how it would work against a predator. Upon picking up a live *M. gracilis* for the first time as a child, one of us (GU) was startled by the "buzz" and dropped the specimen, never to find it again. Perhaps the brief startling effect is sufficient to allow escape. The recent demonstration by Masters (1979) that insect disturbance stridulation does deter predators supports this view. In a related manner, Rovner (1980b) speculates that the defensive web-shaking and leg-vibrating behaviors of araneid and other spiders may produce low frequency, air particle movements as do those of sparassid spiders (see Figure 4.4). These air vibrations may mimic the wingbeat frequencies of predatory wasps and thereby serve to drive predators away.

Legendre (1963) cites several references concerning the "hissing" sound produced by theraphosid tarantulas when cornered. This sound is reported to resemble the hissing of a snake. The "hiss" is apparently created by a stridulatory organ on the chelicerae being rubbed by the pedipalps. Berland (1932) and Legendre (1963) reject the notion that this is a mimicry-based defense mechanism, although Weygoldt (1977) disagrees. Whether or not spider "hissing" noises constitute snake mimicry does not seem to be important. Hissing is a common defensive sound produced by many animals, including both vertebrates and invertebrates. The actual noise is usually startling and may deter predators or at least may allow the animal producing it enough time to escape.

Acoustic communication is also used in aggressive encounters between spiders. The stridulatory organs used for acoustic communication between males and females during courtship are often used in agonistic interactions between males (Weygoldt, 1977). This is the case in *Steatoda bipunctata* and *Teutana grossa*, two theridiid species studied by Gwinner-Hanke (1970). *S. bipunctata* uses stridulation both in courtship and male agonistic encounters, while *T. grossa* uses stridulation only in male-male aggression. Riechert (personal communication) has observed that female *Agelenopsis* spiders produce sounds when charging out on the web to repel intruding conspecifics. These sounds are presumably produced by the tarsi in some way. Rovner (1967) found that *Lycosa rabida* threat display in male-male interactions involved palpal sounds in brief, frequent bursts of pulses occurring at irregular intervals. These sounds were distinctly different from the palp-produced sounds used in courtship display.

Most of the studies of acoustic communication in spiders deal with sounds produced by spiders during courtship. In the review by Weygoldt (1977) mention is made of males of the Linyphiidae, Theridiidae, and Dictynidae using stridulatory organs to produce vibrations while plucking and shaking the webs of females during courtship. For example, the theridiid *Steatoda bipunctata* has a stridulatory organ between the cephalothorax and abdomen, which is used to produce sound with body movements. Other species of web-building spiders appear to rely solely on web vibration and plucking for communication during courtship (Krafft, 1978; Witt, 1975), and in other situations (Burgess, 1979b; Vollrath, 1979b).

Much has been written about the use of acoustic courtship communication by wolf spiders. Males of many species use their palps in drumming on the substratum or in substrate-coupled stridulation as well as percussively (Rovner, 1967, 1975). Several species also produce sounds by tapping the abdomen on the substrate (Bristowe and Locket, 1926; Harrison, 1969; von Helversen in Weygoldt, 1977). Acoustic communication is an important but not essential component of courtship in these spiders. Rovner's experiments with palpless males show that in daylight, females will respond receptively to visual courtship signals. However, since *L. rabida* can be active in the field at night, acoustic communication is an important part of successful courtship under this and other conditions precluding visual communication.

Recent findings by Rovner (1980b) in studies of sound production by sparassid spiders indicate that spiders may also produce sound in courtship by body or appendage vibration (see Figure 4.4). No stridulatory organs are involved, and the sound is a low frequency "buzz" similar to insect flight sounds. These findings raise interesting questions about the observations of other workers who have seen body vibrations of spiders during courtship. Are sounds being produced that are too weak for humans to hear and are they thus going unnoticed? Further research with modern sound recording equipment may reveal this means of acoustic communication to be more common than we have realized (Rovner, 1975).

Evolution of Acoustic Communication in Spiders

The fact that some spiders produce sound without specialized organs merely by vibrating body parts may shed some light on the evolution of acoustic communication in spiders. Weygoldt (1977) suggests that sound production in spiders evolved as a by-product of movements of the palps, legs, or abdomen. Since many such movements are performed by spiders during locomotion, chemoexploratory behavior, etc., as part of courtship, it follows that vibrations thus produced could take on a communicative function as well. The evolution of specialized sound-producing organs and behavior mechanisms would undoubtedly follow. Considering this, the stereotypy inherent in sound production would serve well in producing species-specific courtship sounds, which could function in reproductive isolation of species.

Acoustic Communication and Reproductive Isolation in *Schizocosa*

Our studies of a new species in the genus *Schizocosa* provide an interesting example of how acoustic communication functions in isolating species. The new species, *Schizocosa rovneri* Uetz and

Dondale is named for J. S. Rovner, in recognition of his outstanding contributions to the study of lycosid behavior. It is an interesting species in that it is morphologically identical to *Schizocosa ocreata*, a common wolf spider in the eastern U.S. (formerly known as *S. crassipes*). These species are indistinguishable except that in males of *ocreata*, there is a prominent (although variable) tuft of black bristles on the tibiae of the first pair of legs. These species share similar geographic ranges, and have similar phenologies and habitat preferences (deciduous forest litter). However, their courtship behavior is different (Uetz and Denterlein, 1979).

Pre-copulatory Behavior

The courtship behavior of *S. ocreata* has been described by Dondale and Redner (1978), Kaston (1936), Montgomery (1903), and Rovner (personal communication) as well as observed by us. It consists of tapping with the first pair of legs, accompanied by palpal movements and stridulation. Rovner has suggested that *S. ocreata* may have a "calling song," different in nature from its courtship song. In the later phases of courtship, display consists of alternately raising and arching the first leg pair, then extending the leg forward. Palpal stridulation also is evident at this time.

The courtship behavior of *S. rovneri* differs in many respects from that of *S. ocreata* (Uetz and Denterlein, 1979). The courtship sequence begins with a searching phase, which consists of raising, extending, and lowering the forelegs in a probing fashion. A series of rapid palpal scraping movements occur during searching, followed by alternately raising the palps to the chelicerae. This was characterized as "chemoexploratory" behavior by Tietjen (1977). After a variable number of palpal scraping and raising movements, the male enters a stationary phase wherein the legs are extended and the palps are held nearly perpendicular to and in contact with the substrate. The display phase which follows consists of several bouts of substrate-coupled palpal stridulation (Rovner, 1975), which result in a clearly audible buzzing sound. This is followed by a brief period of inactivity of approximately 5 seconds, after which the spider either goes through another entire courtship sequence beginning at the searching phase, or changes position and resumes stridulation.

Analysis of highspeed films of courtship in *S. rovneri* males has revealed a movement of the entire body associated with stridulation. The spider first raised the cephalothorax and abdomen slightly, then thrust the body downward between the legs, hitting the substrate or nearly hitting it with a "bounce." Rotating movements of the palps were observed at this point, indicating stridulation. Reverberation of the bounce impact was noted in the entire body, and was particularly visible in the abdomen, which continued to vibrate for a few fractions of a second. The onsets of stridulation and reverberation of the bounce impact were simultaneous and appear (when viewed at normal speed) as a "jerk" or "spasm." This may serve to increase the displacement of the tibio-tarsal joint and increase sound output.

The behavior of receptive *S. rovneri* females in response to the male's courtship closely resembles that of *S. saltatrix* (Rovner, personal communication). The female lowers her cephalothorax and extends her forelegs on the substrate, then rises, turns 90°-180° in another direction, and repeats the procedure. Several turns are executed in rapid succession, resulting in approach and mounting by the male.

Tests of Behavioral Isolating Mechanisms

Courtship was displayed by males of *S. rovneri* and *S. ocreata* in most test situations (Tables 4.2 and 4.3), including the presence of a conspecific or heterospecific female or just their silk. However, *S. rovneri* showed quantitative differences in courtship depending on the stimulus (Table 4.4). The stridulation frequency (number of bouts/10 sec) was significantly higher with a conspecific female. The average maximum length of a stridulation bout was nearly twice as long when a male courted a conspecific female as when he courted a heterospecific female, and the interval between bouts of

Table 4.2. Response of male *Schizocosa rovneri* (sp. nov.) to conspecific and heterospecific females or their silk. (+ indicates courtship behavior.)

With female spider			
	+	–	
Conspecific	35	1	$\chi^2 = 1.41$
Heterospecific	28	3	n.s.
With silk (and pheromone)			
	+	–	
Conspecific	10	0	$\chi^2 = 0.0$
Heterospecific	10	0	n.s.

TABLE 4.3. Response of male *Schizocosa ocreata* to conspecific and heterospecific females or their silk. (+ indicates courtship behavior.)

With female spider			
	+	−	
Conspecific	8	1	$\chi^2 = .145$
Heterospecific	14	1	n.s.
With silk (and pheromone)			
	+	−	
Conspecific	10	0	$\chi^2 = .694$
Heterospecific	14	1	n.s.

stridulation with a conspecific female was about half the interval when a heterospecific female was being courted. The male *S. rovneri* clearly shows a higher intensity of courtship when courting a female of his own species.

However, the ultimate arbiter of effective communication is not the measurements of these various parameters, but rather the effect on the female who is receiving the communication. Interestingly, in no case has a *rovneri* female or an *ocreata* female been receptive to a male of the other species. Thirty-five out of 36 *S. rovneri* females responded positively to conspecific males; in 33 trials, there were no responses to heterospecific males. In *S. ocreata*, 24 out of 30 females responded to conspecific males, while none of 30 re-

TABLE 4.4. Quantitative analysis of behavior exhibited by *Schizocosa rovneri* males during courtship. (Means indicated with ± 2 S.D.)

	N	$\bar{X}$ *time in searching mode (min)*	$\bar{X}$ *(no./10 sec) stridulation frequency*	$\bar{X}$ *max. length courtship display (min)*	$\bar{X}$ *interval between bouts of courtship display (min)*
W/conspecific female	20	1.9987 ± 4.02	2.193 ± 1.262	4.03 ± 5.92	6.85 ± 5.82
W/heterospecific female	17	2.44 ± 3.20	1.473 ± 1.060	2.075 ± 2.04	12.407 ± 15.50
Statistic: t-test		n.s.	significant $P < .005$	significant $P < .01$	significant $P < .05$

sponded to heterospecific males. To compare the effectiveness of several modes of communication, we conducted a series of experiments designed to isolate or combine visual, acoustic, and/or chemical communication in these spiders.

Relative Importance of Components of Courtship Communication

The male's response to female silk (containing pheromone) was tested by placing a female cage liner card in a test arena and placing the male on top of it. The male was also tested in the presence of the female plus her silk. In this case, the female was placed under a plastic bubble and her cage card was placed under her (Figure 4.7A). The female's behavior was also noted, particularly any display of receptive behavior. In some cases the male and female were allowed to come into contact with one another. The male's response to visual stimuli from the female was tested by placing the female in a plastic bubble cleaned with alchohol (to remove any trace of pheromone) and sealing her in with vaseline. The test arena was also cleaned with alcohol and fresh arena paper was used.

The female's response to visual stimuli from the male was tested by sealing a female in a plastic bubble with paraffin, and hanging the bubble slightly above the substrate from bricks adjacent to the test arena (Figure 4.7B). The bubble was not touching the substrate paper provided for the male. The female's bubble was also surrounded by a thin sheet of clear plastic to keep the male from touching the bubble. A small amount of airborne sound could have reached the female. However, this set-up effectively eliminated substrate-conducted vibratory communication and chemical communication. If the male did not court with only the visual stimuli, he was provided with a female cage card to induce courtship.

The female's response to vibratory stimuli was tested by placing her in a bubble sealed with paraffin and visually isolating her from the male with white paper. The bubble was then placed on the arena paper on which the male was courting.

Results of these experiments indicate that, for the male, stimuli from the silk of the female are sufficient to trigger courtship behavior. This may involve both chemical and tactile cues. However, the stimulus is not species-specific—the silk of either species will elicit courtship. Visual contact with the female is sufficient to trigger courtship only if the female moves, but is necessary for the male to orient to the female in the absence of vibratory cues. For the female, visual stimuli are not sufficient to release receptive

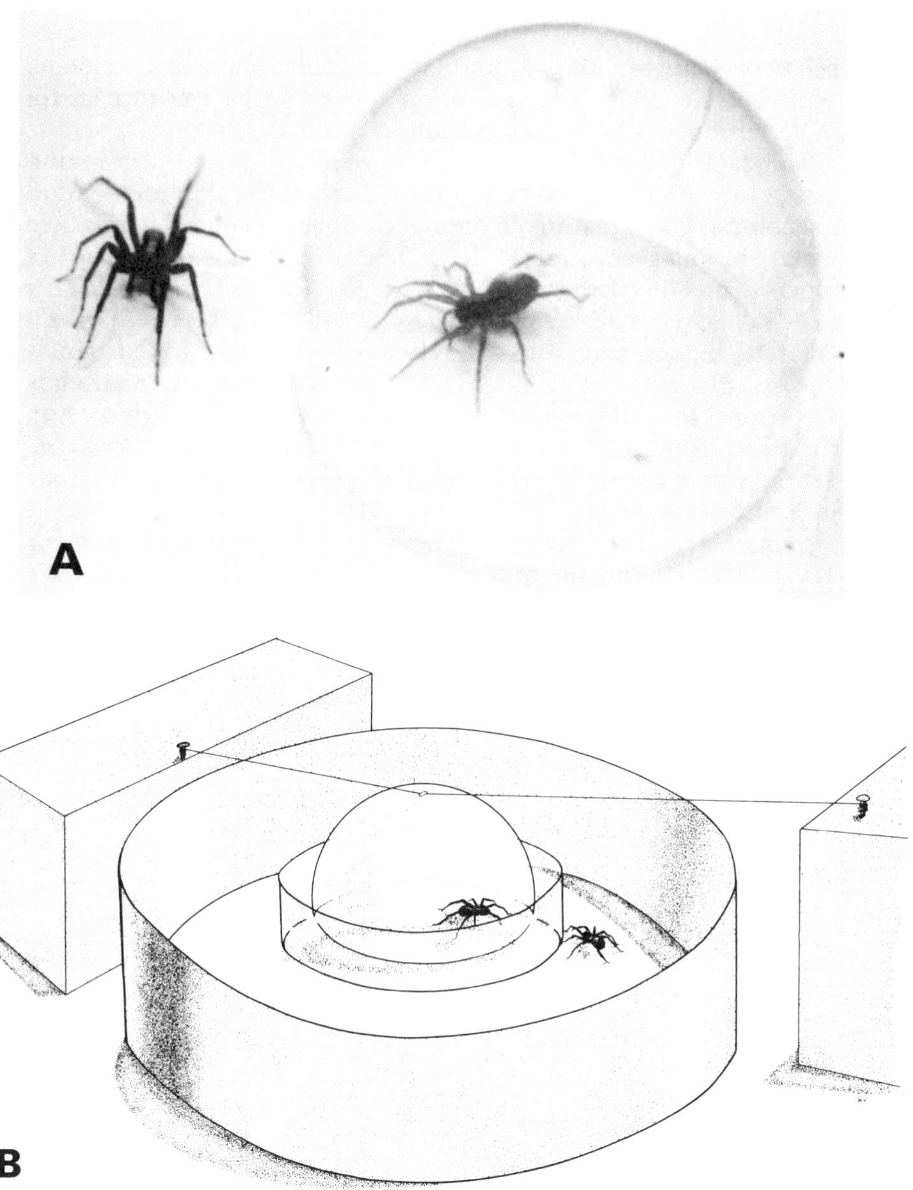

FIGURE 4.7. (A) Photograph of experimental chamber used in study of courtship behavior of *Schizocosa rovneri.* Female (inside plastic bubble) is in visual and auditory contact with the male (outside). The male exhibits courtship in response to the presence of female silk on the paper and/or visual stimuli from the female. (B) Diagram of experimental isolation chamber (used in studies described in text) permitting visual communication between male and female, but no auditory or chemical communication.

behavior. Only vibratory stimuli are capable of inducing female receptivity. Acoustic communication appears to be a critical factor in determining whether copulation will occur.

The data indicate that there is an important chain of communicative interactions between the male and female of *S. rovneri* during courtship and mating. In nature, this chain is probably initiated when the male encounters the female's pheromone. This pheromone is possibly airborne (Tietjen, 1979a), or contained in the female's silk (Hegdekar and Dondale, 1969), or both. This is sufficient to induce courtship stridulation (Table 4.5). If the female "hears" the male (either through the substrate or through airborne vibrations) she will respond with receptive behavior nearly 80% of the time (Table 4.6). Our lab experiments show that if she is prevented from hearing him (by being suspended above the substrate in a closed chamber) but is in visual contact with the male, the response is reduced to 37%. The female's response can also be quantified in terms of the number of receptive display turns she makes per bounce of the male. This number doubled when the female was provided with acoustical signals from the male.

The importance of effective acoustic communication between

TABLE 4.5. Percent response of male *Schizocosa rovneri* to various stimuli from females.

Stimulus	*N*	*Orientation*	*Bounce*	*Jump/Bounce*
Control	39	—	38.5	0.0
Chemical stimulus (silk w/pheromone)	29	0.0	93.1	15.8
Visual stimulus	20	15.0	25.0	0.0
Visual and chemical stimulus	36	88.9	91.6	66.7

TABLE 4.6. Percent receptive response of female *Schizocosa rovneri* and turn/bounce ratio with various stimuli from males.

Stimulus	*N*	*Receptive*	*Number of turns/bounce*
Control	39	0.0	—
Visual stimulus	16	37.5	.221
Auditory stimulus	19	78.9	.456
Auditory and visual stimulus	18	88.9	.547

male and female becomes obvious at this point. The male must "convince" the female that he is a courting male of the same species, lest he waste time and energy with a non-receptive female or end up a victim of the female's predatory instincts. Acoustic communication would seem to be the most appropriate way to do this, because it is effective under conditions in the leaf litter not allowing visual contact and can be perceived at a greater distance than is visual display. Since lycosid spiders use a "sit-and-wait" predatory strategy, hiding among leaves, etc., a male could fall victim to his prospective mate before the message got across. Dead leaves would serve to amplify the effects of vibrations of courting males (Rovner, 1975) and thus acoustic communication by substrate-coupled stridulation would be particularly adaptive in this system.

Visual signals from the female appear to be important only for the orientation of the male to the female. There may be little or no acoustical communication from the female, since the males did not orient to visually shielded females which were still in contact with the substrate; however, the experimental design could have prevented the transmission of weaker signals. In no instance did a male orient to a motionless female when the female was in view, while nearly 71% of the males did orient to a moving female. Of the latter, the largest proportion occurred when the female was on the substrate, receiving acoustical communication from the male, and turning at a higher frequency. This was confirmed with forced mating experiments where females were anesthetized with CO_2. When presented with a non-moving female, the male would often not approach to copulate. However, if the female was moved, even slightly, the male would orient to her and approach. There was a similar increase in intensified courtship behavior when the male was presented with combined chemical and visual stimuli. Visual or chemical stimuli alone were normally not sufficient to trigger this behavior. The movement of the female then appears to be important in triggering the intense courtship phase that normally leads to copulation.

Reproductive Isolation

We found that acoustic communication is not only important in the prevention of cannibalism, but also in the reproductive isolation of these closely related species, whose microhabitat preferences and seasonal occurrence are similar. *S. ocreata* breeds 2-3 weeks earlier, and prefers the complex litter of upland deciduous forests, while *S. rovneri* breeds later and prefers the compressed litter of moist depressions, river bottomlands, and floodplain forests. However, there is much overlap, and individuals could easily come into contact with each other. Differences in courtship behavior are clearly important in maintaining the otherwise incomplete isolation of the two groups, and would obviously be crucial to maintaining species genetic integrity. For these and other reasons we suggest that *S. rovneri* is an ethospecies, morphologically similar to but reproductively isolated from its sibling species by courtship behavior, particularly acoustic communication.

The term "ethospecies" was used by Hollander et al. (1974) in the case of *Pardosa vlijmi*, a species reproductively isolated from a morphologically identical sibling species by courtship behavior. Studies of ethospecies have given some insight into the role of courtship behavior—behavior whose function is sexual communication—in the process of species isolation.

The definition of what is and what is not a species has been discussed often in the biological literature. The species is the only taxonomic category for which there is an ultimate, objective criterion for classification. That criterion is the ability or inability of individuals to interbreed; its origin is the biological species concept of Mayr (1963).

Many animals have been studied with respect to reproductive behavior, and it is assumed that the sequence of behaviors leading to copulation serves as a pre-mating mechanism of reproductive isolation. Although this has been demonstrated only rarely, selection would favor definitive species-separating mechanisms at this point, in that hybrids may be less fit or inviable, and reproductive mistakes would be prevented.

Arthropod systematists rely heavily on morphological characters in making most species decisions—particularly characters of the genitalia. Arthropod genitalia, especially those of spiders, are elaborate structures whose function would appear to be a mechanical reproductive isolation—a "lock and key" system. There are cases of arthropod species with different genital structures which interbreed (Perdeck, 1958), and there is some experimental evidence that surgical alteration of male genitalia in bees may not impair or prevent copulation and fertilization (Sengun, 1944). Conversely, there are species with identical genitalia like *S. rovneri* and *S. ocreata*, which do not interbreed. These are known as cryptic species (Walker, 1964), and it is estimated that one-quarter of all cricket species fall into this classification—undetected by morphological criteria, yet reproductively isolated by the songs they use in courtship and mating.

Clarification of taxonomic problems by means of interbreeding studies or studies of reproductive behavior has been successful with spiders (Dondale, 1967; Hollander and Dijkstra, 1974; Hollander et al., 1973; Rovner, 1973; Taylor and Peck, 1974). It is clear from the behavioral tests we have made that our two species will not freely interbreed (Tables 4.6 and 4.7). Knowing that their genitalia are identical, we wondered if these species might be capable of interbreeding except for the behavioral barrier. We then performed a series of experimental forced copulations to test for interfertility. Females of both species were anesthetized with CO_2 following a variable period of allowing a male of the other species to court. The anesthetized female was placed in front of the male and slowly moved forward with front legs extended. The male would then mount and begin attempting palpal insertion by scraping the side of her abdomen. When females emerged from under the influence of the CO_2, the male was already mounted, and they responded as if in a "normal" conspecific mating.

Offspring were produced in all types of interspecific crosses (Table 4.7), and no significant differences in egg sac production or hatching success were found between heterospecific and conspecific matings. Hybrid males exhibit the distinctive courtship behaviors of each species, frequently "switching" from one parental species pattern to the other. Male hybrids will court hybrid females

TABLE 4.7. Egg sac production and hatching success in conspecific and forced heterospecific matings of two wolf spider species. (+ indicates egg sac produced and/or hatched; − indicates none.)

Egg sac production				
	♀ *S. ocreata*		♀ *S. rovneri*	
	+	−	+	−
W/conspecific	16	7	16	1
W/heterospecific	15	3	14	3
χ^2	1.04		2.00	
Significance level	n.s.		n.s.	
Hatching success				
	♀ *S. ocreata*		♀ *S. rovneri*	
	+	−	+	−
W/conspecific	9	7	13	3
W/heterospecific	8	7	9	5
χ^2	.027		1.099	
Significance level	n.s.		n.s.	

as well as females of either parental species with equal frequency, as will males of both parental species. However, females of parental species rarely respond receptively or copulate with hybrid males (3 in 20 trials with female *S. rovneri*; 1 in 28 trials with female *S. ocreata*). Female hybrids rarely respond to courtship from parental or hybrid males (4 in 45 trials with parentals; 9 in 54 trials with hybrids).

Interspecific matings would probably never occur in the field because of behavioral isolating mechanisms. However, since hybrids can be produced in laboratory matings, there do not appear to be either mechanical barriers or extensive postmating reproductive isolating mechanisms between these two species. (Such mechanisms would presumably involve gamete or zygote mortality.) Data are inconclusive beyond this point concerning other postmating barriers. The data we have, however, underscore the importance of behavioral isolating mechanisms in these species.

Acoustic Signals and Reproductive Isolation

There are three things to be expected if acoustic signals are important in reproductive isolation (Alexander, 1967): (1) Species reproductively active in the same place and time should not have identical acoustical signals; (2) Species which overlap for a portion of their ranges would differ more in regard to acoustical behavior in the area of overlap than where they do not overlap; (3) Closely related species would have similar acoustic signals. These expectations are demonstrated in several studies of reproductive isolation in Orthoptera (Alexander, 1957, 1967; Walker, 1962, 1963, 1964, 1974, and others). Similar patterns were seen in this study of spider acoustic communication. The high probability of overlap in microhabitat during the breeding period of the two species is suggested by Uetz and Denterlein (1979). On the basis of Alexander's (1967) criteria, we would not expect the two species to have identical acoustical signals. However, the two species are apparently closely related. (Based on morphological aspects, they were once considered one species.) We would therefore expect them to have similar acoustic signals. These expectations are realized when we examine recordings of stridulation in these two species.

Sound and vibration recordings were made with a Bruel and Kjaer accelerometer (Type 4366) high-sensitivity vibration pickup leading to a Bruel and Kjaer sound level meter (Type 2203) whose output was recorded by a Teac tape recorder (model 2300 SX). The experimenter could monitor the sounds by headphones as the recordings were being made. Recordings were made at two different speeds:

19.05 cm/sec and 9.52 cm/sec (7.5 ips and 3.75 ips). Equipment other than the accelerometer was placed on a separate table to minimize the effects of machine noise on the animals and on the recordings. The recordings were used to make sonograms and oscillograms.

Sonograms were obtained from a sound spectograph (Kay Electric Co. 6161B Type B/65; 85-16000Hz Spectrum Analyzer). Oscillograms were obtained from a recording oscilloscope (Tektronix 5103N oscilloscope). Pictures were taken of the oscilloscope screen with a camera.

The courtship of male *S. ocreata*, as mentioned above, involves tapping of the first pair of legs, which is accompanied by palpal movements and stridulation. The courtship is active: the male is moving about nearly continuously while tapping and stridulating. Oscillograms of the sounds produced by the courting male *S. ocreata* show complexity without clear temporal patterning (Figure 4.9A). A single episode of sound production can be 10 seconds long, and is followed by a period of silence. There are apparently many

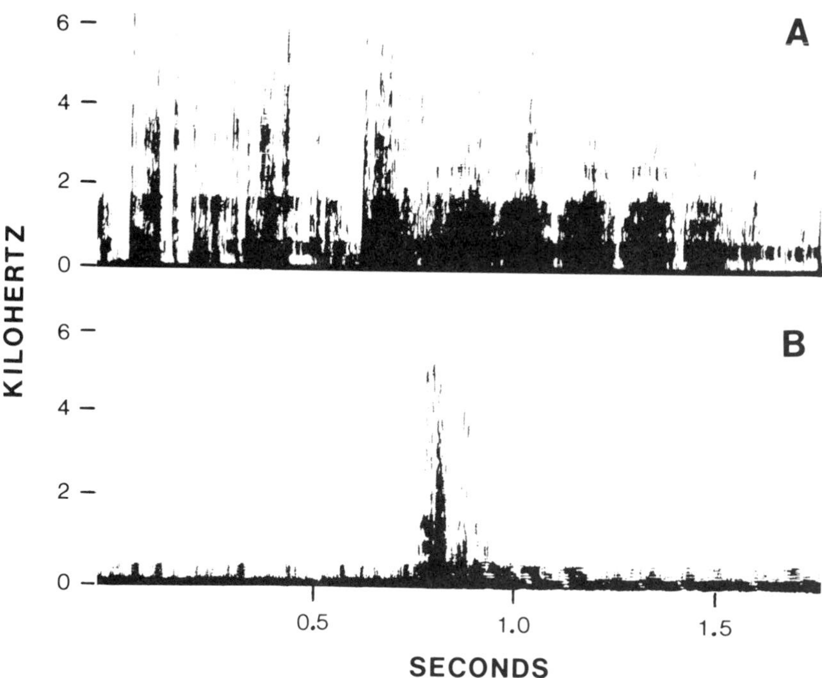

FIGURE 4.8. Sonograms of the courtship sounds produced by male *S. ocreata* (A) and *S. rovneri* (B).

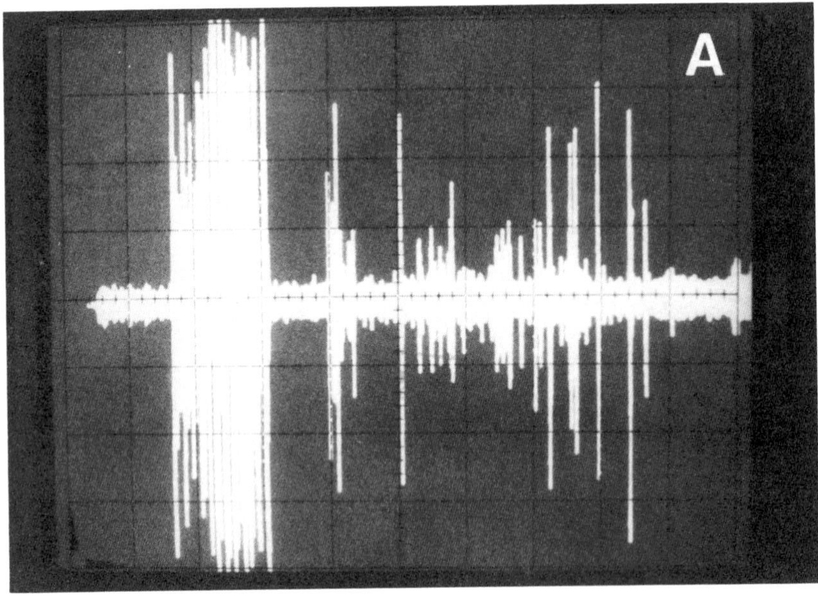

FIGURE 4.9. Oscillograms of the courtship sounds produced by male *S. ocreata.* (A) 1.25 sec/div. (B) 0.125 sec/div. (C) 5 msec/div.

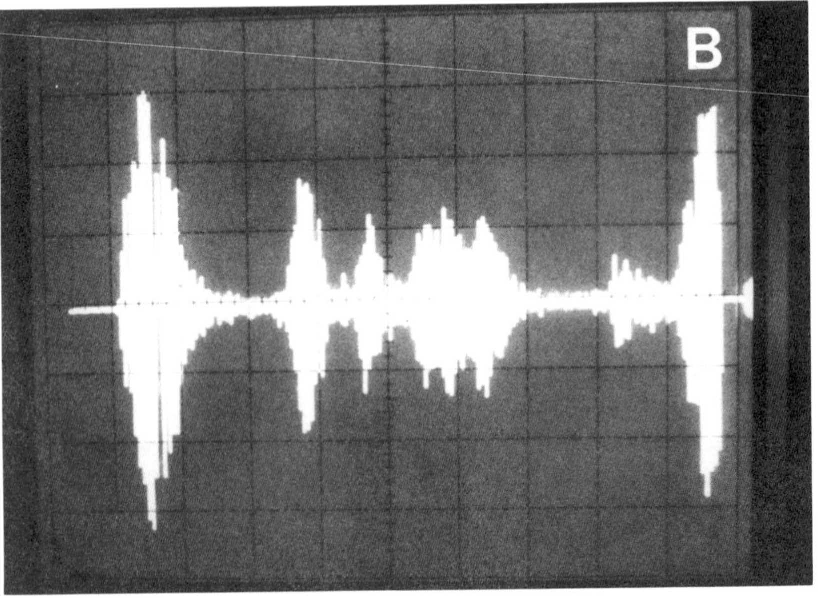

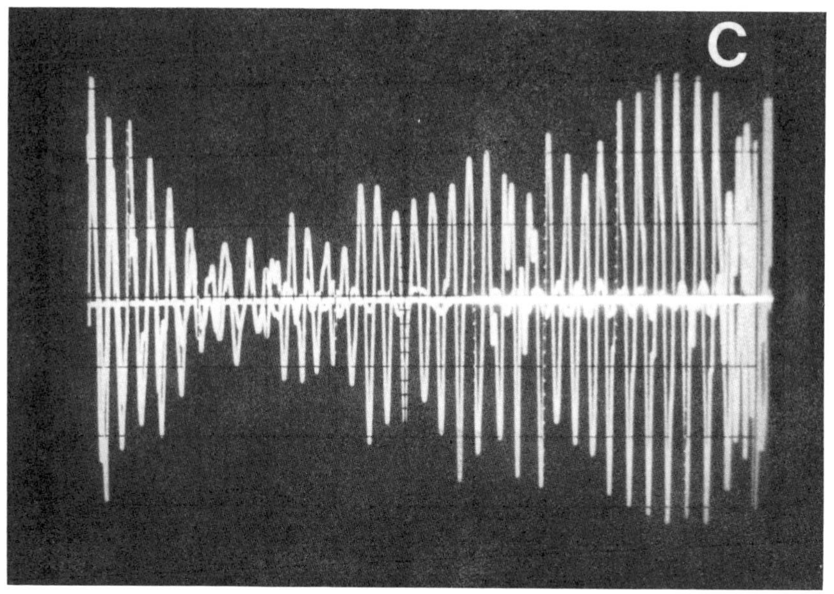

strums or pulses in a single episode (20 or more, Figures 4.9A and B). A single wave-train may last 30 msec (Figure 4.9C). The example given shows 24 waves of variable amplitude. The fundamental frequency can be calculated from this and is about 800 Hz. Analysis of sonograms (Figure 4.8A) shows that most of the energy is below 1.5 kHz.

The courtship of male *S. rovneri* consists of a series of "bounces." The palpal joint appears to rotate several times during a bounce (as observed by highspeed cinematography), but definite passes of the scraper across the file are not easily discernible in either the films or the oscillograms. One bounce occurs about every 3.5 seconds in fairly even intervals (Figure 4.10A). Thus the sounds made by *S. rovneri* are much more regular in their patterning than those made by *S. ocreata*. The sound during a single bounce lasts about 0.25 seconds (Figure 4.10B), and probably includes several pulses or strums. A single wave-train shows high amplitude waves and lasts about 25 msec (Figure 4.10C). As there were 13 high amplitude waves in this particular sequence, the fundamental frequency is about 520 Hz. As with *S. ocreata* the sonogram shows that each strum consists of relatively broad spectrum sounds, with most energy below 1.5 kHz (Figure 4.8B).

It would appear that these two species differ mainly in the temporal patterning of sound production during stridulation. Dondale

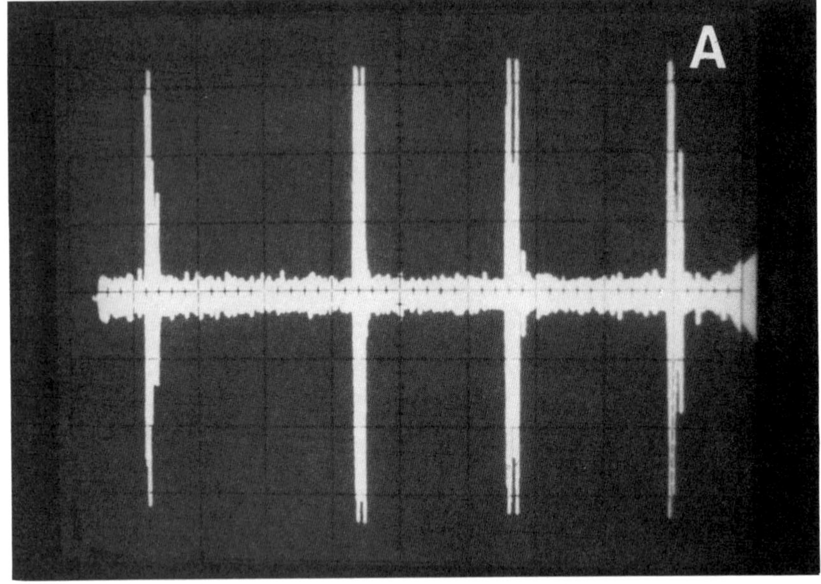

FIGURE 4.10. Oscillograms of the courtship sounds produced by male *S. rovneri*. (A) 1.25 sec/div. (B) 0.125 sec/div. (C) 5 msec/div.

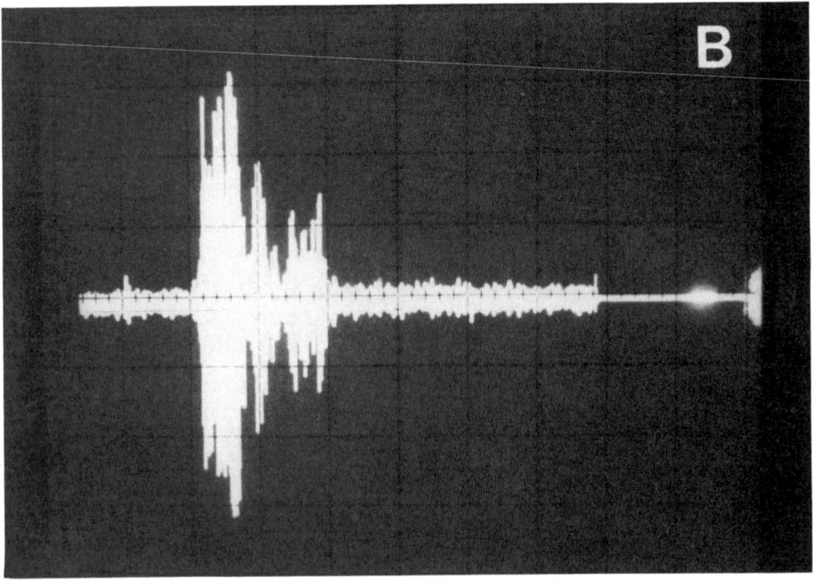

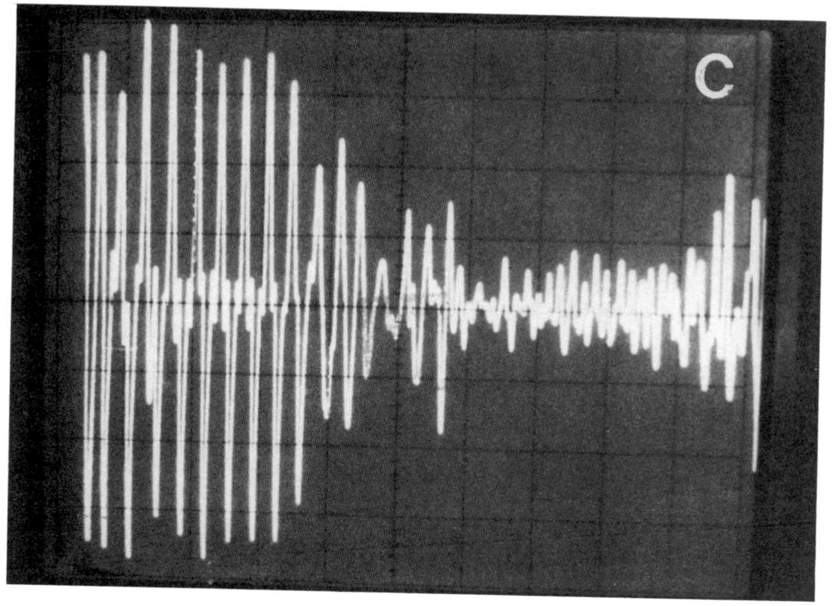

(personal communication) indicated that both species had the same number of ridges (30) on the stridulatory file. However, the number of strums per bout of stridulation is quite different between the two species. *S. ocreata* produces sounds continuously for a much longer period than does *S. rovneri*. This supports the morphological evidence that the two species are using the same device to make different sounds. These recordings add credence to the hypothesis that these closely related species have recently diverged.

The use of substratum-coupled sound production by wolf spiders (Rovner, 1975) raises some interesting questions about the linking of behavior and ecology. How important is the structure of the environment in transmitting or amplifying sound? We have speculated elsewhere (Uetz and Denterlein, 1979) that the differences in courtship behaviors of *S. rovneri* and *S. ocreata* may be related to the preferred litter substrate of each species. The prolonged, low amplitude "purring" of *S. ocreata* may be more adaptive in the complex, dry leaf litter of upland deciduous forests than in the lowland areas preferred by *S. rovneri*. The brief but louder burst of stridulation accompanied by a body "bounce" demonstrated by *S. rovneri* may be transmitted better in the open areas with compressed leaf litter inhabited by this species. We are currently investigating this problem, and our results will undoubtedly raise further questions about the connection of behavior and habitat. Are spiders adapted to communicate acoustically using a particular substrate? Harrison (1969) says that *L. gulosa* does not attempt to strum when it pauses on a different substrate (i.e., a substrate other than dead leaves). Von Helversen (personal communication to J. S. Rovner) reports that *Hygrolycosa rubrofasciata* would court in the lab only on suitable substrates. If so, are they at a disadvantage on other substrates and thus restricted in habitat choice? If the microhabitat influences the nature of sound produced during courtship, might that interaction also play a role in the reproductive isolation of species? Are hybrids between species capable of communicating effectively on both substrates, or neither? We do not yet know the answers to these questions, but hope to find them soon. There are implications for other species as well: how is the mode of acoustic communication (percussion, stridulation, vibration) related to the substrate used? Is the behavior adapted to the substrate, or is the preference for the substrate dictated by the behavior?

Concluding Remarks: Future Research

As is obvious from this review, the study of acoustic communication in spiders is in an early stage. There is certainly a great deal to be learned, and many questions to answer. With the advent of high-sensitivity vibration and sound recording instruments and computer analysis of sound and frequency patterns, we will likely see many advances in the next few years. In concluding this chapter, we wish to suggest several areas of inquiry that may prove fruitful to future researchers.

Earlier in this chapter, we noted that there were many types of sound generators among the approximately 26 families of spiders capable of sound production. Acoustic communication obviously is important in the order Araneae, the diversity of mechanisms indicating that it has arisen independently in unrelated groups over evolutionary time. Extensive surveys of acoustic communication in spiders, including recording and analysis of sounds produced, and assessment of their biological function, may reveal and/or substantiate phylogenetic relationships and evolutionary patterns within and between groups. Considering the amount of groundwork already done within the family Lycosidae, the opportunity exists for comparative studies of genera and species. The results of such research would be invaluable to taxonomists as well as ethologists.

Variability is an important aspect of acoustic communication in need of much clarification. Are the sounds produced by individuals within a species identical in all respects? In what ways do they differ, and what independent variables influence them? Is temperature an important variable in regulating temporal patterning in these poikilothermic animals, as it is in various insects (Busnel, 1963)? Does age or behavioral state modify the nature or pattern of the sounds produced? Does sound production in courting spiders vary with respect to stimulus intensity, as was suggested earlier for *Schizocosa*? We have shown that spider acoustic communication is important in maintaining the reproductive isolation of species. It would be valuable to know how much of the variability of sound pattern is due to heredity or to experience. Is the species sound pattern genetically "fixed"? Do hybrids between species show a combination of acoustic elements found in the sounds of their parents? These questions need to be answered before definitive comparisons can be made.

Another area of potential research is the organization of spider acoustic communication. We know that some species use sound in both courtship and agonistic interaction and that the sounds produced in each case are different. Do many spider species have "calling songs," "rivalry songs," and "courtship songs" as crickets do? Can the pattern of sound production be varied to create different "messages"?

As yet, there are no answers to these very basic questions. We hope that our future research and that of our colleagues and others unknown to us will provide them.

Acknowledgments

The research reported in this chapter was supported by grants from the University of Cincinnati Research Council and the Society of Sigma Xi, and the Department of Biological Sciences Graduate Research Funds. Figure 4.7B was prepared by Kevina Vulinec. The authors thank all those who have played important roles in the research and writing, and their assistance is gratefully acknowledged.

Chapter 5

VISUAL COMMUNICATION IN JUMPING SPIDERS (SALTICIDAE)

Lyn Forster
Otago Museum
Gt. King Street
Dunedin, New Zealand

Introduction

One of the most elegant examples of visual communication in invertebrates is the display behavior of the jumping spiders. These sometimes brightly colored and distinctively patterned spiders have been renowned for their spectacular courtship displays ever since the Peckhams (1889, 1890, 1894) first drew attention to them. Indeed, it was the Peckhams who originally proposed a communicatory role for such behavior, arguing that the fine eyesight of these spiders permitted them to observe each other very closely and in particular, enabled the females to appraise the relative attractiveness of male suitors and so select a suitable mate.

Although successive workers have postulated other functions for courtship behavior (See Crane, 1949a,b), the present evidence, to be reviewed in this chapter, is that jumping spiders have exceptionally acute eyesight (Drees, 1952; Dzimirski, 1959; Forster, 1979a; Homann, 1928; Kaestner, 1950; Land, 1969a,b, 1972b, 1974) and that, from both qualitative and quantitative analyses, visual communication undoubtedly occurs during courtship interactions (Forster, 1979b). These findings are supported by other studies which show that jumping spiders are highly visual animals, and that much of their daylight behavior is mediated by three of their four pairs of eyes (Crane, 1949b; Forster, 1977a; Heil, 1936; Homann, 1928; Land, 1971; Manly and Forster, 1979). Of the two pairs which face forward, the middle eyes are especially large and conspicuous (Figure 5.1).

In addition to courtship, other stereotyped pattens of display take place during interactions between conspecifics. When two juvenile *Trite planiceps* face each other, for instance, they exhibit a variety of postures and movements, many of which depend on the relative sizes of the participants, as well as on their patterns of locomotion and the distance separating them (Forster, in prep.). Most male spi-

FIGURE 5.1. Peering out from beneath a leaf, a jumping spider (*Trite bimaculosa*) waits for prey. The two large principal eyes in the center are flanked by a smaller pair. Photograph by courtesy of R. R. Forster.

ders adopt ritualized patterns of posture and approach when they meet (Bristowe, 1958; Crane, 1949b; Forster, 1979b); in some species, females exhibit similar behavior under rather more specific circumstances (Forster, 1979b). Mirror-image displays, elicited from males of many species, testify to the visual nature of such agonistic sequences (Crane, 1949b; Forster and Forster, 1973; Hutchinson, 1879; Peckham and Peckham, 1894), and suggest that these behaviors may also function in the exchange of information.

There is evidence too, of the occurrence of interspecific visual communication when two salticid species live in habitats which overlap (Forster, unpubl.). Since this overlap is apparently a common phenomenon among larger salticid species (Blest, personal communication; Enders, 1975b), it is obvious that confrontations must occur and that such jumping spiders require, within their behavioral repertoire, visual reactions which promote their mutual survival.

Interest in animal communication has burgeoned in recent years, and the reader is referred to Hinde (1972), E. O. Wilson (1975), Sebeok (1977), and Smith (1977) for some excellent discussions of the theoretical issues involved. This chapter adopts the view that communication requires a sender, a signal, and a receiver, and that communication has in fact taken place when a signal from the sender alters the probability of response by the receiver (Wilson, 1975). To this we will add Burghardt's (1970) proviso that the signal should also confer some advantage to the signaler.

Jumping spiders are excellent subjects upon which to undertake both qualitative and quantitative analyses of visual communication, as this chapter hopes to show. They are small terrestrial animals often readily observed in the wild (see Bristowe, 1929, 1958; Gertsch, 1979; McKeown, 1936; Peckham and Peckham, 1894, for some delightful descriptions of salticid behavior) and easily kept in the laboratory (Jackson, 1978c). Much of their behavior is manifested by highly predictable "sets of responses" largely elicited by visual stimulation, while sexual dimorphism and epigamic decorations are common at maturity, characters which lend themselves to description and quantification.

But while the behavior of jumping spiders appears to be dominated by visual skills, this does not preclude the use of other sensory modalities. Crane (1949b) suggested that olfaction had a role in courtship and mating in some species, while more recently Jackson (1977b) has demonstrated that *Phidippus johnsoni* employs two types of courtship, one tactic being visual and the other vibratory and perhaps chemotactile. In the light of this finding, and in view of Platnick's (1971) earlier comment that courtship is a conspicuous activity which places the participants in considerable danger, then the question is raised as to why visual forms of courtship and agonistic behavior have evolved and why indeed they persist, if these spiders have safer and more reproductively successful alternatives at their disposal (Jackson, 1980b). Such questions make studies of visual communication in salticids all the more important; some of the answers, when they are obtained, may have much wider implications, for they may help us to understand the evolution of visual signaling systems in other animals.

Courtship and Agonistic Displays

Anyone who casually observes a male salticid courting a prospective mate will realize very quickly why the Peckhams (q.v.)

were so enthralled by these fascinating spiders. In many parts of New Zealand, for instance, there is an active little jumper, *Euophrys parvula* (generic placement according to Bryant, 1935; it is possible that this species is not congeneric with those of the Northern Hemisphere), commonly found on the walls of houses. Upon maturity, the male features a dark red patch of hairs on his forehead, augmenting the creamy white head-band and faintly patterned abdomen typical of both the juvenile and the adult female spider. When the opportunity comes for these spiders to mate, the male begins his courtship dance. With his third pair of legs lifted high in the air, he struts and sidles in front of the docile female in a dazzling display of color and movement (Forster and Forster, 1976).

All the elements of the male's performance, the prominent red forehead, the spectacular postures, the exaggerated movements, suggest that they are involved in presenting visual information to the female. But if we are to regard the male's behavior as communicatory, there should of course be visual information deriving from the female and perceived by the male. Insufficient attention has been given in the past to this particular point. Hence the first step toward an analysis of the communicatory nature of courtship in *E. parvula* involves a breakdown of both male and female behavior into smaller components. Once we know the way these events are ordered we can examine this sequence in relation to the significant non-behavioral visual feature that distinguishes the sexes—the red forehead of the male. In the present instance, events which dominate agonistic interactions between male spiders can also assist in the analysis.

Sequential Organization of Behavior

For convenience the smaller components (acts or events) of courtship and agonistic displays of *E. parvula* are grouped into three stages (Figure 5.2). Stage 1 consists of orientation of the spiders toward each other and the subsequent posture and/or movement (primary events) adopted by both spiders. Stage 2 consists of those postures and movements which usually close the gap between them, while Stage 3 events are those pre-mounting postures which occur when the spiders are in close proximity. As far as is known, Stage 3 postures are common to all salticids. It is anticipated that the ritualized courtship and agonistic behaviors of many salticid spiders will fit within this broad framework.

A more detailed account of courtship events in *E. parvula*, and

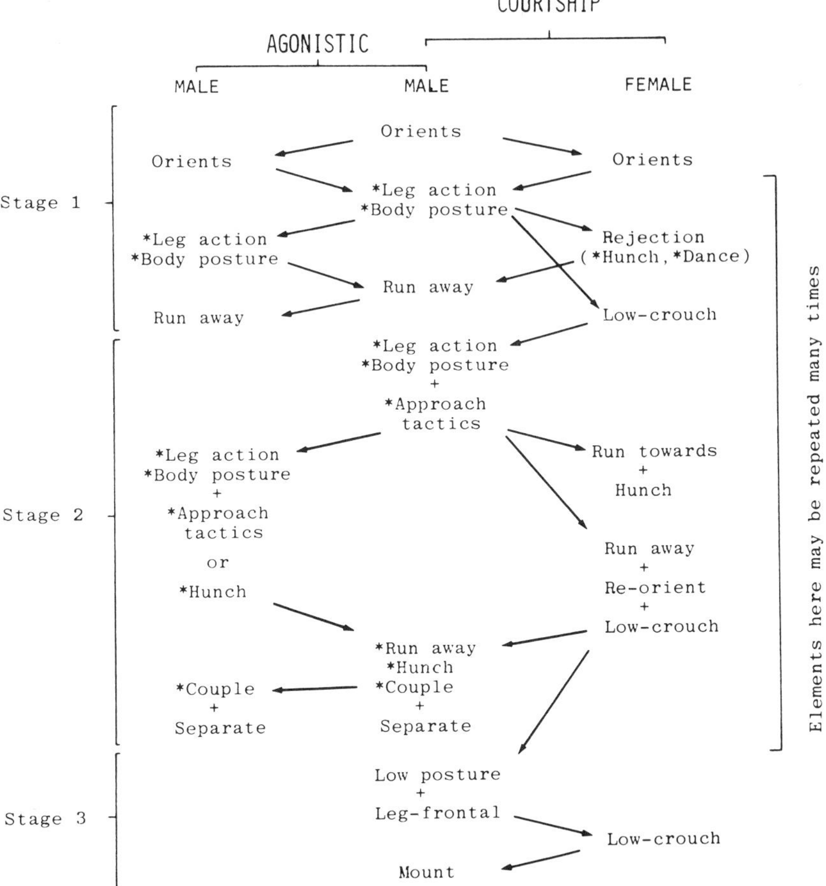

FIGURE 5.2. Order of events which occur in agonistic and courtship interactions in salticid spiders. Sequences are divided into 3 stages (see text). Antecedent events are shown in vertical columns and arrows point to subsequent events in the order of occurrence. Events marked with asterisks are optional; where several occur within a set, one or more of them may occur at that stage of the sequence. + indicates events which may occur simultaneously. In an agonistic encounter in one species, for instance, the first male orients (center column), then the second male orients (left column). The first male leg-raises and zig-zags, the second male does likewise, and so on. In a courtship encounter in another species, the male orients (center column), then the female orients (right column). The male leg-raises and circles, the female may perform a rejection dance, or she may adopt a low-crouch posture. The male then runs away or repeats his first performance, and so on.

their subdivision into three stages is as follows. When a male spider turns (orients) to face a female, he rises up on tiptoes (stilts) and with his abdomen lifted and swaying from side to side (body posture) and his third pair of legs high in the air (leg-raise), he prances and sidles in wide circling arcs from one side of the female to the other (zig-zags). The female turns (orients) toward the male and if receptive, she sits passively but attentively with her body close to the ground (low crouch). This is Stage 1 courtship.

As the male zig-zags, in conjunction with body posturing and leg-raising, he gradually gets closer to the female. Then quite suddenly, the legs are lowered to the ground; the zig-zagging arcs, still accompanied by body posturing, become smaller. Sometimes the female maintains her docile posture until the male's performance brings him close enough to adopt pre-mounting postures. At other times, perhaps when she has already mated or is not receptive for some other reason, the female mimics the male's body posturing and leg-raising patterns. When this happens, the male very soon ceases to display and runs away. These events represent Stage 2 courtship.

When the male is almost within touching distance of the female (2-3cm), Stage 3 courtship begins. If the female's low body posture persists, the male ceases zig-zagging, lowers his abdomen, and shifts his forelegs into a forward parallel position, pointing toward the female (leg-frontal). He creeps slowly toward her until these legs touch and tap the female's head, immediately prior to mounting. We will assume that subsequent events do not contribute to a communicatory system based on vision.

Perhaps it should be pointed out here that the "warning away" or mimicking dance performed by *E. parvula* females is an unusual type of rejection display as far as is known. In most other species, rejection displays are as follows: spiders (males or females) raise themselves up as high as possible and bend their legs sharply (hunch) (Figure 5.3), creating in effect an extra-large silhouette. From this position they make shuffling, lunging, or striking motions toward the other spider. In some species, running sharply toward, jumping at, or running away may occur.

When two male *E. parvula* spiders meet, they mirror each other's performance, zig-zagging and posturing with both abdomen and the third pair of legs, gradually closing the gap until they touch (Figure 5.4); subsequently both spiders cease displaying and separate. But at no stage during this intermale exhibition is the third pair of legs lowered as in the courtship sequence. These dances and the differences between them suggest that males and females are able to recognize each other.

Support for the assumption that both the discriminatory process

FIGURE 5.3. Mountain spider in hunched position.

FIGURE 5.4. Two male *Euophrys parvula* jumping spiders side-step and circle toward each other with legs lifted and abdomen raised. They mirror each other's performance as they approach.

and sequence of events are based on visual information can be demonstrated in a very simple yet convincing manner. When a male spider is confronted with his own image in a mirror, he performs in exactly the same way as he does to a male conspecific. In contrast, female spiders do not display to their mirror images but sit passively and watch intently, sometimes for as long as 10 minutes. This is consistent with their behavior in intrasexual encounters. Apparently they reserve their brief leg and abdominal displays to "warn away" would-be suitors. What should we conclude?

Table 5.1 shows that the sexually dimorphic characters are sufficient to account for the major differences in behavior. If this is so, then what is the function of the behavior itself?

TABLE 5.1. Summary of the major display components which typify interactions between mature male and female *Euophrys parvula* according to the presence or absence of a red forehead.

		2nd Spider	
		Red forehead	*No red forehead*
1st Spider	*Male*	3rd leg-raise, abdomen waving, zig-zag	abdomen waving, zig-zag
	Mated female	3rd leg-raise, abdomen waving	no display
	Unmated female	attentive, passive	no display

Perhaps, in anthropomorphic terms, we can analyze it as follows. Males announce their sex and state of maturity by means of a red forehead; they declare their intentions by raised legs and swaying abdomen. They repeat this information as they approach, principally to keep the female's attention firmly on the ultimate objective (mating) and to facilitate the male's safe approach (see section on approach tactics, this chapter). Courtship may, of course, have other functions as well, but the concern here is with communicatory aspects.

If this information is transmitted to another male, he responds in kind; the approach ends in contact, and spiders separate without injury to each other. Since agonistic events during Stage 2 are specific to males, this behavior must have some function, otherwise the sequence would surely be abandoned. There are two possibilities which warrant investigation. One is that agonistic behavior operates as a spacing mechanism to preserve interindividual distances, and the other is that it separates spiders into winners (those that stand their ground) and losers (those that run away). Winners may be more likely to display to females at the next encounter and so approach and mate. Similar mechanisms were demonstrated in a study of agonistic display in the wolf spider *Schizocosa crassipes* (= *ocreata*), behavior shown by Aspey (1977a,b) to be a complex male-male communication system based primarily on vision.

However, we cannot conclude that any particular set of observations such as those described for *E. parvula* provides a satisfactory answer to the question of the function of similar behavior in all salticids, or indeed that sexual dimorphism is solely responsible for the spiders' capacity to discriminate between the sexes. In the first place, despite the seemingly stereotyped nature of such performances in jumping spiders, there is considerable variability at both intraspecific and interspecific levels. Secondly, the generalized description given here for *E. parvula*, for example, obscures a number of factors which may influence the course of displays. These are: distance separating spiders upon initiation of courtship, size of participants, vigor of male displays, initial readiness of female to mate, frequency of behavioral components, space available for zig-zagging dances, duration of display, the presence of obstacles or other animals, and prior experience (both immediate and long-term), to name the more obvious aspects. Perhaps the sexually dimorphic characters or "visual badges" (term from Smith, 1977) account for the initial acts performed by spiders, while the subsequent behavior reinforces or supplements this information; there may of course be other functions as well. We will return to this idea later.

Transition Analysis of Display

The well-defined behavioral distinctions exhibited by *Euophrys parvula* are by no means universal. In *Trite auricoma*, for example, a male employs exactly the same primary postures and movements throughout Stages 1 and 2, whether confronting a female, another male, or looking at his image in a mirror (Figure 5.5). Clearly, visual information influences the male's behavior, but it is less easy to decide, from the sequence of events, at which stage the male recognizes the female, and what particular characters lead to this decision.

Courtship behavior in *T. auricoma* can be regarded as composed of thirteen distinctly recognizable events (Table 5.2). But since we

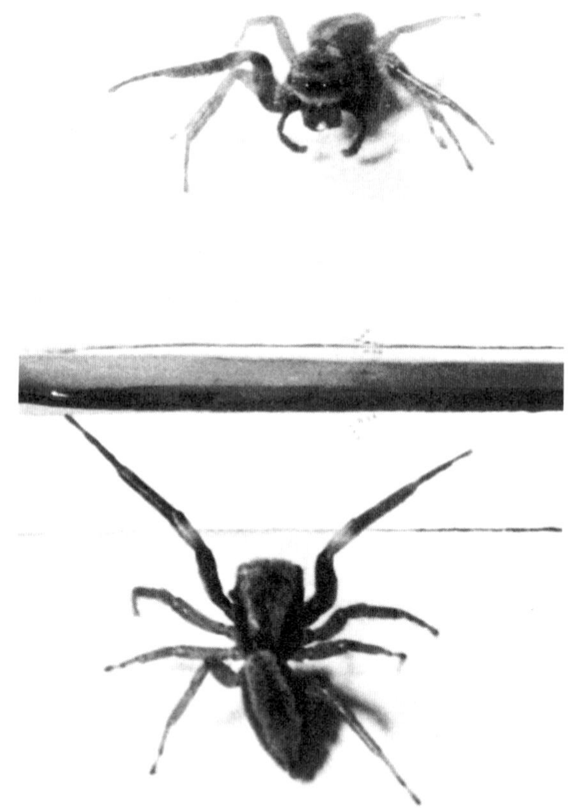

FIGURE 5.5. When a male *T. auricoma* catches sight of his reflection in a mirror, he raises his legs and zig-zags toward the "other spider."

cannot be certain, merely from qualitative "feelings," that any of these events are actually signals with a high probability of eliciting specific responses in recipients, we must find a quantitative approach which will allow us to distinguish real "signals" and appropriate responses to them from accidental sequences of events.

One approach is to consider what would happen if none of the thirteen events was a signal. In that case, any event performed by one spider would have the same chance of being followed by each of the events available in the repertoire of the other spider. If the thirteen acts are arranged in a two-way table, all interactions can be disposed as two-act sequences, or dyads, following the method

TABLE 5.2. A brief description of events comprising behavior of male and female *T. auricoma* spiders.

Act	*Symbol*	*Description*	*Used by*
Orientation	0	a turn to face the target	Male, female
Leg-raise	L-R	forelegs raised above head, diagonally	Male
Leg-wave	L-W	forelegs waved, diagonally	Male, female
Leg-lower	L-L	forelegs lowered, frontally	Male
Zig-zag	Z-Z	approach, side-to-side stepping	Male
Leg-frontal	L-F	forelegs pointed forward, parallel	Male
Mount	M	climbing on female, tapping, drumming	Male
Low-crouch	L-Cr	passive, attentive, low stance	Female
Hunch	H	high stance, all legs bent sharply	Female
Couple	C	1st and 2nd legs, palps, out-stretched	Female, male

of transition analysis described by Slater and Ollason (1972). If no communication is taking place, dyadic frequencies from a large sample of observations will be scattered at random with respect to the cells of the table; if any of these events are indeed influencing the performance of other events, then some of the cells will be heavily occupied compared to others. A goodness-of-fit test can then be used to determine whether cells contain more observations or fewer observations than might be expected to arise by chance.

From interactions recorded for twenty pairs of adult *T. auricoma*, dyads were extracted as follows. If, for all spiders tested, male leg-raises were followed by 36 low-crouches and 23 hunches, then the frequency of the "leg-raise to low-crouch" dyad is 36, the "leg-raise to hunch" dyad is 23, and so on. Very often, if a male leg-raised and the female did nothing, the male would leg-raise again, a situation recorded as a "leg-raise to leg-raise" dyad. Hence intraindividual as well as interindividual sequences of behavior were taken into account, an approach adopted by Baylis (1976) in his study of interactions in cichlid fish.

A schematic representation of the 13 × 13 frequency matrix used to set out these data is shown in Table 5.3. The thirteen possible antecedent events performed by either male or female spiders correspond to the horizontal rows and the thirteen possible subsequent events correspond to the vertical columns. Each cell contains three numbers: the observed frequency, the expected frequency (calculated according to an iterative method devised by Goodman, 1968) and a goodness-of-fit statistic determined as follows.

If, in isolated dyads, the difference between the observed and ex-

TABLE 5.3. Schematic representation of a frequency matrix. Rows (r) are antecedent events and Columns (c) are subsequent events. Thirteen events, characteristic of courtship in male and female *Trite auricoma* (see text) are listed from r1 to r13 and c1 to c13 (complete table in Forster, 1979b). Thus all possible combinations of two-act sequences within and between sexes may be accounted for in this matrix. Values in the cells, of which there are 169, represent the frequency of occurrence of a pair of events for all spiders observed.

			Subsequent events: Male and female events		
			c1	...	c13
Antecedent events	Male and female events	r1	f11	...	f113
		⋮	⋮		⋮
		r13	f131	...	f1313

pected frequencies is greater than three times the square root of the expected value, the linkage is highly significant ($P < 0.001$) (Wilson and Kleiman, 1974). The complete matrix is given in Forster (1979b); the following account is a modification and brief summary of the findings presented there.

Transition analysis of the 13 × 13 matrix of behavioral frequencies shows that twenty-six dyads are linked significantly ($0.05 > P > 0.001$). Twelve of these dyads occur less often than expected (inhibitive) and fourteen occur more often than expected (directive) (terms from Hazlett and Bossert, 1965); only the nine highly significant ($P < 0.001$) and directive dyads (Table 5.4) will be discussed here.

A diagram constructed from these results (Figure 5.7) shows that all relationships are based on interactions between male and female spiders, rather than on intraindividual sequences. Of particular interest is the finding that orientations are not linked to the main

FIGURE 5.6. The principal courtship postures in *T. auricoma*. These illustrate the change in male stance from a high stilting position to a low posture as he nears the female shown here in the typical docile, low-crouch position (left to right): (a) Male—leg-raise posture; (b) Male—leg-wave posture (sometimes used by females); (c) Male—leg-frontal posture; (d) Female—low-crouch posture. The male spider leg-raises with a high body stance (Stage 1) which becomes progressively lower as he approaches the docile female (Stage 2). Forelegs are also lowered until they are pointing toward the female (Stage 3). The female may leg-wave, but not while in a receptive stance. See Table 5.2 for description of other courtship events.

sequence of events; other studies show that this act initiates all organized sequences of behavior in jumping spiders (Forster, 1977a, 1979a). Since such sequences differ markedly (see section on visual mechanisms of communication, this chapter), very specific information must govern the performance of succeeding acts. The pres-

TABLE 5.4. Nine highly significant transitions ($P < 0.001$) in which the observed frequency of the two-act sequence exceeded the expected frequency. In 5 cases, female events followed male events, and in the other 4 cases, male events followed female events. Observed (ϕ) and expected (E) frequencies for each of these transitions is shown, with differences (d) in all cases greater than three times the square root of the expected value ($3\sqrt{E}$). See Table 5.2 for abbreviations. Modified from Forster, 1979b.

♂ *Event followed by* ♀ *event*			ϕ	E	d	$3\sqrt{E}$
O	→	O	60	27.4	32.6	15.7
L-R	→	L-Cr	36	10.0	26.0	9.6
L-R	→	H	23	7.3	15.7	8.1
Z-Z	→	L-Cr	16	6.2	9.8	7.5
Z-Z	→	H	14	4.5	9.5	6.4
♀ *Event followed by* ♂ *event*						
L-Cr	→	Z-Z	78	51.2	26.8	21.5
L-Cr	→	L-F	65	29.2	35.8	16.2
L-Cr	→	M	30	13.4	16.6	11.0
C	→	C	40	11.4	28.6	10.2

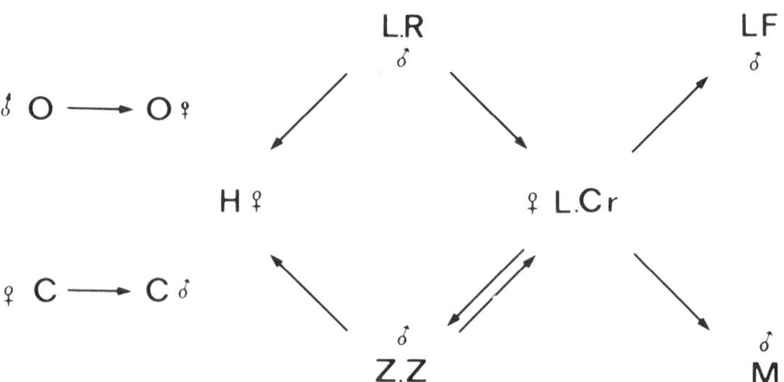

FIGURE 5.7. Relationships between events which occur in courtship encounters between male and female *Trite auricoma* spiders. Male or female acts are indicated in the diagram. All dyads shown are significant ($P < 0.001$), as demonstrated by transition analysis.

ent findings suggest that a visual badge, a chemical, or even an auditory signal, not included among the event categories, forms the necessary linkage between orientation and the specific courtship events. Note also that the male, by orienting, brings about an equivalent response in the female, indicating that males are the most likely initiators of courtship.

Coupling also forms an independent dyad. This is because such behavior may occur after any one of a number of events; hence we say it is randomly distributed. Always initiated by the female in courtship situations, perhaps coupling represents the ultimate means by which a female can repulse a male, given that he has failed to react appropriately to an earlier signal such as hunch, for example. Supernumerary observations show that males often back or run away when females hunch, as depicted in the schematic representation of behavioral patterning in Figure 5.2.

The remaining events, relevant to courtship, are linked together. The female's low-crouch and the male's leg-raise form focal points, and several other events depend on these postures. Their importance in this context was demonstrated by Drees (1952), whose model tests support the view that these postural configurations are visual signals. Leg-raises invariably initiate courtship, in the true sense, but it is not until the female responds with a docile posture that zig-zag is performed on a regular basis. If she hunches, then the courtship does not proceed. Low-crouch and zig-zag appear to reinforce each other; the former stance, however, influences leg-frontal and mounting once the male is close enough.

Only those events which reliably lead to other events, such as orientation, leg-raise, and zig-zag in males, low-crouch, hunch (from observations), and coupling in females, qualify as signals. With the exception of coupling, which is a contact posture, the other five events are presumably visual signals, taking into account the evidence from mirror-image reactions. Relatively few signals, therefore, are implicated in successful courtship; apparently other events included in the frequency matrix are irrelevant.

Perhaps "irrelevant" acts have other functions. For example, leg-waving may provide a means whereby spiders periodically sample environmental stimuli such as vibration while their other sensory systems are preoccupied with the perception and processing of courtship information. Alternatively, this action by male spiders may serve to pick up chemical signals from the female, as suggested by Crane (1949b). If, eventually, a chemical is implicated in the recognition of females in this species, then perhaps pheromones provide the fine tuning in courtship interactions while visual signals control the gross components. Some support for this view comes from preliminary studies (Forster, unpubl.) in which *T. auricoma* females were sealed in clear plastic containers: under such circumstances male courtship behavior was not only more variable and intermittent, but was also abandoned much more readily.

In *E. parvula*, the presence or absence of a red forehead explains behavioral differences between the sexes, while in *T. auricoma* the quantitative analysis demonstrates linkages in the event sequence without recourse to sexual dimorphism, suggesting that behavioral information supplements or reinforces non-behavioral input. It also supposes considerable redundancy in the system, that is, much more information is available to spiders than is strictly necessary for successful courtship to occur.

The method of analysis used here does have some drawbacks, however, since it does not take into account the duration of acts, or the effects of delayed or indirect relationships, for example. The findings reported above should be regarded as preliminary, but I hope they will lead to the use of other quantitative techniques. Advances in this area of ethological assessment are reviewed in a publication edited by Hazlett (1977).

The examples presented here are but two of the many variations in appearance and behavior exhibited by this family. Some idea of the extent of these differences and the extraordinary manner in which various epigamic characters and displays may be combined and paraded is given by Jackson (this volume). Because of this variability and the fact that relatively few species have been studied in detail, extrapolation from the studies presented here should be done with caution.

Juvenile Interactions

Many analyses of communication in animals have been confined to interactions between adult members of a species, and relatively little attention has been paid to the ontogeny of display behavior and its function in the juvenile stages. Undoubtedly, in some studies, there has been an implicit recognition of communicatory interactions among juveniles (Andrew, 1972; Fine et al., 1977); other studies have been concerned with adult-juvenile, especially parent-offspring patterns of behavior which clearly involve an exchange of information (Lorenz, 1970). In general, though, these investigations have concerned themselves with social animals, particularly with those birds and mammals which live together in groups for at least some period of their lives.

Most spiders are regarded as solitary animals (except during mating activities), but studies by Aspey (1977a,b) show that, in at least one species of wolf spider, *Schizocosa crassipes* (= *ocreata*), adult males exhibit complex patterns of socially organized behavior. Aspey (1975) also found that young *S. crassipes* exhibit leg-waving movements which become progressively more stereotyped and regular as spiders mature. He concluded that leg-waving is a visual signal and that its purpose is to space the juveniles and to minimize cannibalism.

In salticids, brief displays involving immature spiders were dismissed as abortive events or rudimentary threats by Crane (1949b). These descriptions, however, do not fit the stereotyped patterns of behavior observed when juvenile *Trite planiceps* spiders encounter each other (Forster, in prep.). Unlike *S. crassipes* (Aspey, 1975) or *Pardosa* spp. (Koomans et al., 1974), these young spiders possess a large repertoire of distinctive postures and movements, most of which also appear in adult male-male interactions.

Juvenile spiders used in these investigations ranged from third-instar individuals to subadults. They were ranked according to size and classified according to whether they won or lost an encounter. A spider was specified as a loser if, either at an early or late stage, it retreated (backed) and ran away from the other spider. Hence the winner was the spider which "stood its ground." The overall results show, quite convincingly, that in 64% of encounters the larger spider emerged as the winner ($\chi^2 = 6.9$, $P < 0.01$). For encounters where size disparities were greatest, large spiders won in 77% of encounters ($\chi^2 = 12.8$, $P < 0.001$); but as size difference decreased, winning or losing became less predictable ($\chi^2 = 1.06$, $P > 0.05$).

All test spiders exhibited some form of display during encounters, ranging from a simple foreleg-side-stretch, usually accompanied by several vigorous taps on the ground, to very complex patterns of response (Figure 5.8). However, the greater the disparity in size between interacting spiders, the briefer the period of confrontation. In two instances (2% of encounters), both involving the highest and lowest size ranking, the larger spider switched to predatory behavior (see Forster, 1977a), and quickly seized its erstwhile opponent as soon as the opponent turned and ran. Since a running target induces chasing and capture by *T. planiceps* (Forster, 1979a), it seems that unless spiders engage conspecifics in display activities until an appropriate time to retreat and run away occurs, they may be treated as prey. Prolonged display sequences invariably ended with both spiders retreating and running away at much the same time. In only 2% of the total remaining interactions were attack sequences initiated, and none was fatal. There was no significant difference in the size of spiders spurred into jumping at their opponents ($\chi^2 = 0.56$, $P > 0.20$). Since attack behavior apparently plays only a minor role in interactive situations, it seems likely that one function of the stereotyped juvenile displays in this species is to reduce the risk of cannibalism, as Aspey (1975) also found.

Unlike Aspey's studies, however, no assessment of individual distances separating a group of spiders was undertaken, so it is not possible to say whether these behaviors also function as spatial density control mechanisms. However, the present investigations included an examination of the reactions of spiders to their own mirror images and these results are described briefly below.

The previous findings suggest that spiders can judge the size of an opponent relative to their own; hence it must be concluded that because a mirror image presents an opponent of similar size, the "viewing" spider is able to make an approximate assessment of this by some visual means. That this is so is confirmed by the fact that all spiders reacted to their mirror-image challengers with a relatively complex sequence of responses, consistent with those displayed by the equally or closely matched pairs of spiders involved in the earlier interactions. However, all spiders eventually ran away from their own image, but "running away" invariably occurred at the conclusion of either of two distinct tactics employed in this confrontational situation.

The two tactics are referred to as "defense" and "attack"; the

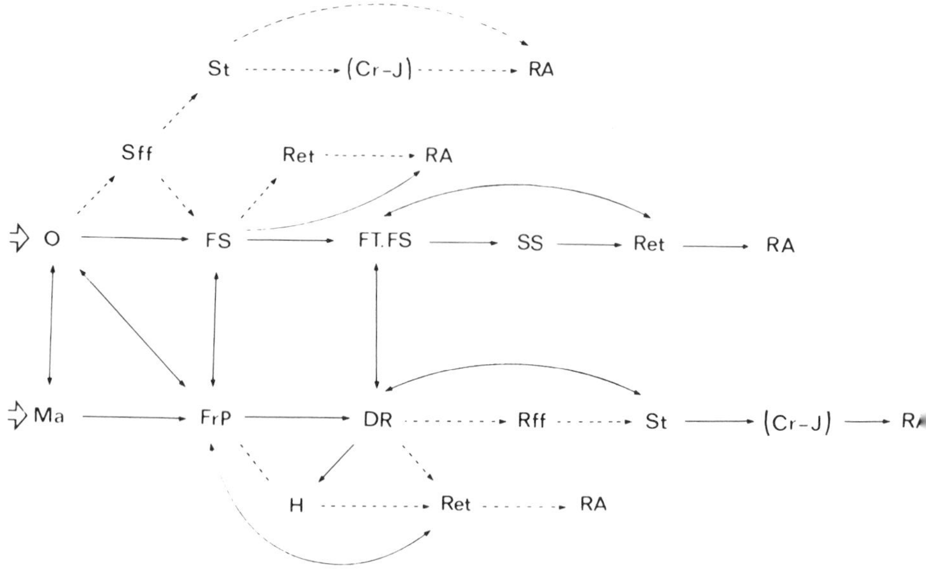

FIGURE 5.8. Flow diagram depicting actual event sequences observed in a study of juvenile interactions in *Trite planiceps*. Sequences represent individual patterns of activity and disregard the intervening acts performed by the second participant. Solid lines represent linkages most commonly observed, while broken lines indicate those only occasionally noted. The probability of the occurrence of a particular sequence is not shown here, since this changes according to various stimulus parameters such as size, distance, age, duration of performance, and prior experience of both spiders (Forster, in prep.).

Symbols for Figures 5.8 and 5.9

(Cr-J):	crouch and jump	O:	orientation
DR:	direct run	RA:	runs away
FrP:	frontally positioned	Ret:	retreat
FS:	forelegs sideways	Rff:	rounded forelegs forward
FT.FS:	forelegs sideways with forelegs tapping	Sff:	stiff forelegs forward
H:	hunch	SS:	side-stepping
Ma:	moving about	St:	stalk

sequence of events commonly exhibited in mirror-image encounters is depicted in Figure 5.9. Attack events are similar but not identical to those shown in predatory situations, while defense tactics lack some of the components of the two-spider confrontations. The mirror-image situation in itself provides an explanation of such differences, since the imaginary spider always "stands its ground," does not initiate acts, maintains eye contact during a frontal orientation, provides no movement or postural antecedents, and is never "seen" to run away.

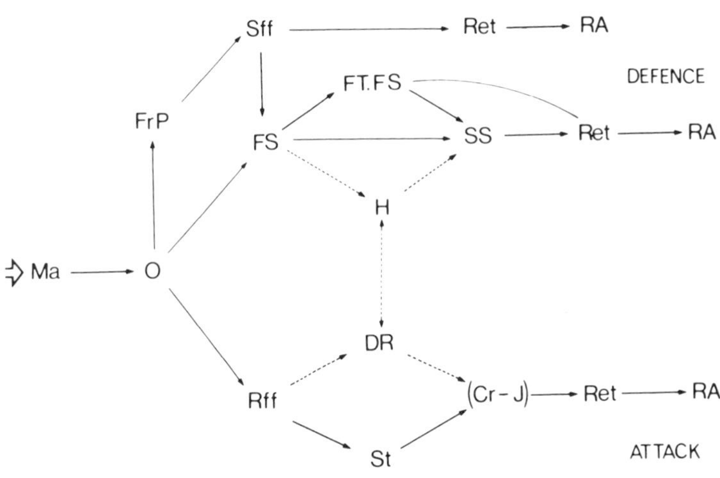

FIGURE 5.9. Flow diagram of actual event sequences performed by juvenile *Trite planiceps* in response to their mirror image. Two tactics were observed, attack and defense, which may occur quite independently. In contrast, the same events are highly intermingled in the two-spider situation (see Figure 5.8), suggesting that spiders "choose" postures according to the needs of the moment. The mirror-image situation is more stable; spiders react in a more organized fashion and sequences are of much shorter duration.

Other than the foreleg-side-stretch, acts occurring in juvenile interactions have never been observed in isolated individuals of *T. planiceps*. If, for example, a lone spider is disturbed in its container either by an unidentifiable movement within its visual ambit or a sudden noise, it may react by swiveling in the direction of the stimulus and instantly extending its forelegs sideways. Most likely this leg action is a conditioned response, but clearly this does not prevent it also functioning as a visual signal.

The behavior of spiders toward their mirror image indicates that the signals to which they are reacting are visual and that the accompanying sequences are based on a stimulus-response principle. Comparison with the two-spider situation suggests that visual signals contain information relating not only to the size/age of the other spider but also to its class. This is demonstrated by the fact that the spider equates its own acts reflected in a mirror with those of another spider; and since imaginary performances and real interactions are fundamentally similar, it is reasonable to suggest that such behavior promotes intraspecific recognition. There seems little doubt, therefore, that interactive displays among juvenile spiders operate as a visual communicatory system.

Interspecific Communication

When two salticid species occur in the same habitat, it is likely that they possess a number of adaptations designed to reduce competition and hostility. Many species of *Phidippus*, for example, occur in sympatric pairs, but they mature at quite different times, so that a considerable size difference is maintained throughout the year (Enders, 1975b). As a result each species retains considerable independence in the size of prey hunted.

Two New Zealand species of jumping spiders, *Trite planiceps* and *Trite auricoma* (these two species are probably not congeneric; see Forster, 1977a), are commonly found living together in flax bushes (*Phormium tenax*). There is no evidence of differential maturation, but competition is minimized because hunting activities appear to be scheduled for different times of the day (Forster, 1977b). Inevitably some overlap occurs, so that members of these species must meet each other from time to time. What happens when they do?

If male, both spiders orient toward each other and immediately adopt their primary species-specific posture (Figure 5.10). After a long pause, *T. planiceps* taps sharply with its outstretched forelegs and runs briskly forward. Sometimes *T. auricoma* side-steps and waves its forelegs before turning and running away; at other times its retreat is more immediate. However, *T. auricoma* is always the spider which retreats. The forward run and outstretched tapping legs of *T. planiceps* have but one message for *T. auricoma*. For a conspecific, in another context, these signals must have a quite different meaning.

For comparison, a male *T. auricoma* was paired with another male salticid (sp. *A*) of similar size from an entirely different habitat. *T. auricoma* is found almost exclusively at sea level but sp. *A* is an unidentified montane spider living among the stones at an altitude of some 2,000 m (Forster and Forster, 1970). Unlike other salticids, male mountain spiders do not acquire extra adornments at maturity; instead they lose their most distinctive juvenile feature, a prominent band of orange hairs beneath the eyes. Females, in contrast, actually retain this character. Males, therefore, are almost entirely black, except for some faint markings on the dorsum of the abdomen, and are quite unlike male *T. auricoma*, which have a light covering of yellow hairs over a brown body and a prominent band of yellow hairs below the front eyes.

When two males of these apparently unrelated species encounter each other at a distance of about 15 cm, an interesting behavioral

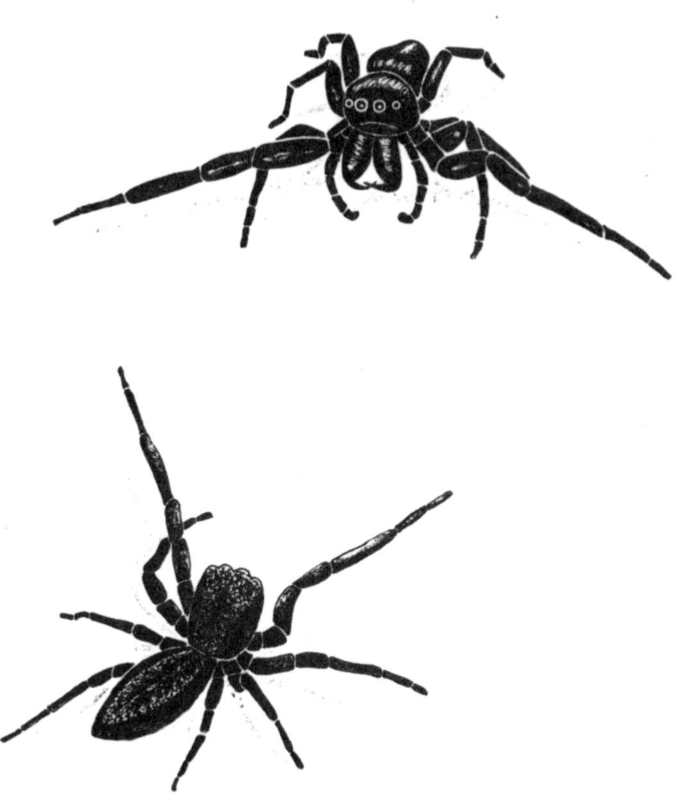

FIGURE 5.10. A male *Trite auricoma* raises his forelegs when he is confronted by *Trite planiceps. T. planiceps,* on the other hand, stretches his forelegs sideways and taps vigorously on the ground.

sequence occurs. After orientation, both spiders raise their forelegs above their heads in an almost identical posture. *T. auricoma* advances by zig-zagging and leg-raising. The second spider approaches in a very different fashion, running foward a few steps, then backward a few steps. Once within touching distance, both spiders may stretch their legs sideways, couple, and then separate. Sometimes, however, the mountain spider adopts a hunched leg posture, in which case *T. auricoma* usually takes flight. In one courtship trial, a male *T. auricoma* made a successful approach to a female mountain spider that watched him attentively and passively while he postured and sidled. Only when he touched her with his forelegs did she reject him.

In the inter-male interaction both spiders apparently performed with their normal, agonistic "set of responses," although never, in nature, would they encounter each other. Using the same postures, but different movements, *T. auricoma* was able to approach the female mountain spider with impunity. The most interesting conclusion to be drawn from these observations is that there are some basic "salticid" signals, recognizable in any context. Drees's (1952) studies support this principle, since the models he used to elicit courtship were quite unspecific. The primary posture, foreleg-raise, apparently conveys a common message, and because this posture is fundamentally similar in both species, males are induced to approach each other. When primary postures are different, as in the earlier encounters, one spider retreats at once.

These findings are of a preliminary nature, but they point to a potentially valuable area of research, and strengthen Wickler's (1968) suggestion that knowledge of interspecific signals may throw light on the evolution of intraspecific communication.

Systems of communication involving salticids are not necessarily restricted to intra- or interspecific relationships. Some of the most remarkable communication systems are found in certain species and even genera of jumping spiders which have gone to considerable lengths to mimic the signals of another group of animals, the ants.

Salticid ant-mimics fall into two categories: those that lead solitary lives, and those that live with ants. Both groups have been studied by Reiskind (1977), who found that the relationship between ants and their spider mimics may be quite complex. The advantages of mimicry to either the spider or the ant are not thoroughly understood, the most favored explanation being that mimicry by the spider enables it to avoid potential predators (Reiskind, 1977). However, in situations where a spider mimic and an ant model encounter one another, Reiskind notes that both exhibit avoidance reactions. Since intraspecific avoidance reactions in juvenile *Trite planiceps* are communicatory, there is a good case for pursuing this possibility among ant-mimics. An extraordinary example of interphyletic communication, for example, has been shown to occur in a behavioral, mutualistic symbiosis between the goby *Psilogobius mainlandi* and two species of shrimp, *Alpheus rapax* and *A. rapacida* (Preston, 1978). The extent to which ant-mimicry has developed suggests that there have been strong selective pressures at work; and this seems hardly likely unless there are considerable benefits entailed.

Visual System

In order to perceive and react to visual stimulation, jumping spiders need specialized optical and neural equipment. Several studies, notably those of DeVoe (1975), Eakin and Brandenburger (1971), Hardie and Duelli (1978), Homann (1928, 1971), Land (1969a,b), and Yamashita and Tateda (1976) show that the eyes of these spiders do indeed have the expected sophistication. A summary of the main features of the visual system is given below. This section is largely derived from Land (1969a,b), and measurements refer to the species discussed in those papers unless a statement is made to the contrary. The reader should remember that there is likely to be adaptive variation of eye design, so that it will differ between species, quite possibly in major ways.

Anatomy of the Eyes

The arrangement of the four pairs of simple eyes on the anterior part of the head is shown in Figure 5.11. The AM (principal) eyes have evolved remarkable features which distinguish them from the lateral (AL and PL) eyes. The tiny PM eyes, whose function is not known, are vestigial versions of the other lateral eyes and will not be discussed here.

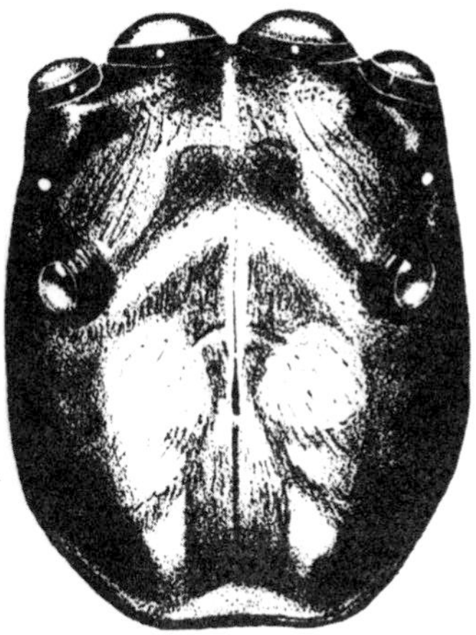

FIGURE 5.11. Diagram of 4.2.2 arrangement of eyes in a salticid spider. In the first row, the large central anterior-median (AM) eyes are flanked by the somewhat smaller anterior-lateral (AL) eyes. Behind them lie the very small pair of posterior-median (PM) eyes while the third row consists of the medium-sized posterior-lateral (PL) eyes. From R. R. Forster (1967).

Each of the two main kinds of eye consists of a dioptric apparatus (a cornea continuous with the cuticle of the carapace, a lens, and a vitreous anterior compartment composed of the cells which made the lens) and a retina which consists of a mosaic of receptors. Light focused on a retina by the lens is absorbed by the receptive segment of each receptor and converted into an electrical signal.

In the AL and PL eyes each receptive segment is surrounded by screening pigment. The PL eye consists of a broad but shallow cup-shaped capsule with an almost hemispherical retina, while the retina of the AL eye is smaller and flatter and placed at the rear of an elongated capsule (Figure 5.12). In both pairs of eyes, the spacing between receptors (in terms of their angular subtense) lies between 1° and 1.5°, except that in the AL eyes it is reduced to 30 mins. at the region which regards the spider's axis (see Figure 5.12).

The front lenses of the AM eyes have very long focal lengths, and this dictates the elongated, tubular form of the eye capsule (Figure 5.12). Each retina is boomerang-shaped and lies in a narrow trough consisting of screening pigment which does not invade the layers of receptive segments. The latter consist of four tiers which are discussed in a later section. They are embedded in a dense matrix composed of processes from the pigment cells; and the interface between the matrix and the anterior vitreous chamber of the eye is a cone whose rounded apex is positioned at the optical axis. The construction of the AM retina as a slender strip of receptors oriented dorso-ventrally means that it has an extremely narrow horizontal subtense of approximately 2°, and a vertical subtense of some 21°. Some components of the receptor layer nearest to the lens, Layer 4, can be seen from outside the eyes of the living spider by means of an ophthalmoscope (Land, 1969a). The images of the two retinae thus obtained are separately inverted and reversed, and therefore indicate to an observer how, acting in unison, they sample the visual world. The most extensive receptor Layers, 1 and 2, cannot be seen through an ophthalmoscope, partly because they reflect no light, and partly for optical reasons which will not be discussed here. However, it can be deduced that the fields they look at have the form of a diagonal cross of the kind indicated in Figure 5.13a. Receptors in the third layer occupy a smaller region in the central area; and Layer 4 receptors, closest to the lens, are scattered (Figure 5.13b).

Fields of View of the Eyes (Figure 5.14)

The four main lateral eyes between them encompass a horizontal visual ambit of almost 360°, and their primary task is to detect

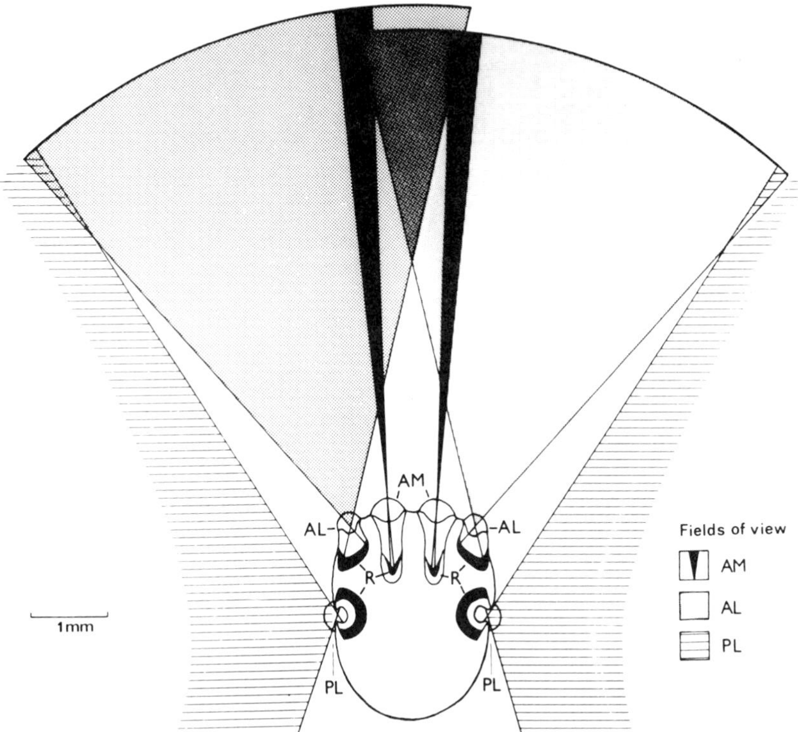

FIGURE 5.12. Diagram of horizontal-frontal section through the ocular region of a jumping spider. The AM retinae (R) are situated in a U-shaped depression at the rear of long tubular eye capsules; lateral movements increase their fields of view. Other types of eye movements subserve special functions (Land, 1969b) (see text for details). The AL retinae (R) are situated at the rear of elongate eye capsules, while the PL retinae (R), almost semi-circular in shape, provide for extensive fields of view (See Figure 5.14). Adapted from Homann (1928).

moving objects in the environment (Crane, 1949b; Homann, 1928; Land, 1971). Having done so, the spider swivels (Forster, 1977a), a turning movement which brings the source of stimulation into the fields of view of the principal eyes (Land, 1971, 1972a). In *Trite planiceps* there is some binocular overlap of the fields of the AL eyes, each of which also shares a portion of its field with the AM eyes.

Once the spider has swiveled toward its target, subsequent strategies depend on the identity of the target and the rate at which it moves (Forster, 1979a). For stationary targets, identification is achieved by the AM eyes. Objects entering the fields of the AM retinae are fixated by their central areas and scanned; for this purpose the retinae are so constructed as to be motile.

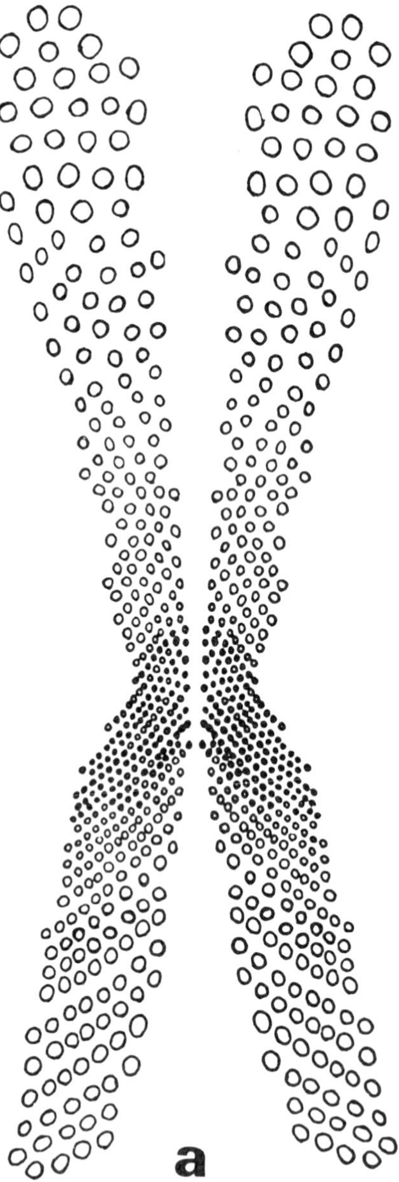

FIGURE 5.13. (a) Receptors of Layer 1 of both AM retinae of *Metaphidippus aeneolus*, each separately inverted and reversed so as to indicate the geometry of the fields they regard in object space. Adapted from Land (1969a). (b) Layers 1-4 (left to right) of the right AM retinae of *M. aeneolus*. Copied from Land (1969a).

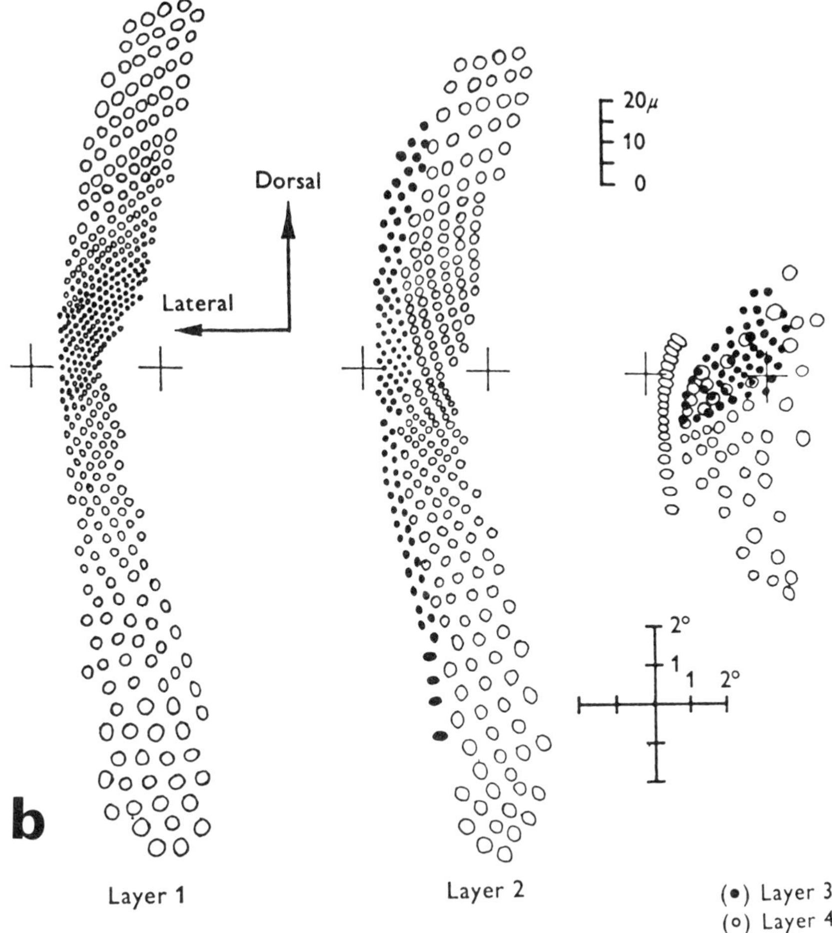

MOVEMENTS OF THE RETINAE OF THE PRINCIPAL EYES

The lenses of all the eyes are fixed. The retinal strips of the AM eyes, therefore, can only be made to survey their fields by movements of the tubular eye capsules, which are provided with muscles for that purpose. Scheuring (1914) described the muscular harness by which each tube is secured, and noted the lateral movements; Dzimirski (1959) found that there are vertical and rotational components as well. Later, Land (1969b) constructed an ophthalmoscope which allowed him simultaneously to monitor movements of both retinae by observing components of Layer 4 of the receptors through the lens, while presenting targets in the visual fields of the same eye. Four types of movement were identified (Figure 5.15).

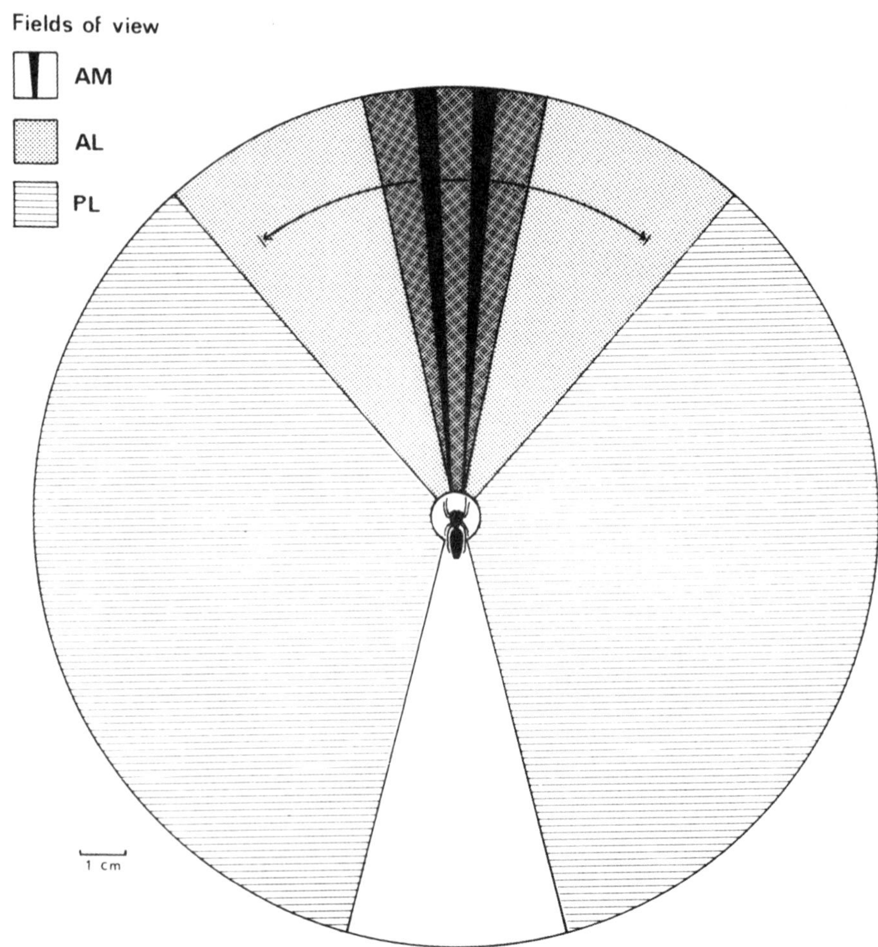

FIGURE 5.14. Fields of view of the AM, AL, and PL eyes in a jumping spider (modified from Homann, 1928, and Land, 1969a). Each PL eye encompasses a field of 120°; each AL eye covers a field of 55° with a frontal overlap of 25° (double hatched). The AM eyes have narrow inherent fields of view extending to 35° on either side of the midline (arrowed) as a result of movements of the retinae (see text for details).

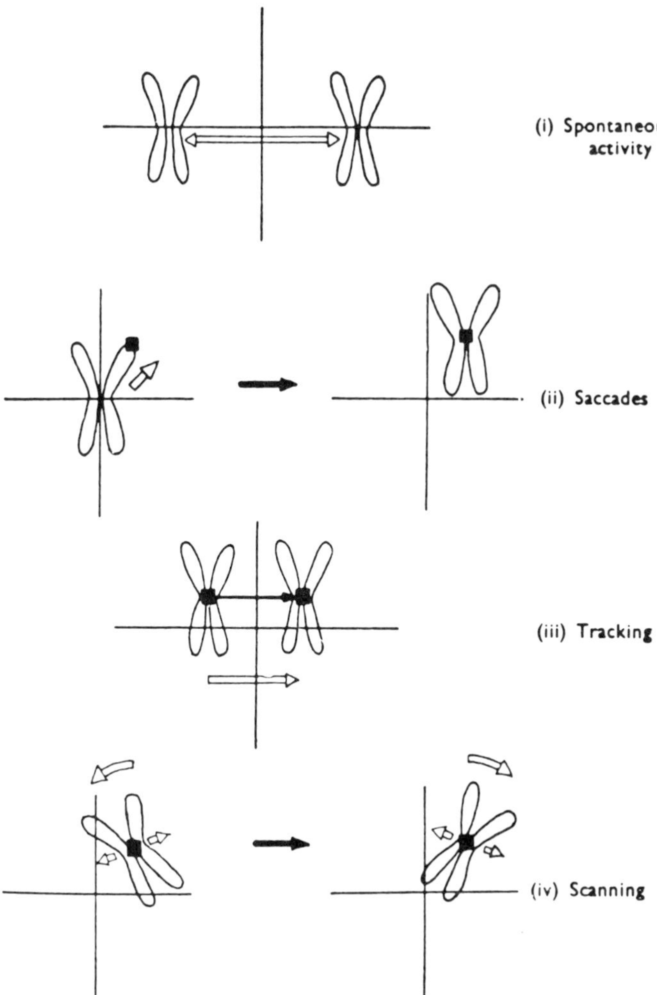

FIGURE 5.15. Diagram illustrating the four kinds of eye movements. Movements of the retinae are indicated by open arrows; movements of the stimulus are indicated by solid arrows. See text for explanation. Copied from Land (1969b).

1. *Spontaneous activity*: Variable, side-to-side movements which occur when the spider is alert but which do not depend on the presence of a target in the field of view.
2. *Saccade*: A swift movement of the retinae which centers and fixates them on the target.

3. *Tracking*: Both retinae track (follow very closely) the movement of a target in the visual field in such a way as to maintain its fixation by the central region of the retinae.
4. *Scanning*: A regular side-to-side displacement which moves the central region of the retina backwards and forwards over the image of the stationary target, upon whose presence it depends. Scanning generally follows a saccade; and within limits its amplitude is related to the size of the target. It is combined with a rotation of the retinae about the optical axis through an angle of approximately 50° in each direction.

Dzimirski (1959) believed that the eye capsules move independently of one another and that convergence might increase the binocular overlap. However, Land (1969b) found that they always move in tandem except during spontaneous activity when one eye sometimes leads the other by about 25° in the absence of a target. This means that, although the musculature would permit it, convergence and divergence of the eye capsules are not used as devices for the binocular fixation of objects in the visual field. In addition, the eyes cannot accommodate (Homann, 1928; Land, 1969b) and their apertures are fixed (Land, 1969a,b). Retinal movements ensure that each eye surveys a larger area than is provided by the retinal subtense alone. An AM eye can survey a horizontal angle of around 70° as a result of movement, while its horizontal retinal subtense in the central region is only some 2°. Why have the AM eyes evolved in this unique manner?

First, it is with these eyes that the spider performs its most important visual discriminations. For physical reasons, receptors cannot form a mosaic of less than a certain grain (Kirschfeld, 1976), so that at the limit thus imposed, further acuity becomes a function of the magnification provided by the lens. High magnification requires a long focal length. If this was combined with a retina which accepted a conventionally large visual angle, there would be no room in the spider's carapace to house it. Reduction of the retina to a strip and the evolution of large-scale retinal movements are ways of dealing with this problem of construction. They offer a second advantage: arthropod visual systems are designed to perceive the consequence of movement. A jumping spider which has started to stalk its prey must often need to assess the nature of something which is motionless. In an absence of retinal movements, the whole spider would have to move in order to scan its target. Selection probably favored the evolution of the motile retina as a means of allowing the spider to remain still until an immobile target has been identified.

Land (1969b, 1972b) first suggested that scanning mediates recognition of targets. How the process might work in terms of real visual discriminations will be discussed later; but an account of his hypothesis in outline is needed at this point because it will influence the way in which we approach the organization of the AM retina. The hypothesis supposes that linear arrays of receptors are sampled by higher-order neurons in such a way that one collector responds when a sufficient number of its receptors "in line" with respect to a retinal image are stimulated. Such a high-order neuron can be termed a "line-detector." This method of analysis of contours is known to play an important part in the early stages of image-processing in the mammalian visual cortex. It is assumed that in the case of the jumping spider, line-detectors will be stimulated by appropriate features of a target being scanned, and that the rotatory movements of the retinae are performed so that particular linear arrays of receptors lock (or fail to lock) onto significantly oriented boundaries. So direct a system of feature recognition is highly attractive, because it achieves a substantial amount of classification of input at a peripheral level of the nervous system, and thus, again, conserves space.

Function of Retinal Stratification

The photoreceptors in the eyes of animals in several phyla are arranged in more than one layer; the purposes that such arrangements may serve are discussed by Miller and Snyder (1977). In jumping spiders the presence of four layers of receptors in the principal eyes is one of the most extreme examples of retinal tiering, and one for which there is currently no completely adequate explanation. Following his discovery of this arrangement, Land (1969a) considered three hypotheses which might explain it. Before these are outlined, it must be pointed out that his calculations for the principal eye optics of two Californian species (*Phidippus* and *Metaphidippus*) rest on the assumption that the only major refracting interfaces of the optical system of the eye are those of the front lens. It will be explained below that this may not be correct. However, in terms of his 1969 data, Land considered the following alternatives.

1. Tiered receptors increase the light capture potential of the retina, successive layers capturing light transmitted, and therefore lost, by their predecessors. Although Land was wrong to dismiss this hypothesis merely because he believed jumping spiders to be wholly diurnal—many species live in such dimly lit habitats as the litter layer of rain-forests—other reasons adduced by him are com-

pelling. First, the neural anatomy is such that the pooling of signals which the hypothesis requires could only take place at a location in the brain at least two synapses from the receptors; and there are functional disadvantages to such an arrangement. Second, if either of his two further hypotheses were also right (and neither wholly excluded the present one), images of different degrees of sharpness would be combined by pooling; and this would appear to be nonsensical in terms of the way the eye as a whole seems to be designed to achieve maximal acuity. It may also be pointed out that even if all layers receive focused images, pooling would result in considerable image degradation, unless the receptors of successive layers were homotopic; and in the central areas of several species examined ultrastructurally it is now known that they are not (Blest and Williams, in prep.). Two more attractive hypotheses are both suggested by Land in terms of compensatory devices to allow optimal exploitation of focused images.

2. If Layers 1 and 2 of the receptors have both spectral maxima in the blue-green, Layer 1 will receive its best-focused images from a plane about 3 body-lengths in front of the lens, and Layer 2 from infinity. Thus, this model assumes that, in part, the tiering exists to compensate for the inability of the eye to accommodate. It does not explain the functions of Layers 3 and 4; and for that reason Land preferred the next hypothesis.

3. Tiering compensates for longitudinal chromatic aberration of the lens, such that Layer 1 receives sharply focused images of distant objects at long wavelengths (i.e., red) and Layer 2 of distant objects in the blue-green. Layer 3 could exploit a sharp ultraviolet image, but Layer 4 cannot receive a focused image at all, and Land suggested that it might be a polarization analyzer. Eakin and Brandenburger (1971) subsequently showed that the structure of Layer 4 receptors would allow them to perform this role, although there is as yet no evidence that jumping spiders navigate by polarized light. For this model to work, therefore, the receptors require three different photopigments in Layers 1-3. Layer 4 could share its photopigment with one of the other layers, most plausibly Layer 3.

Land (1969a) stressed that no real test can be made of these hypotheses until intracellular recordings have been obtained from marked cells so that spectral attributes can be linked to layers. Three attempts have been made to record from AM eyes, with varying degrees of success. DeVoe (1975) found no evidence for long wavelength receptors for electroretinograms, although negative evidence from ERGs can never be treated as more than suggestive. Intracellular recordings disclosed a preponderance of green cells, a smaller number of cells peaking solely in the ultraviolet, and a still smaller number with peaks of variable relative heights in both the UV and the green. The difficulties inherent in making recordings from AM eye preparations are well indicated by the fact that only 15 cells were fully analyzed, from a total population of 170, the majority of which were held by the electrode for periods of time too short for full spectral information to be obtained from them. Yamashita and Tateda (1976) obtained only seven cells from *Menemerus*, one peaking in the UV (360 nm), two in the blue (480-500 nm), three in the green (520-540 nm), and one in the yellow (580 nm) region of the spectrum. In a study in progress at the time of writing, Hardie has obtained full spectral data from some 20 cells from *Plexippus validus*, and marked the location of five cells by the intracellular injection of Lucifer Yellow (Hardie and Blest, in prep.). All but two cells were mid-spectral green receptors with spectral properties similar to the receptors of the PL eyes of the same species (Hardie and Duelli, 1978). Three green receptors were marked, one in Layer 1 and two in Layer 2. The two UV receptors were located in peripheral and central Layer 4 respectively.

Taken together, these results do not encourage us to believe that Layer 1 is likely to consist of red receptors in any species. They also suggest that there are differences between species in the spectral distributions of their AM receptors. It should be remembered, however, that a retina which possesses receptors for UV and green light is able to sustain dichromatic color vision. DeVoe (1975) has pointed out that some jumping spider adornments may have significance to other jumping spiders because of striking patterns in the UV which the human observer is unable to perceive.

Spectral data, then, do not support Land's third hypothesis; and we might be tempted to conclude that his second hypothesis, which supposes that tiering in part compensates for inability of the eye to accommodate, is proved. Unfortunately, the issue is complicated by a further factor: no analyses of the optics of the principal eyes have taken into account the matrix in which the receptors are embedded. Figure 5.16 shows a longitudinal section through the AM eye of *Plexippus validus* embedded in resin after fixation for electron microscopy, and stained with toluidine blue. There is a sharply demarcated interface between the darkly stained matrix and the anterior vitreous chamber. It is now known that this interface, which has precise and species-specific shapes, acts as a refracting surface whose rather complex optical properties are currently under examination (D. S. Williams and P. McIntyre, 1980). One inference which is inescapable is that previous calculations of the

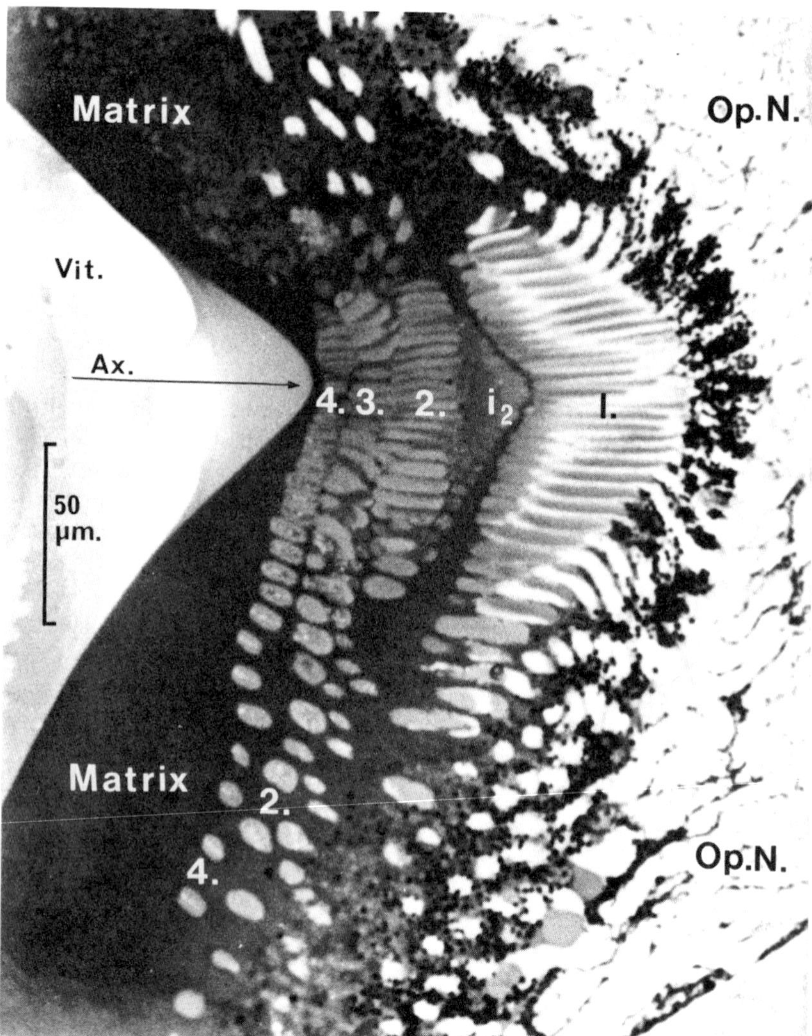

FIGURE 5.16. Vertical longitudinal section through the AM eyecup of *Plexippus validus*, passing through the central region and one arm of the boomerang-shaped retina. Receptor Layer 1 is separated from Layer 2 by intermediate segments of Layer 2 receptors (i_2). The thin arrow (Ax.) indicates the approximate optical axis of the eye. The interface between the vitreous compartment (Vit.) and the matrix within which the receptors are embedded is of complex shape, and the tip of the conical interface is filled with amorphous material of higher refractive index than the vitreous. In *Plexippus validus*, ophthalmoscopy suggests that there is unlikely to be significant image magnification derived from this interface (D. S. Williams, unpublished observations). Photomicrograph by courtesy of A. D. Blest.

focal planes for objects at different distances from the eye must be reexamined, and the optical geometry of the receptor mosaics reconsidered in terms of the results. This has important consequences for estimates of the acuity of the AM eyes discussed in the next section.

Visual Acuity

Both Homann (1928) and Land (1969a, 1972b) likened the organization of the jumping spider's complement of eyes to that of the individual human eye. The spider's lateral eyes, with low resolution, are primarily concerned with detecting movement, and hence resemble the human peripheral retina. The principal eyes, with fine receptor grain at the centers of the retinae, are involved in the task of identifying targets, and hence resemble the human foveal region. What order of acuity do the principal eyes achieve?

Receptor separations in the central region of the retina, expressed as angular subtenses in minutes, have been measured for a number of species: 12′ (*Evarcha blancardi*, Homann, 1928); 12′ (*Evarcha falcata*, Kaestner, 1950); and 11′ (*Phidippus johnsoni*) and 9′ (*Metaphidippus aeneolus*, both Land, 1969a). These values make the acuities of the central regions of the retinae at least three times better than is achieved by the best insect compound eye, and only about one magnitude worse than in man (Land, 1972b, 1974). Acuities estimated from behavioral data by Dzimirski (1959) for *Sitticus truncorum* (8′) and by Forster (1979b) for *Trite planiceps* (9-11′) are in the same range.

Given the jumping spider's life-style, one might suppose the principal eyes to be under strong selection pressure to attain the maximum acuity possible. Some of the limits to acuity can be calculated from optical parameters which are readily accessible. The performance of a simple eye is limited in two particularly important ways: (1) the diameter of the light-trapping component of a receptor (the rhabdom) is limited by wave-guide properties such that the minimum width permitted to it is roughly 1 μm (Kirschfeld, 1976) and (2) the lens has a diffraction limit which constrains the amount of image detail it can pass (see Land, 1969a, for an account of how this limit can be estimated for the principal eyes). It would be interesting, therefore, to determine how closely the principal eyes are operating to these two limits.

Eakin and Brandenburger (1971) examined the ultrastructure of the principal eyes of the same two species studied by Land; measurements from their figures show that the rhabdoms have widths substantially greater than 1 μm. Land (1969a) considered the diffraction limit for the same species. From the measured optical parameters of the lenses he calculated the period of a sinusoidal grating which the system would just fail to resolve, assuming the system to be free from aberrations. If the latter qualification were true, then the receptor spacing would be half this spatial cut-off value if the receptor mosaic were maximally to exploit the image provided to it. In fact, Land found that the receptor spacing at the central regions was twice the spatial cut-off, that is to say, *four times* the optimal spacing which theory predicts.

This discrepancy is large if we assume that lens performance and receptor spacing are evolved to match each other, and remains so even if we extrapolate from less sophisticated spider eyes, where it is known that the retinae undersample the images which they receive (Blest, 1978; Laughlin, Blest, and Stowe, 1980). Land (1969a) offered an explanation: aberrations, and especially the spherical aberration estimated for these lenses, degrade image quality so that the receptor spacing is, in fact, matched to what it can usefully sample. While this may be true, we now know that the lenses of spider eyes can be constructed ingeniously to correct the spherical aberration inherent in their geometries (Blest and Land, 1977; Blest, Williams, and Kao, 1980; Williams, 1979).

Mention has already been made of the conical interface between the retinal matrix and the vitreous anterior chamber. At the time of writing it is established that at the rounded tip of the cone the interface acts as a minuscule diverging lens to produce magnification of images at the central region of the retina by virtue of a telephoto effect. Image magnification in one case has been estimated to approach × 1.6. The effects of the interface on the peripheral retinal image are more complicated and are not yet fully understood (Williams, in prep.). However, the central magnification effect could at least bring the receptor spacing closer to the diffraction limit set by the lens; and perhaps the nature of the matrix surface interface makes it an aberration-prone system which accounts for the remaining discrepancy.

The receptor mosaics of Layer 1 are of great regularity: rigorously transverse sections through retinae which have been prepared for electron microscopy by procedures which obviate tissue distortion reveal this (Figure 5.17) in a way which light microscope reconstruction cannot. It becomes apparent that a line detector sampling the central array of mosaic such as this could be literally responsive to a continuous straight line imaged on the retina rather than to its approximation. This should further encourage us to believe that analysis of the optical constraints on the performance of the prin-

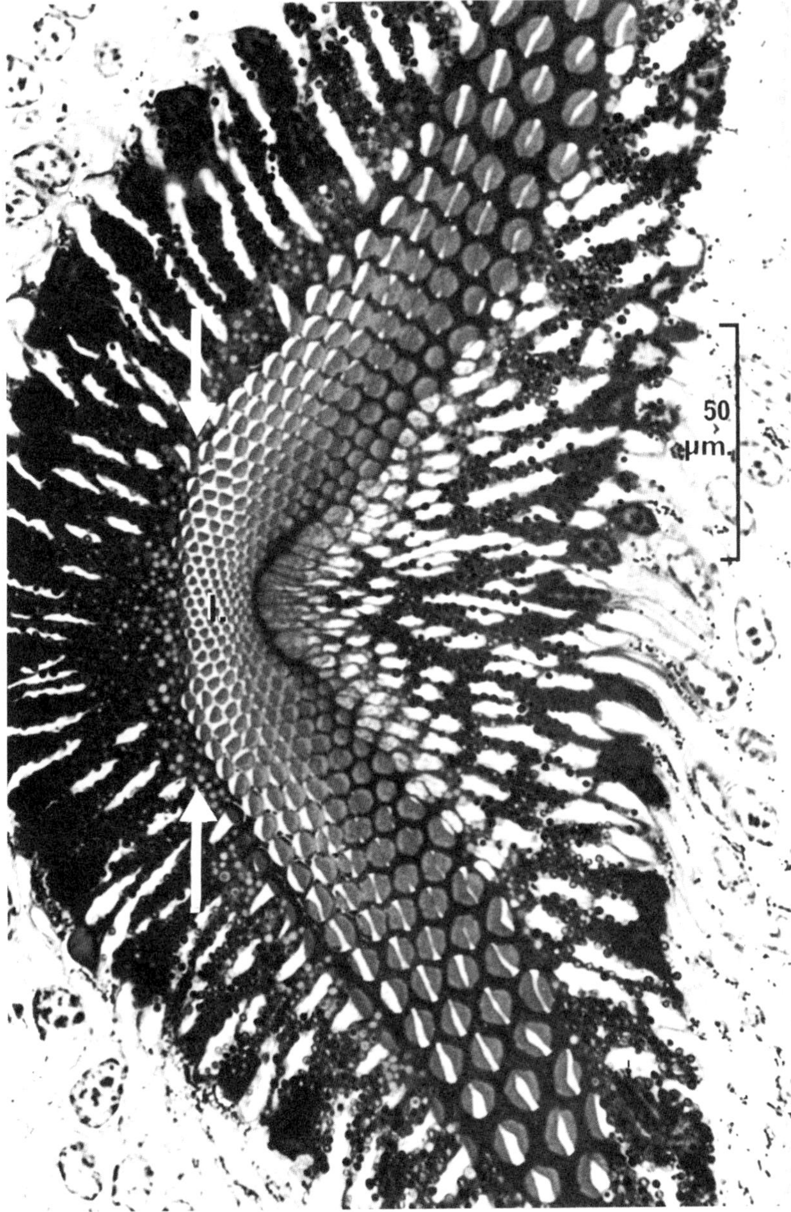

FIGURE 5.17. Transverse section through the AM retina of *Plexippus validus* at the level of Layer 1 of the receptors, taken close to the tips of the rhabdoms in the central area. The two white arrows indicate the approximate plane of longitudinal section shown in Figure 5.16. Photomicrograph by courtesy of A. D. Blest.

cipal eyes is worthwhile, for the precision achieved by the retinal mosaics implies that they are in receipt of images worth handling with care.

A brief account of the visual system is necessarily inadequate, and this section has been largely about the challenging and enigmatic principal eyes. There is much left to discover. The optics are imperfectly understood, and we do not know why the retinae are tiered. The originality and elegance of Land's study of the visual system of jumping spiders seems, regrettably, to have deterred arachnologists from further work on it. But it is apparent that progress in the broad field of vision research during the last decade opens up many new possibilities. Tiering might be related to optimization of sensitivities so that different layers work best at different light levels; some of the four classes of receptor may have distinctive adaptational properties, as has been shown for the duplex retina of a slug (Suzuki et al., 1979); linkage to line or boundary detectors might be arranged so that each layer and its associated neuropile is used for the recognition of a single class of releasing stimuli; or, the receptors of each layer may have spatial frequency transfer functions optimized for images which move across the retina at different speeds. None of these suggestions is implausible, and evidence has yet to be sought for any one of them.

Further progress along these lines is unlikely until a physiological preparation is designed which allows intracellular recordings to be made from the receptors of principal eyes whose optics are unimpaired. Less ambitiously, once the tiered receptors have been fully classified in terms of their spectral properties, the optical construction of the eyes can be reassessed. It has already been pointed out that the popular assumption that jumping spiders uniformly enjoy bright sunlight is mistaken, and that many live in very dimly lit habitats. This range of preferred environmental illuminances should be reflected by the retinal design of the principal eyes in distinctive and informative ways. A preliminary study (A. D. Blest and D. S. Williams) shows that this is, indeed, the case. One conclusion from this comparison is that Layer 1 is always preserved as a high-resolution mosaic, whatever the adaptive fate of more distal layers. This should color our thoughts about functions that might be attributed to each of the four layers, but much more information is needed before reliable conclusions can be drawn.

Pattern Recognition

Beginning with the Peckhams (q.v.), almost all observers of courtship in salticids have described both adornments and displays in rich and imaginative terms. In addition to detailing their spectacular postures and movements, authors have stressed the significance of their brilliant hues, prominent hairs and fringes, and strikingly patterned bodies. If, now, there is reason to suppose that some spiders have dichromatic color vision, what other evidence is there that they can actually see those postures and fringes and patterns?

What one jumping spider looks at when it confronts another is a wide forehead, a row of prominent eyes, legs and palps positioned in various ways, and the often large, basal segments of the chelicerae. The relevance of some of these characters in eliciting displays was examined by Crane (1949b) during the course of a long series of investigations into the behavioral biology of a large number of salticid species found at Rancho Grande, Venezuela.

During these studies, Crane tested the responses of male *Corythalia xanthopa* to several cardboard models simulating a variety of frontal perspectives (Figure 5.18). She found that a model equivalent in size to the viewing spider, having a well-defined forehead above a yellow patch with legs retracted slightly, was more successful in eliciting a threat (agonistic) behavior from males than a model with eyes, or models with thick, prominent legs or a large, dark head and chelicerae, for instance. In further tests, she obscured or altered the appearance of dead males until other males failed to respond to them, even when they were moved mechanically. She examined the relevance of facial characters, normally enhanced in the mature male with splashes of color and additional features such as tufts and fringes, by obliterating them or by painting hairs, or attaching replicas of these features to the female. In this way she was able to establish the relative value of such adornments in courtship behavior.

To summarize, Crane's most important conclusions concerning the external releasers of display are that:

1. Visual stimuli alone are sufficient to release display.
2. Motion, shape, and suitable size are the most effective stimuli while legs are also important releasers.
3. Patterns and colors play little part in releasing courtship.

Crane, however, did accept that spiders could see both patterns and colors, and this led her to conclude that sexual decorations are very important "reinforcers" of display. Furthermore, she emphasized the enormous variability among species, more particularly with respect to the specific sign stimuli that release displays, noting also that olfaction seemed important in some cases. It is worth pointing out, however, that Crane's work concentrated on ascer-

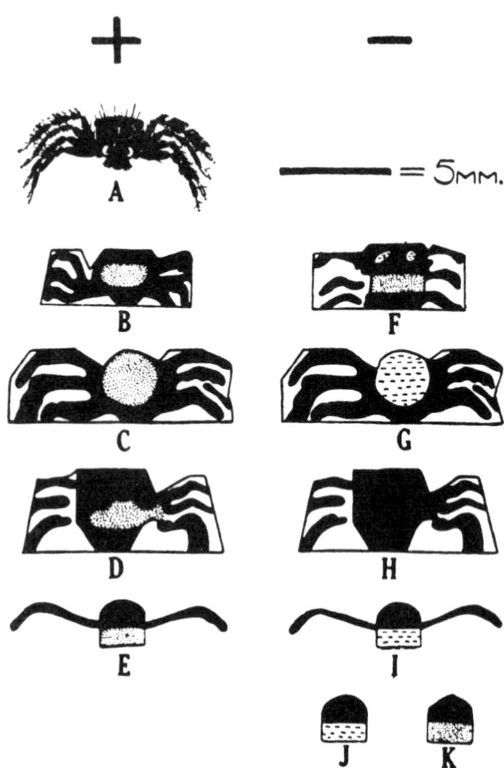

FIGURE 5.18. Examples of models used by Crane (1949b) in testing form-pattern-color perception in *Corythalia xanthopa*. Stippled areas were painted yellow, dashed areas white, backgrounds light green. (A) detailed drawing on cardboard of *C. xanthopa* in threat position; (B-E) drawings of cardboard models which drew threat responses, in order of success; (F-K) unsuccessful models. In some cases, the only difference between successful and unsuccessful models was the presence or absence of a yellow median area. Copied from Crane (1949b).

taining the nature of the characters that initiate courtship and that her studies did not examine the detailed behavioral interactions that occur between male and female spiders.

Drees (1952) carried out more specific tests than Crane, using just one species, *Epiblemum scenicum (Salticus scenicus)*. He too found that size and the thickness and position of legs were important in eliciting displays (Figure 5.19). Spiders that had been trained to jump at a variety of shapes and later avoided those which had been paired with aversive stimulation provided evidence that shapes could be perceived and distinguished (Figure 5.19b). Tests with other models led Drees to claim that the black and white striped abdomen of the female, as well as the relative proportions

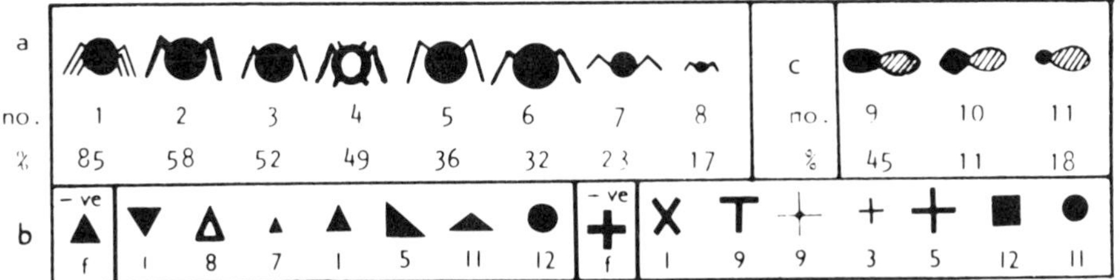

FIGURE 5.19. (a) Models (1-8) used by Drees (1952) to elicit responses from *Epiblemum scenicum*. Percentage of response shown beneath each form. (b) Spiders are trained to respond to ▲ and ✚, which were then paired with an electric shock (aversion stimulation). The models depicted here were then presented. Spiders showed, by avoidance behavior, that they recognized models similar to those from which they received a shock. Frequency (f) of response is shown beneath each shape. (c) "Female" models with black cephalothorax and striped abdomens. Note that each cephalothorax is of a different size and shape. From Drees (1952).

of the abdomen and cephalothorax, influence the male's response. (Figure 5.19c).

What Drees's (1952) studies show, quite convincingly, is that jumping spiders are able to discriminate between certain geometric configurations and patterns. We should, however, regard some of his other conclusions with caution. For example, test models used to show that male *Epiblemum scenicum* spiders discriminate between the relative proportions of different parts of the body are inadequate chiefly because they are not all the same size (Figure 5.19c). Spiders may well have been responding to the "total size" of the models, which diminish in size as proportions change, a variability which Gardner (1966) showed could affect the probability of response. In addition, the shape of the cephalothorax varies, another potential source of response differences.

In another set of experiments Drees (1952) showed that female *E. scenicum* spiders ceased hunting a small moving dot as soon as two wire "legs" were attached and waggled in an upraised position. Males, on the other hand, promptly courted models consisting of a black disk (body) with several lines placed in leg-like positions (see Figure 5.19a, no. 1). Obviously these models, which have an immediate effect on both participants, represent quite fundamental signals. These studies support the view that jumping spiders discriminate between targets on the basis of their form or shape, although the mechanisms are as yet not fully understood. Since there is apparently considerable variation in the organization of the principal retinae in different species (Blest, personal communication) the capacity for pattern recognition probably varies accordingly.

Visual Mechanisms of Communication

Catching prey, avoiding predators, and finding a suitable mate are the main tasks which confront a jumping spider. Success in any of these tasks depends primarily on the spider's capacity to perceive and process the appropriate visual input.

Detection and Localization

Jumping spiders swivel toward a moving object which has stimulated a region of the lateral retina (Bennet and Lewis, 1979; Dill, 1975; Drees, 1952; Duelli, 1978; Dzimirski, 1959; Forster, 1977a; Gardner, 1964, 1966; Giulio, 1979; Land, 1971). This orientation movement may mark the beginning of an escape reaction, a predatory sequence, an agonistic exchange, or a courtship interaction (Forster, 1977a). Moreover, swiveling by one jumping spider frequently alerts another to its presence, in which case it is also a communicatory act.

During the swivel, further visual input is suppressed (Duelli, 1978), so that once a spider has reacted by way of its lateral eyes to a particular stimulus, it is blind to all other environmental information. The advantage is that, at the completion of a turn, the spider's attention is concentrated upon the original source of stimulation which is now centered on the principal retinae. At the same time, the precise direction from which the movement originated has been established. Now, with its four anterior eyes directed at the target, the spider's next assignment is to decide upon its identity.

Target Discrimination

A jumping spider's immediate evaluation of the target is based on three factors: size, distance, and rate of movement. If, for example, the spider sees a target more than twice its own size, it backs and runs away (Crane, 1949b; Drees, 1952; Forster, 1979b; Gardner, 1965, 1966); if the right-sized target is within range (ca. 20 cm for *Trite planiceps*) and moving at more than 4°/second, then the spider treats it as prey and chases it, behavior which is guided by the AL eyes (Forster, 1979a). If, however, the target is stationary or moving at less than 4°/second, then the spider must decide whether it is prey or a conspecific (Forster, 1979a), a decision mediated by the foveal regions of the principal eyes (Land, 1969b).

Properties of the retinae which may be involved have already been discussed (see section on the Visual System).

The decision occurs promptly. If the target is prey, the spider begins to stalk it, but if it is a conspecific then the ensuing actions depend upon the sex of the spider itself, as well as the sex and maturity of the other spider (Forster, 1979a). It has already been suggested that, in some species, spiders cannot always make this latter judgment at the outset. Nevertheless, salticids possess a variety of visual signals which allow them to cope with this situation and to make the necessary identification at some stage in the sequence.

Once orientation has occurred, all intraspecific exchanges are mediated by the principal eyes, a conclusion derived from experiments in which interacting spiders were observed under various conditions of visual impairment (Crane, 1949b; Forster, 1979a; Homann, 1928). This means that the perception and processing of signals involved in visual communication are the prerogative of the principal eyes.

Having embarked upon a course of action which specifies the target as a conspecific, male spiders then have to close the gap between them. Receptive females, on the other hand, just wait and watch.

Approach Tactics

Two approach tactics are common to males of many species. One of these consists of side-to-side patterns of locomotion which may be fairly rigidly prescribed, zig-zag sidles such as those seen in *Trite auricoma*, or wide, rather variable, circling motions as practiced by *Euophrys parvula*. The second tactic involves more direct methods of approach. For example, in *T. planiceps* the male periodically performs short runs toward the female, often preceded or followed by a vigorous tapping of his outstretched forelegs. *Phidippus johnsoni*, on the other hand, progresses slowly forward in a series of smooth, deliberate steps (Jackson, 1977a).

For all their flamboyance, these two tactics could be based on very practical considerations. They may be designed chiefly to minimize the probability of releasing hunting behavior in the female, by reducing visual exposure in those spatial zones likely to provoke pursuit. The basis for this idea is as follows.

In some jumping spiders, hunting behavior consists of two, post-detection strategies during which prey is either chased or stalked

(Forster, 1979a). Both strategies result in the spider moving along its medial axis in the direction of the target. Therefore the parts of the retinae which are directed toward this axis have the potential to initiate hunting behavior.

Chasing is mediated by the AL eyes and is only activated toward a moving target (Forster, 1979a); hence the central region of the binocular trident monitored by these eyes corresponds to the medial axis of the spider (see Figure 5.14). For an approaching male, therefore, this axis is the most dangerous path toward a female. Stalking, on the other hand, is mediated solely by the principal eyes, presumably while they are scanning the target, and is directed toward stationary or slow-moving targets (Forster, 1979a).

To diminish the chances of being chased, the male must either avoid the medial axis of the female or move along it in a novel way. Moreover, his next objective presents him with something of a dilemma, since he needs to be "scanned" by the female in order to convey other relevant items of information, but at the same time he must avoid being stalked. How have these problems been solved?

A male *Trite auricoma* raises its forelegs as soon as it has turned toward the female, and then commences to zig-zag. At each zig-zag end-point he pauses briefly, jerks his body, and waves his forelegs. He then sidles across to the other side of the female and repeats his previous performance. This continues as the spider follows the path to the female depicted in Figure 5.20. This route overlaps the triangular area in front of the female which is monitored by the binocular overlap of her AL eyes and the retinal motions of her principal eyes. By crossing back and forth in ever-decreasing arcs, he may avoid activating the "chase" and "stalk" receptors directed toward the medial zone. Perhaps the male's jerking and the waving display on either side serve to attract attention to himself so that the female then directs her principal eyes toward him. Thus as he approaches, she "scans" his displays and "tracks" his zig-zags. In this way, essential information is transmitted to the female, and at the same time the male is able to approach safely.

A male *Trite planiceps* makes use of the second tactic when approaching a female. Although he may side-step from time to time, generally he runs in very brief spurts down the midline. Presumably this would be hazardous, were it not for the fact that it is associated with vigorous tapping movements by the outstretched forelegs. Perhaps the female initially watches and "scans" the male during his short runs but immediately her attention is distracted by the sharp movements at about 2 cm on either side of her. This

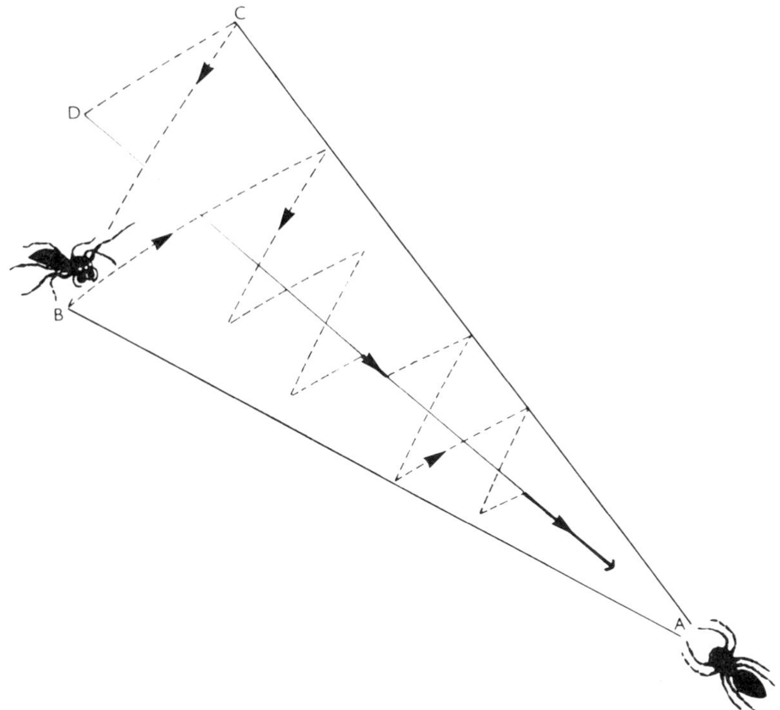

FIGURE 5.20. Approach path of male *T. auricoma* as he zig-zags toward the female (diagrammatic). During these maneuvers he always faces the receptive female. The area bounded by ABC corresponds to the visual trident monitored by the AL and AM eyes of the female. AD is the midline which the male spider criss-crosses as he approaches.

might inhibit hunting behavior, as suggested by experiments which show that *Trite planiceps* cease chasing a previously attractive target when two or three of these targets are presented together (Forster, 1979b). Singleness appears to be an important stimulus parameter. If the displaying male is producing three simultaneous sources of stimulation, arising from the body in the medial field of view and leg movements on either side, this may explain why the female refrains from pursuing him.

The function ascribed to the tactics outlined here is hypothetical, but it does offer a reasonable explanation for the male's safe conduct during approach. But do these locomotory tactics also have have a communicatory role? They can be seen in this light only if the male's behavior elicits responses from the female which otherwise would not have occurred. It is suggested here that these take the form of tracking and scanning eye movements. Unquestionably, from Land's (1969b) studies, eye movements occur as a result of the male's approach, although they are not ritualized events enhanced for the purpose of communication. Moreover, if the female monitors the male's activity in this manner, then she must also be receiving information from this source. Significantly, Crane (1949b) observed a striking use of retinal movements in *Lyssomanes* sp. during courtship, and accordingly she designated them as a display. Perhaps this covert form of response by females should be more widely recognized as a display, taking into consideration a proposal that the male's approach tactics promote an unfamiliar pattern of visual stimulation, of a kind not normally associated with prey-catching, for instance.

The Evolution of Diurnal Strategies

The benefits of a diurnal life-style to jumping spiders should be viewed in terms of both the predatory and reproductive strategies that have evolved. To begin with, hunting for prey in daylight opened up a whole new food resource, one that is utilized only sparingly by other spiders. Within this daylight regime, jumping spiders have exploited a light intensity gradient ranging from the dim recesses of a leaf litter habitat, through the fluctuating luminances of an arboreal life-style, to the exigencies of a bright, sunlit existence.

Concomitant with these benefits came a need for new hunting strategies that would allow them to take the fullest advantage of a capacity to see their prey, rather than just feel or taste it, or sense its vibratory signals. A variety of predatory strategies evolved, associated with different habitats and life-styles and correlated with adaptive modifications to the visual system (Forster, in press).

At the same time, prospective mates became visible, paving the way for a proliferation of visual signals by which they could be distinguished, first perhaps, from other species which happened to run around in daylight. More importantly, though, these visual signals provided them with mechanisms of prey discrimination, species and sexual distinctions, and safe conduct during approach to a prospective mate. Therefore, once jumping spiders had begun to depend on vision in their day-to-day activities, a system of visual communication became inevitable.

Such advantages are not without their shortcomings; ways of attracting the attention of a mate are equally likely to attract the attention of a predator, for example. Those with good eyesight such as spider wasps, praying mantids, robber flies, lizards, frogs, and birds are probable offenders. Countermeasures by salticids have not been seriously investigated except perhaps for ant-mimicry, which has developed, according to Reiskind (1977), as a protection against predation.

In some species, procryptic colors and patterns on the carapace and abdomen suggest that they are devices for deceiving predators (Forster and Forster, 1976), while "warning signals" are a potentially useful but as yet unreported practice. By and large, displays are conspicuous occasions and apparently place the participants in considerable danger (Platnick, 1971) unless they are less eye-catching to the potential predator than is generally supposed, a possibility which warrants investigation. Alternatively, the hazards of conspicuousness might be offset by the fact that ritualized sequences based on a rapid exchange of visual information might serve to minimize the time required for courtship and, by the same token, maximize success. Consider, for example, the absence of such distinctive features and movements, together with more haphazard ways of locating and approaching mates, and it is not difficult to see that, even in visually sophisticated animals, reproductive success might be less assured.

Without doubt, the degree of dependence on vision varies from species to species (Crane, 1949b) and hence the extent to which visual communication has developed may be correspondingly limited. For instance, not all salticids rely on visual interactions prior to mating; Jackson (1980b) reports that, in *Phidippus johnsoni*, mating frequently takes place inside the female's nest after a specialized non-visual type of courtship, and that the net benefit to the male is greatest if this occurs when the female molts after cohabitation.

My observations suggest, however, that in *T. auricoma* the two tactics described by Jackson are simply part of one overall strategy. When males encounter females out in the open, while both presumably are on predatory excursions, visual courtship occurs. The female, after watching the preliminaries with a docile stance, turns and runs away for a few steps with the male following. She reorients toward the male whereupon he begins leg-raising and zigzagging once more (Forster, 1977c). In a simulated environment, where the female has already established a nest, she leads the way back to it by a series of short runs and reorientations. The path she follows is quite direct and not consistent with the outward path, indicating that she makes use of some navigational capacity. Upon arrival at the nest she enters it at once, followed shortly after by the male. However, the male does not engage in any particularly specialized activity prior to entry, as Jackson (1977a) found for *P. johnsoni*.

Clearly, these two species (*T. auricoma* and *P. johnsoni*) behave quite differently, and it has yet to be shown which of these situations is the more prevalent. The strategy employed by *T. auricoma*, however, suggests that visual courtship has arisen because of the great probability that the two sexes will meet when searching for prey. Perhaps the deployment of a particular courtship strategy relates to the "hunting manner" of that particular species. Hunting manner refers to the distinctions made (along a continuum) between "searchers" on the one hand and "pursuers" on the other, determined by the way in which a predator's time or energy budget is allocated (Enders, 1975b). Searchers, for instance, spend most of their foraging time in actively looking for prey, whereas pursuers spend relatively more time in seizing (handling) prey, once it has been noticed. Based on this principle, those salticids designated as searchers might be more likely to encounter conspecifics and consequently to be more dependent upon visual courtship strategies. If it can be shown that *T. auricoma* is a searcher and *P. johnsoni* is a pursuer, then this may explain the major differences in the courtship strategies they each employ. Moreover, it should be borne in mind that there may be much more variability in the behavior of jumping spiders than is currently known, so that more conclusive analyses must await further investigations.

It is probable, therefore, that visual courtship has a multiplicity of functions which includes not only those mentioned earlier but also the inhibition of predatory behavior, a notion empirically demonstrated and discussed by Drees (1952), and a mechanism whereby the female invites the male to retreat with her to the relative seclusion of her nest. The behavior exhibited by *T. auricoma* has the virtue of providing a reasonable hypothesis as to how visual communication may have arisen and a logical explanation as to why it persists. It may be, of course, that *T. auricoma* represents the more primitive situation whereas *P. johnsoni* has advanced to the point of being able to utilize either of two tactics, depending on the circumstances of the moment.

Use of Vision in Other Spiders

By far the greatest contribution to our knowledge of the eyes of spiders has come from the detailed studies of Homann (1928, 1971), who examined both optics and properties of the retina in all the most widely recognized families. Homann's work, however, has not been matched by behavioral studies, so that today we still know little of the way in which most spiders' eyes are used.

There are, of course, some notable exceptions. As this chapter shows, the most sophisticated use of vision is found in salticids to an extent that far outstrips any other family. Perhaps the lycosids (wolf spiders) are next, although their visual capacities vary from species to species. According to Bristowe (1958), males of several species of *Lycosa* and *Trochosa*, for example, are distinctly marked and wave their legs and palps in a conspicuous manner during courtship; this led him to suggest that visual recognition is an essential component of display.

This conclusion was supported by Rovner (1968b) who found that pre-copulatory behavior in *Lycosa rabida* was influenced by visual signals as well as acoustic and chemical information. Because of Homann's (1931) claim that lycosid vision depends chiefly on movement perception, Rovner agreed that movements of the conspicuous black forelegs of the male and the leg-waving display of the female undoubtedly contribute to visual communication in this species. More recently, Aspey (1976) showed that intermale display in *Schizocosa crassipes* (= *ocreata*) is a visually mediated communication system functioning to preserve interindividual space within a socially organized population. Observations also suggest that *Lycosa hilaris* makes use of the visual perception of movement in the capture of prey (Forster and Forster, 1973), although this ability has not been empirically verified. However, Lloyd (1977) provides an astonishing example of the way in which one wolf spider, *Lycosa rabida*, capitalizes on an unusual source of visual information. Apparently these spiders are important predators of *Photuris* fireflies, which they probably locate by either their illumination self-signals, predatory false-signals, or mating signals.

Yet another lycosid species, *Arctosa variana*, navigates by means of polarized light. Its capacity to do so was discovered by Papi (1955), while later studies (Magni et al., 1964) showed that it is mainly the principal eyes which are responsible. Tretzel (1961) found that *Coelotes*, an agelenid, makes use of polarized light in choosing a nest site while Görner (1962) showed the the funnelweb spider, *Agelena labyrinthica*, also orients with the aid of polarized light.

Peucetia viridans is an oxyopid with the curious habit of mating while dangling from silk threads (Whitcomb and Eason, 1965). During the preliminaries, the male apparently recognizes the female by sight, subsequently vibrating his abdomen, drumming his palps, and waving his first and second pairs of legs up and down. This can last as long as eleven minutes before the female suddenly moves to the edge of a leaf and drops from a thread; the male follows shortly afterwards.

Probably the largest simple eyes found in any arthropod are the posterior median eyes of *Dinopis subrufus*, the ogre-faced spider. The eyes have very great visual sensitivity and have been modified for night vision (Blest and Land, 1977). After dusk, *Dinopis* constructs a small "sticky" web and then hangs downwards, holding the net with the tips of its first two pairs of legs. Robinson and Robinson (1971) demonstrated that when the spider sees something moving beneath it, the net is stretched, the spider lunges toward it, and the net is thrown at the "source of movement" even if this is just a shadow. Capture of prey is thereby brought about by the perception of movement.

With the exception of the salticids and to a lesser extent those species mentioned above, few spiders have specialized in the use of vision. Some species of lycosids and oxyopids may make limited use of visual communication: only in the salticids has it reached a degree of sophistication equivalent perhaps to that found in many fishes and birds.

Conclusions

The behavior of jumping spiders is dominated by visual skills, to an extent which makes a system of visual communication inevitable. However, there is considerable variability with regard to the reliance on vision within this family, so that until more species have been studied in detail, generalizations should be treated with caution.

Courtship and agonistic sequences are generally conspicuous affairs, and it is shown here that much of this behavior is based on a stimulus-response system of communication. Many events, typically performed during agonistic interactions, are also characteristic of courtship behavior in the species discussed here; the relationship between these behaviors in other species warrants further investigation. Visual badges, postures, and movements all participate in the exchange of information, although it is likely that other sensory input contributes to the interaction at certain stages.

Sequences are divided into three stages, the first stage consisting of primary events, the second stage consisting of those events which promote the approach of the male to the female, or prevent it, and the third stage involving pre-copulatory postures and movements. The particular events of the third stage are believed to be universal among salticids.

In *Trite planiceps*, juvenile spiders exhibit a wide variety of stereotyped postures and movements when they encounter each other. These displays, shown by mirror-image stimulation to be visually mediated, appear to have a role in the reduction of cannibalism among conspecifics.

Interspecific interactions between *T. planiceps* and *T. auricoma* males suggest that their primary signals serve to keep them apart. However, a *T. auricoma* male was able to approach (Stage 2) an "alien" species with impunity, although he was repulsed at the Stage 3 level. Males of both these species have comparable primary and secondary postures (although approach tactics differ); hence it appears that there are some basic salticid signals which are universally recognized.

Jumping spiders have a highly specialized visual system, one which is eminently capable of perceiving the visual signals described here and hence mediating the appropriate behavior. Mechanisms of communication relate to the detection and localization of a source of stimulation, the discrimination of a target, the approach tactics of male spiders, and the "sit and watch" or "runaway" behavior of females.

When stimulated, the lateral eyes induce orientation of the spider; post-detection strategies depend on the identity of the target, its size and distance, and its rate of movement. If the target is a conspecific, given that it is of the right size and within range, then the sexes of the interacting pair of spiders dictate the form of behavioral strategies employed; such strategies are mediated by the principal eyes.

The evolution of a diurnal life-style in jumping spiders has led to the development of visual strategies involving predator avoidance, prey-capture, and conspecific recognition, as well as elaborate forms of courtship behavior. It is suggested that the need to hunt for prey in daylight some distance from a secluded nest gave rise to visual mechanisms designed to reduce cannibalism, space individuals (unverified as yet), promote intra- and interspecific recognition, and, in the case of *T. auricoma*, provide a means whereby the female leads a male back to her nest. It is argued that the disadvantages of conspicuous behavior may be offset by strategies which minimize the time or energy required for detection, localization, and approach (as opposed to more haphazard procedures), and at the same time maximize success.

The use of vision by other spiders is discussed briefly. While there is evidence that some lycosid species exchange visual information during encounters, it is concluded that salticids, above all other families of spiders, make use of remarkably sophisticated, visually mediated, communication systems.

Acknowledgments

The author would like to extend her thanks to Dr. H. Levi and Dr. M. F. Land for reading earlier drafts of the manuscript; Dr. A. D. Blest for assistance in writing up the section on the visual system; Mrs. Denise Hession, Mrs. Ruth Newlove, and Miss Fiona Kennedy for valuable technical assistance.

The following publications appeared or will appear after completion of this chapter:

Forster, L. M. Prey-catching strategies in jumping spiders. Amer. Scientist (in press).

Laughlin, S., A. D. Blest, and S. Stowe. The sensitivity of receptors in the posterior median eye of the nocturnal spider, *Dinopis*. J. comp. Physiol. *141*: 53-65, 1980.

Williams, D. S. and P. McIntyre. The principal eyes of a jumping spider have a telephoto component. Nature *288*: 578-580, 1980.

Chapter 6

THE BEHAVIOR OF COMMUNICATING IN JUMPING SPIDERS (SALTICIDAE)

Robert R. Jackson
Department of Zoology
University of Canterbury
Christchurch 1, New Zealand

Introduction

If it is proper to judge animals on a scale of beauty and elegance, the Salticidae must surely be rivaled by few other groups of animals. Often they are ornamented with colors ranging from brilliant red, yellow, and orange to metallic green and iridescent blue. Their faces may be decorated with pigmented scales, their palps clothed with snow-white setae, and their heads fringed with multicolored hairs, to name but a few variations.

Male salticids tend to be more extravagantly marked than the females, and mating is usually preceded by a courtship in which the male postures and prances in front of the watching female. The first detailed descriptions of salticid behavior were made at the turn of the century by George and Elizabeth Peckham (1889, 1890, 1894, 1909); they argued that the ornamentation of the male is important in communication, and specifically in sexual selection. This conclusion stirred up considerable controversy, and other arguments have been proposed. However, discussion has frequently been hampered by confusion concerning the important questions.

In this chapter I shall attempt to clarify some of the questions about salticid behavior by using a framework derived from Smith's (1977) semiotic approach. His approach makes distinctions not prevalent in more traditional studies.

The basic components of the communicatory process, according to this framework, are two individuals (the sender and the receiver) and a signal (attributes of the sender which are adaptations for information transfer). The signal's message is the information it conveys, and its meaning is its effect on the receiver's behavior. By communicating, the sender gains a certain degree of control over the actions of the receiver, and the adaptive significance to the sender of this type of control is the function of the signal.

As implicated by their family and common names, the Salticidae are one of the few groups of spiders which move about by jumping; but the characteristic which most distinguishes the salticids from other spiders is their highly sophisticated and, in many respects, unique visual system (Forster, this volume; Land, 1972b). Much of the communicatory behavior of these spiders is visually mediated. However, the evolution of sophisticated visual abilities in the salticids has not precluded communication involving other sensory modalities.

The salticids are a conspicuous component of most terrestrial faunas, living in habitats from the intertidal to more than 6,000 m above sea level on Mount Everest, from deserts to tropical rain forests, on oceanic islands, and on every continent except Antarctica. Study of this large family (approximately 4,000 described species; Prószyński, 1971), which is one of the major groups in the animal kingdom to have evolved visually mediated communication, holds great potential for advancing our understanding of animal communication. However, their importance in this respect is often overlooked. For example, Hailman's (1977a) review of communication by reflected light failed to mention the salticids.

Non-Visual Communication

Use of visual displays does not preclude simultaneous communication by means of other sensory modalities. For example, both visual and auditory displays are used by *Phidippus mystaceus* (Edwards, in press), and there is preliminary evidence in several species of olfactory stimuli that alert males to the presence of females and lower the males' thresholds for visual displays (Crane, 1949b; Richman, 1977a).

Multichannel communication is not necessarily limited to simultaneous usage, however. Probably it is often segregated into phases within the normal courtship sequence as in *Phidippus johnsoni* (Bristowe, 1941; Crane, 1949b; Jackson, 1977a; also see comments by Savory, 1928) during courtship outside of nests (Figure 6.1). Initially the male employs a visual display, but upon mounting the female he engages in specialized tapping and stroking behavior during which tactile and chemical information may be exchanged.

Salticids build silken nests in which they molt, ovipost, and often reside at night, during inclement weather, or during other periods of inactivity (Jackson, 1979a). In *P. johnsoni* communication by different sensory channels is segregated in a manner related to the nest (Jackson, 1977b). Each individual male may use any of three distinct mating tactics depending on the type of female he encounters and whether she is inside or outside her nest (Figure 6.1). If he encounters an adult female outside her nest, he performs vision-dependent displays (type 1 courtship) in front of the facing female. If he encounters an adult female inside her nest, he employs a different type of courtship (type 2) consisting of various tugging, probing, and vibrating movements made on the silk of the nest and not dependent on vision. If he encounters a subadult female (one that will mature at her next molt) inside her nest, he initially performs type 2 courtship and later spins a second chamber on the female's nest and cohabits until she molts and matures. There is evidence to suggest that this kind of courtship versatility and use of alternative tactics is widespread in the salticids (Bristowe, 1958; Forster and Forster, 1973; Hill, 1977a; Jackson, 1978b).

Crane (1949b) tentatively proposed classifying salticids into three

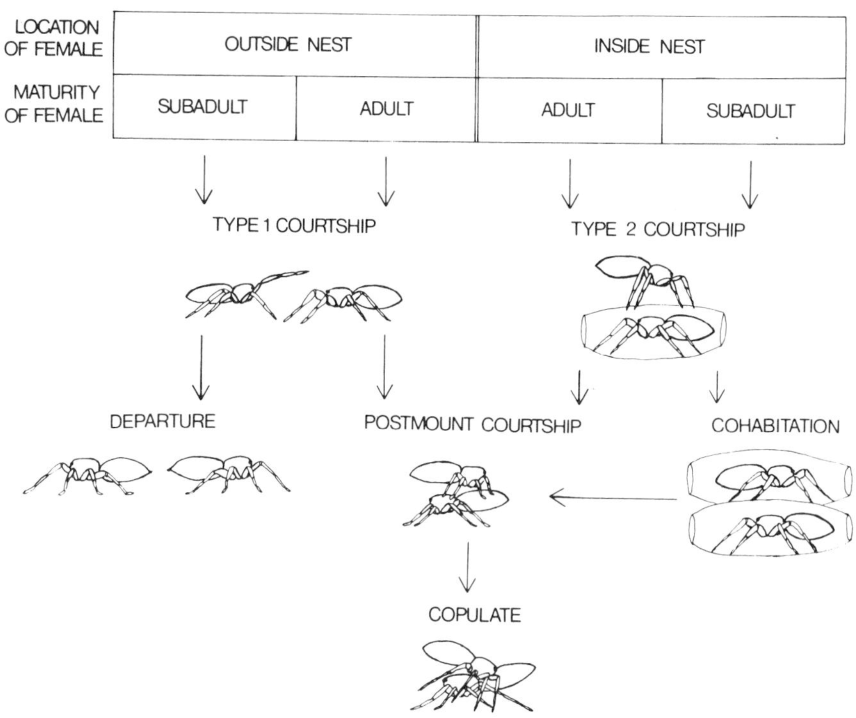

FIGURE 6.1. Alternative mating tactics of the adult males of *Phidippus johnsoni*.

groups with respect to jumping and other characteristics (Table 6.1). The runners, presumed to be closer to the primitive salticid stock, rely more on chemical stimuli during both communicatory and non-communicatory behavior, whereas the hoppers are more vision-dependent spiders. Intermediates between these two types make important use of both olfaction and vision. It is tempting to propose that nest-related, non-visual courtship follows a similar pattern, with the runners relying more on vibratory forms of courtship than the hoppers. However, it is likely that ecological determinants are more important than phylogenetic ones in both the trends discussed by Crane (Forster, 1977a) and in trends related to courtship versatility (Jackson, 1977b).

Syntactics

Basic Forms of Displays

The diversity of displays that occurs in the Salticidae is at first overwhelming. Palps, chelicerae, abdomen, cephalothorax, and each of the four pairs of legs may be positioned, waved, jerked, or rotated in special and varied ways. Yet there seem to be some basic themes.

It is useful to distinguish between aspects of display that involve movement patterns and those that involve postures. Postures are static, with the spider holding parts of its body in a particular stance for a sustained period. When referring to legs of the spider, legs I are the most anterior pair; legs II, the next most anterior pair, etc.

Elevated-leg displays. While elevated, the legs may perform specialized movement patterns, or be held in specific postures (Figures 6.2 and 6.3). Elevated-leg displays involving legs I seem the most prevalent. Use of legs II and III is less common but not necessarily rare, while legs IV are seldom used in this form of display (Crane, 1948, 1949b). Corresponding legs on the left and right sides of the body tend to be used in unison; but the phasing of such movements may be parallel or alternate (Figure 6.2B and 6.2C), with the former possibly being more common.

Palpal displays. Although displaying salticids frequently perform specialized movements and especially postures with their palps, this type of display may be frequently overlooked in less detailed descriptive studies.

Cheliceral displays. The chelicerae may be held in special pos-

Table 6.1. Behavioral characteristics of Crane's (1949b) three categories of salticids.

	Courtship	*Inter-Male Display*	*Mirror-Image Display*	*Visual Acuity*	*Reliance on Chemo-reception*	*Tendency to Wave Palps*	*Posture of Legs I during Locomotion*	*Use of Hopping during Locomotion*
Runners	Simple	Practically absent	Rarely elicited	Least developed	Chemotaxis and olfaction important	Strong	Elevated	Hop only to cross gaps and capture prey
Inter-mediates	Inter-mediate	Similar to courtship	Inter-mediate	Inter-mediate	Olfaction relatively important	Inter-mediate	Waved inter-mittently	Hop in absence of gaps and prey
Hoppers	Complex	Distinct from courtship	Readily elicited	Most developed	Relatively unimportant	Weak	Kept on substrate	Hop frequently in absence of gaps and prey

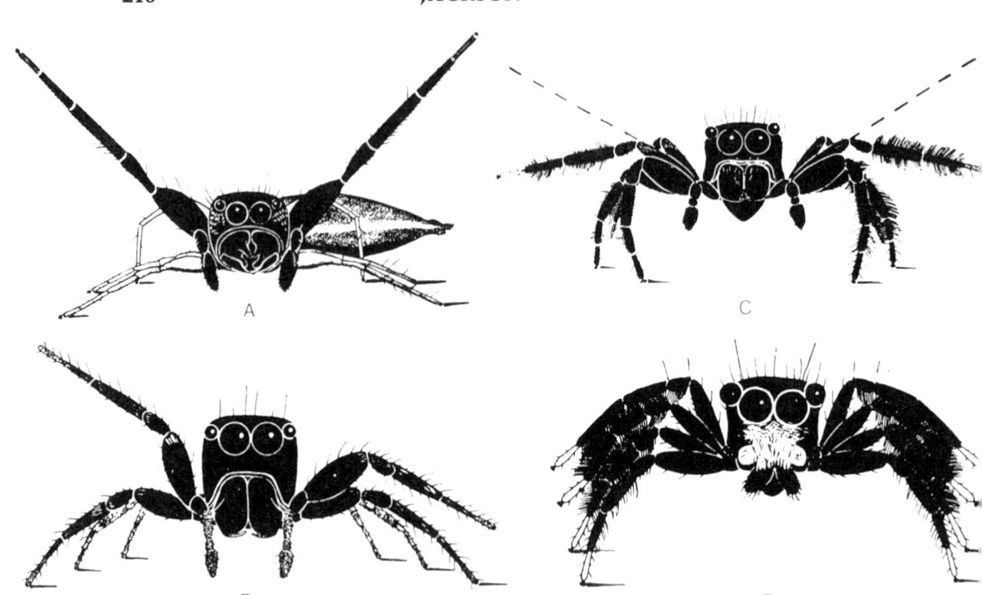

FIGURE 6.2. Elevated-leg displays. (A) Adult male of *Ashtabula furcillata*. Courtship. Legs I 45° to sagittal and frontal planes of cephalothorax. Palps spread to side. Abdomen bent to spider's left. Legs II and III extended outward resulting in lowering of the cephalothorax. (B) Adult male of *Mago dentichelis*. Courtship. Legs I moved alternately. Palps held downward. (C) Adult male of *Corythalia chalcea* displaying to another male. Dotted lines indicate peak positions of legs III during waving. (D) Adult male of *Corythalia xanthopa* displaying to another male. Fan display: successively waves II, III, and IV. Note markings on face. From Crane (1948, 1949b).

tures which may involve the spreading apart of the paturons (basal segments) and/or extending of the fangs downward from the paturons. Movement patterns, as opposed to postures, of the chelicerae have not been reported.

Abdominal displays. The spider may hold its abdomen in a special posture or move it in some specialized way. Twisting it to the side is especially common (Figure 6.2A). Abdominal flaps are expanded during the courtship of a few salticids, and a number of species raise the abdomen (Figure 6.4). Twitching of the abdomen (rapid, low amplitude up-and-down or back-and-forth movements) is a very common element. It seems unlikely as a visual display since the spider's cephalothorax and legs usually obscure the abdomen from the view of another facing spider, nor is there evidence at present of stridulatory organs or other means of sound production associated with these movements.

Cephalothoracic displays. The spider may raise or lower its

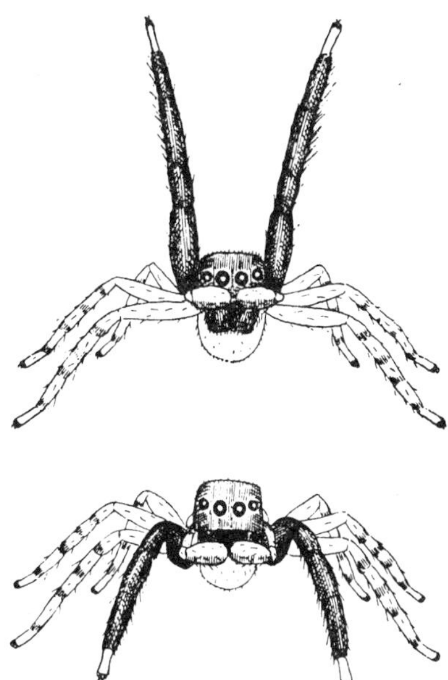

FIGURE 6.3. Adult male of *Euophrys frontalis*. Courtship. Legs I repeatedly raised from substrate and jerked to almost vertical, then lowered slowly before repeating. Palps bent inward and vibrated up and down. From Bristowe (1958).

cephalothorax as it performs elevated-leg displays, dancing, or rocking; and it is common for the female to lower her body before the male mounts (Crane, 1949b).

Dance. Dancing is a specialized pattern of locomotion in which the spider repeatedly alters its direction of stepping. The dance is normally oriented toward another spider, and two types have been discerned: (1) zig-zag: side-to-side arcing by the male in front of the female with the arcs becoming smaller as the distances between the spiders decrease (Figure 6.5); (2) linear: alternation of forward and backward locomotion along a straight line (Figure 6.6).

Sway, rock, and lunge. These are displays in which the spider moves its body from side to side (sway and rock) or forward and backward (lunge) without stepping. The legs extend and flex without their tarsi leaving the ground. During rocking (Figure 6.7), as opposed to swaying, the body tilts with respect to the substrate.

Charge and truncated leap. These movement patterns differ from sway, rock, and lunge in that the tarsi leave the ground. During a charge, the spider runs a short distance toward another spider, then abruptly stops. A truncated leap carries the spider forward only a few millimeters, after which it may immediately run away. Leaps

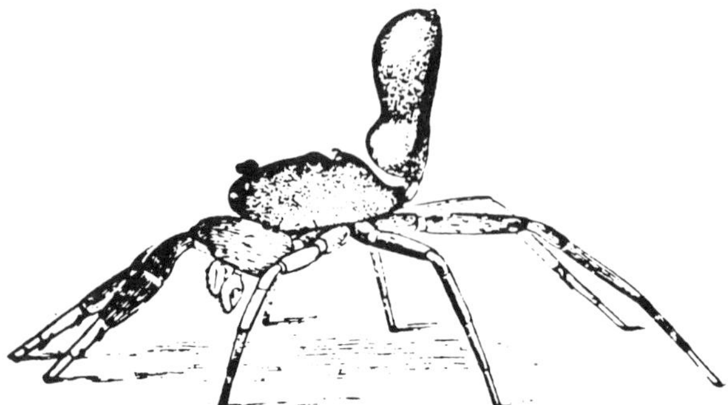

FIGURE 6.4. Adult male of *Peckhamia (Synageles) picata*. Courtship. With abdomen elevated, male rocks from side to side. Legs I remain on substrate. Tarsi of legs I flattened and ornamented. From Peckham and Peckham (1889).

during predation and normal locomotion tend to be of considerably greater amplitude.

Strike. To strike, the spider raises its legs I, then rapidly moves them forward and downward against the substrate or another spider (Figure 6.8). Sometimes the spider steps toward its target during the strike.

Prod. Prodding sometimes occurs during male-male interactions after one spider decamps. The prodding male brings its face into

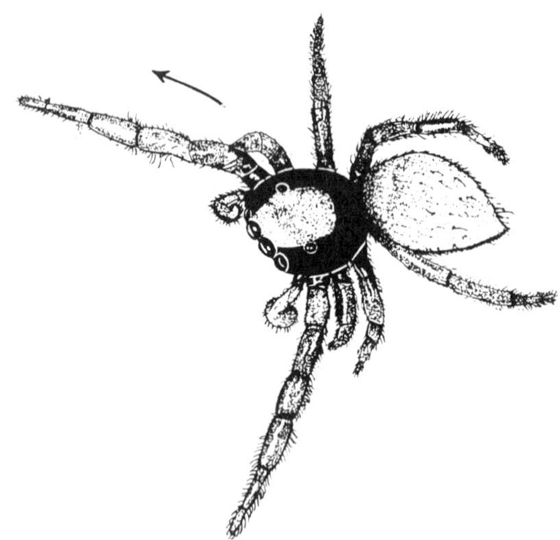

FIGURE 6.5. Adult male of *Phidippus johnsoni*. Courtship. Zig-zag dance. Stepping in direction of arrow. Abdomen bent to side. Gestures with legs I. Legs I move repeatedly from position in this figure (45° to sagittal and frontal planes) to position in Figure 6.6.

FIGURE 6.6. *Phidippus johnsoni*. Courtship. Adult male (left) performs linear dance. Stepping in direction of arrow. Gestures with legs I. Legs I almost parallel with sagittal plane and raised nearly 60° from frontal plane. Female waves palps; abdomen bent to side.

contact, sometimes quite forcibly, with the abdomen of the decamping male.

Embrace. Two spiders stand face to face with their chelicerae and/or legs I touching when embracing (Figure 6.9). Sometimes the palps and legs II may also make contact. While embracing, the spiders may remain essentially motionless, or they may push and grapple. Embracing seems quite common in the salticids, but apparently it has mostly been referred to as "fighting," an unfortunate expression because it seems to suggest that the spiders have resorted to physical force and there is little need for further consid-

FIGURE 6.7. Adult male of *Ballus depressus*. Rocks during courtship. Cephalothorax tilts alternately to left and right. Extends legs on side opposite movement. Palps spread to side. From Bristowe (1941).

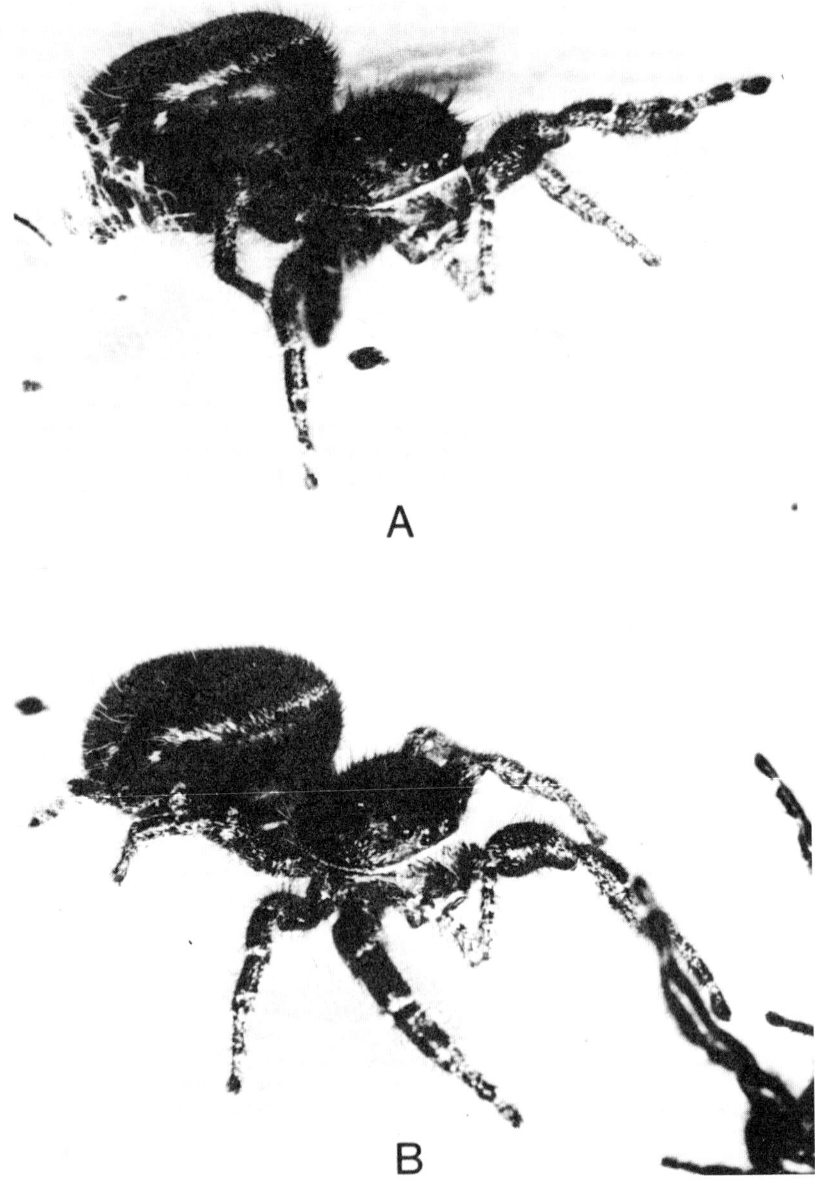

FIGURE 6.8. Striking by adult female *Phidippus johnsoni*. (A) Legs I move upward and medially as female departs her nest (left) to begin strike. (B) At completion of strike, legs I on substrate. Note erected legs of male on right. Note basal band on abdomen. From Jackson (1977b).

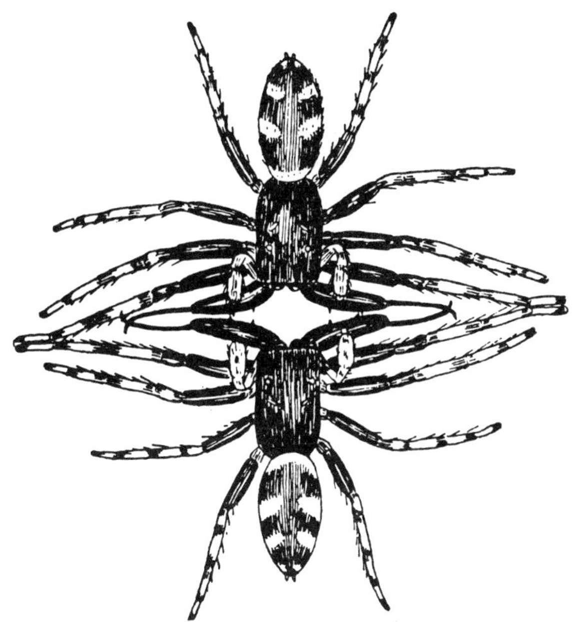

FIGURE 6.9. Adult males of *Salticus scenicus*. Embracing. Note large chelicerae. Legs I and chelicerae in contact. Palps spread to side. Note markings on abdomen. From Bristowe (1941).

eration of communication. More likely, these specialized postures and movements are displays, although this does not rule out the involvement of physical force as well.

Mount. As a necessary preliminary to mating, males mount females; but the manner in which this is achieved may also be an adaptation related to information exchange. In most cases, just before mounting the male walks up to the female as he displays with his legs I forward. Commonly the male taps the female with his legs and palps as he climbs onto her. Males of *P. johnsoni* sometimes also mount directly from the embraced posture; and when mating inside nests, males often mount while spinning (Jackson, 1977a). Males of some species mount by leaping onto the female (Bristowe, 1929; L. Forster, personal communication; Homann, 1928).

Postmount displays. After mounting, males of *P. johnsoni* perform various tapping, scraping, and stroking movements with their legs and palps on the female's abdomen (Figure 6.10); the female's behavior includes turning, walking, and lifting the forelegs, as well as rotation of the abdomen (Jackson, 1977a). Males of *Pellenes*

FIGURE 6.10. *Phidippus johnsoni.* Postmount courtship. Male (upper) taps female's abdomen (lower) with his legs and palps as he stands on her cephalothorax. Note distinct black band on female's abdomen and more uniform coloration of male's abdomen.

umatillus spin silk over the female after mounting (Griswold, 1977).

Nest-associated displays. These displays have been discussed in detail for *P. johnsoni* (Jackson, 1977a). While probing (Figure 6.11), which is especially common at the door of the nest, the male moves his forelegs alternately backward and forward with his tarsi in contact with the silk. During tugging, the male grips the silk with his chelicerae and moves his cephalothorax up and down several times. Vibration, perhaps the most distinctive feature of type 2 courtship, consists of a volley of very rapid (ca. 10 per sec), low amplitude, up-and-down movements of the spider's legs and body which cause the male and the silk in his vicinity to take on a blurred appearance.

The female in turn may perform various behaviors from within the nest, such as stabbing at the male with her legs, bumping against the nest with her cephalothorax, and pulling on the silk with her legs. It is particularly common for the female to hold the

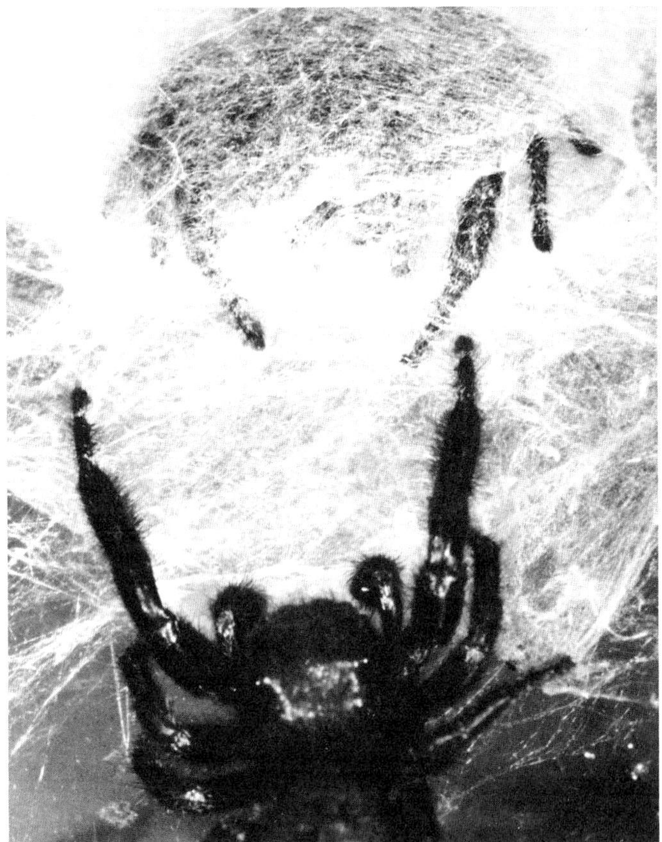

FIGURE 6.11. Adult male (lower) of *Phidippus johnsoni* probing near the door of the female's nest. Female (upper) inside nest and holding door down.

nest door closed by pulling on the silk while the male is probing at the door (Figure 6.11).

BADGES

Displays are signals that are behavioral in character, i.e., adaptations entailing movement. Badges, in contrast, are static properties of the spiders' morphology. In practice it may be difficult to determine the relative importance of the two since they tend to act in concert.

Peckham and Peckham (1889, 1890) emphasized the distribution on the spiders' bodies of coloration, markings, and other structural peculiarities, especially those which are sexually dimorphic. As a rule, these are located in positions that are likely to be noticed by another interacting salticid, especially during display behavior. As with displays, the number and diversity of badges in salticids seem at first overwhelming, although there are some basic themes.

Leg and palp badges. Very often the forelegs of male salticids are longer and thicker than those of the female (Figure 6.4). Other modifications include such things as spurs, tufts of hair, and coloration. Palps also are frequently decorated.

Cheliceral badges. Some of the most spectacular coloration in salticids occurs on the paturons of the chelicerae. In addition to brightly colored scales (Hill, 1979a) and setae, there may be distinctive iridescence due to the structure of the cuticle.

Cephalothoracic badges. The face is an area that is especially often decorated with colored scales and tufts of hair (Figure 6.12). Other regions of the cephalothorax may also be decorated.

Abdominal badges. In many species, the abdomen is conspicuously patterned, particularly in those which bring the abdomen into view during display.

Chemical badges. This type of badge (see Tietjen and Rovner, this volume) might be important during embracing and postmount behavior in the salticids, although this possibility has not been investigated in depth. Olfactory pheromones produced by the female are another potential type of chemical badge mentioned earlier.

Construction Signals

Unoccupied nests of the females of *P. johnsoni* elicit type 2 courtship from conspecific males. Males discriminate between empty female-nests and empty male-nests, as well as between those of a predatory gnaphosid spider and those of conspecifics (Jackson, 1976b, 1981a). Perhaps the chemical and/or physical properties of the nests have been evolutionarily modified to convey this type of information.

Ritualization

Ritualization refers to evolutionary changes that enhance the communicatory nature of certain elements of behavior. During routine activity, many salticids tend to wave their palps and legs (especially legs I) in the air or tap with them on the substrate (see Hill, 1978; Savory, 1935). Generally, leg and palp movements of these types have been referred to as chemosensory, which seems a reasonable hypothesis since Foelix (1970) and Hill (1977b) have

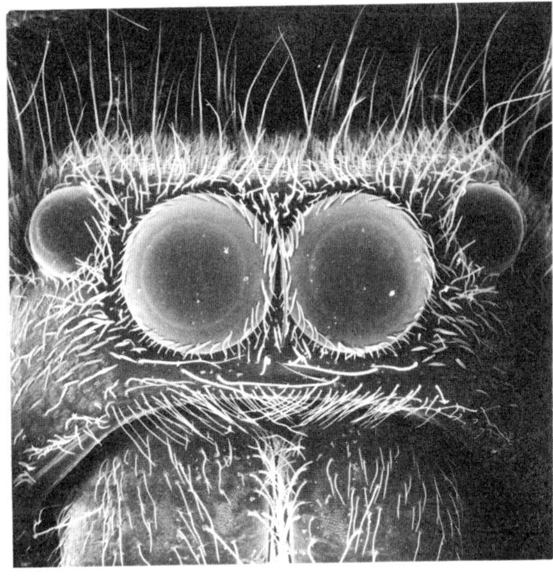

A

B

FIGURE 6.12. Face of *Phidippus johnsoni* (S.E.M. from shed exoskeleton). (A) Scales and setae, responsible for color pattern, surround eyes and clothe clypeus. AM eye diameter: 500 μ. (B) Enlarged view of scales. Diameter at widest point of scale at bottom left: 10 μ.

pointed out chemoreceptors on the legs of the spiders. Waving of the legs and palps is often especially pronounced just after a potential prey or mate comes into the spider's visual field, which is consistent with a chemosensory function. Movements of these types would seem conspicuous to a facing conspecific spider. Bristowe (1941), Bristowe and Locket (1926), and Crane (1949b) suggested that many salticid displays involving the palps and legs are the result of ritualization of such routine palp and leg movements.

Salticids often groom their antero-medial eyes by slowly moving one palp at a time downward across the cornea (Hill, 1977c). Perhaps palp-waving displays, such as those of *Phidippus femoratus* (Jackson, 1981b) and *P. putnami* (G. B. Edwards, personal communication), are the result of ritualization of such grooming behavior. Similarly, dancing displays may have resulted from the ritualization of non-display walking; and displays involving open fangs, from predatory behavior (for additional hypotheses, see Crane, 1949b).

Consideration of neuromuscular economy (Barlow, 1977) might prove valuable for understanding the particular types of modifications that have taken place during ritualization. For example, one of the displays performed by males of *P. johnsoni* (Jackson, 1977a) is gesturing, in which the legs move up and forward, then back and down (Figure 6.13). Gesturing would seem to entail alterations superimposed on the same basic patterns of movement used by the spider during routine leg waving. During gesturing, the legs are stiffly erected instead of simply raised, and their movements are more jerky and result in different positional end-points compared with routine waving. It would not seem that major reorganization of existing neuromuscular pathways would be required to derive these differences. Perhaps little more than alterations in thresholds have taken place (see Manning, 1965).

Stereotypy

Some degree of stereotypy seems necessary in order for displays to be distinguishable from other stimuli competing for the attention of the receiver. However, questions about just how much stereotypy is typical of displays is an issue under increasing discussion in ethology (Schleidt, 1974). Displays have often been referred to as fixed action patterns (Thorpe, 1951), a term which has a number of connotations; but the one that has gained preeminence is "extreme stereotypy." Barlow's (1977) alternative term, "modal action pattern," is preferable in that it leaves open the question of degree of

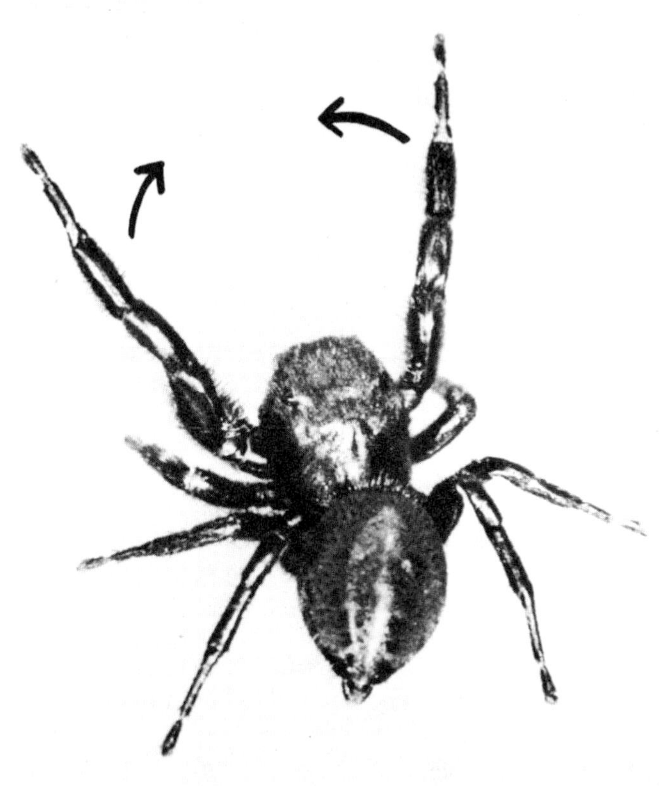

FIGURE 6.13. Adult male of *Phidippus johnsoni.* Gesturing during courtship. Arrows indicate upward and forward movement. Abdomen bent to side. Note stiff appearance of legs I. Posterior-dorsal view.

stereotypy. Stereotypy is an important aspect of salticid displays, but there has probably been a tendency to overemphasize it to the point of overlooking important variability. For example, there seems to be considerable variability in velocity of stepping during dancing, degree of extension of the fangs during threat displays, velocity and amplitude of the raising of legs during displays, etc.

Various factors may favor variability in the characteristics of displays. For example, variable aspects may be anti-monotony devices (Hartshorne, 1973), since too much stereotypy may cause the receiver to habituate to the signal before responding in a manner appropriate for the sender.

A given display may have numerous physical characteristics, such as the velocity, amplitude, and duration of movements, the sequence in which different components occur, and so on. Great stereotypy in one characteristic is not always mirrored by great stereotypy in another; and this might be the key to understanding how animals cope with conflicting, simultaneous demands for stereotypy and for variability (Barlow, 1977). One or more characteristics might be highly stereotyped, enhancing detectability of the signal, while other features may be variable, counteracting monotony. As a possible example in *P. johnsoni*, the physical form of stepping movements during zig-zag dancing seems relatively stereotyped; but the duration of each bout, the distance of the male from the female, and whether or not he gestures with his elevated legs I seem quite variable.

Complexity

Salticid displays apparently differ greatly in complexity. For example, one of the most ornate of the Australian salticids is *Saitis pavanis*. Its courtship display is one of the most complex in the Salticidae, although Dunn's (1957) description of a single male-female interaction is apparently the only published record (also quoted by Main, 1976). The male displays to the relatively drab female with his abdomen and legs III elevated nearly vertically and his abdominal flaps expanded. With his body tilted forward, he performs a zig-zag dance as he vibrates his abdomen, palps, and legs III. As he moves to the right, his right leg III is held more nearly vertical (perpendicular to the substrate). The left leg III remains at a more acute angle and vibrates more strongly, creating the illusion that it is propelling the spider to the right. The roles of the two legs III switch as the spider steps to the left. As he nears the female, the male extends his palps and gradually lowers them as he mounts.

At the other extreme there are some apparently very simple male courtship displays. For example, males of *Menemerus (Marpissa) melanognathus* elevate their legs I and extend them forward as they perform what seem from Bhattacharya's (1936) descriptions to be linear dances; and the males of *Metaphidippus (Icius) sexamaculatus* seem to do little more than elevate their legs I and walk directly to the female (Richman, 1977a).

A related but distinct issue concerns the complexity of the species' repertoire of displays. Although individual displays of other species seem frequently more complex than those of *P. johnsoni*, the repertoire of this species is more complex than those described so far for other salticids. The number of "major displays" (Moynihan, 1970) has been estimated as 24, and the rules describing how and when they are used do not seem simple. Some are performed by adult males only, others by adult females only, and still others by adults and subadults of both sexes. Some of the displays performed by males occur only in interactions with other males, some only in interactions with adult females, some in both. Some occur only during nest-associated interactions, and some only outside nests with relatively high levels of ambient light. Still other displays occur only after the male has mounted the female. For a given type of interaction, some displays always occur, some occur frequently but not always, and some occur quite rarely (see Jackson, 1977a).

There is a need for an objective method to measure the complexity of displays and of repertoires. Designating the number of components in individual displays and the number of distinct displays in a repertoire is a start, but this is inadequate because the relationships between components within the display or the repertoire need to be taken into account also. Possibly a measure could be devised which is based on the number of yes/no questions that have to be answered (bits of information required) in order to characterize adequately individual displays and display repertoires. This should include questions about which sex/age classes use each type of display, the circumstances in which each display is used, and so on. Another problem is that the displays of different salticids have not been described with equal thoroughness. Very likely, many of the seemingly simple displays and display repertoires would be found to be more complex if they were to be more exhaustively examined.

Yet another form of complexity in communication needs consideration in the Salticidae. In *Maevia (Astia) vittata* males occur in two morphological forms which are so different that originally they were considered to belong to different genera. Each morph performs a distinctly different courtship (Figure 6.14) (Painter, 1913; Peckham and Peckham, 1889). This extraordinary spider has hardly received the attention it would seem to deserve from students of salticid behavior, but similar examples of intraspecific variation in other animal groups are currently a topic of considerable interest (e.g., Cade, 1979; Gadgil, 1972). Apparently intraspecific variation in morphological traits occurs in the males of numerous salticid

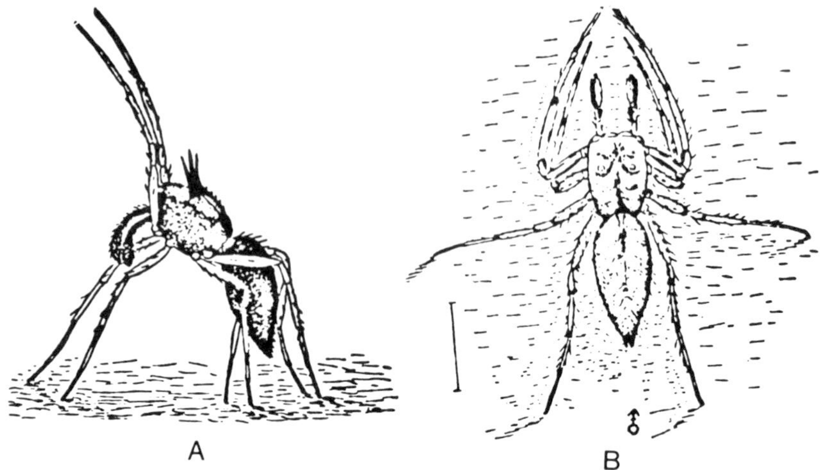

FIGURE 6.14. Courtship displays of the dimorphic males of *Maevia (Astia) vittata*. (A) Niger form. Darker coloration. Tufts of hairs on carapace. Coloration on ventral abdomen. Legs I elevated and held nearly parallel to sagittal plane of body. Palps held downward. Cephalothorax elevated. (B) Vittata form. Lighter coloration. Body held close to substrate. Palps extended forward. From Peckham and Peckham (1889).

species (Peckham and Peckham, 1889, 1909), and careful comparison of the communicatory behavior of these would probably prove very rewarding.

Messages

Smith (1977) specified four categories of messages: selectional, supplemental, identifier, and locational. A given signal will generally have multiple messages, including some from each category.

Selectional messages provide information predicting the sender's future behavior; supplemental messages provide adverbial information concerning how likely and in what form the future behavior will be performed. For example, elements of male courtship may often predict copulatory behavior. Females of some species lower their bodies while being courted, and this seems to provide information about their receptivity to mounting by the male (Crane, 1949b; Forster, this volume; Jackson 1977a). The messages during intermale displays probably include information concerning the likelihood that the sender will "attack" (i.e., perform behavior which can potentially injure the receiver) and perhaps information concerning how capable the sender is of inflicting such injury.

FIGURE 6.15. Adult male of *Phidippus johnsoni* wags in response to another male. Arrow indicates downward movement of leg I, from femur-patella joint. Femora remain almost stationary. Legs held out at approximately 90° to the sagittal plane of the body. Left leg I not in focus. *Pivot point.

Larger individuals will often be more capable of causing physical injury, and threat displays frequently enhance the sender's apparent size (Crane, 1949b). The hunched-legs displays of *P. johnsoni*, which are performed almost exclusively during male-male interactions, seem aptly described in this way (Figure 6.15).

Specification of species and sex/age class would seem the most common identifier messages in salticids, since males of different species commonly have different badges and displays, and these are also frequently sexually dimorphic.

Locational messages specify the spatial position of the sender relative to the receiver. Crane (1949b) pointed out a specialized locational badge. The females of many species have a basal band on the anterior dorsal abdomen (Figure 6.8) which seems, in some instances, to inform the mounted male of where to position himself in order to locate her epigynum before copulation.

Meanings

Careful observation of interacting spiders can provide some indication of the meanings of displays. For example, the first responses of females to male courtship usually include arresting locomotion and watching the male. Receptive females eventually respond by permitting the male to mount and copulate. However, methods such as transition analysis and information theory provide a more precise means of assessing responses (Hazlett, 1977). Although increasingly common in ethological studies, analyses of these types have apparently been employed only once in a study of salticids (Forster, this volume).

Despite their considerable utility, the limitations of such methods must be realized. They are most appropriate for releaser systems, in which displays elicit specific and more or less immediate responses from the receiver. The concept of releasers had an important place in early ethological studies, and there was often a tendency to view communication as largely or completely a releaser system (see Platnick, 1971). However, much of animal communication does not seem aptly viewed in this way. Our concept of salticid communication needs to be broad enough to encompass cumulative, tonic, delayed, and primer effects (see Cullen, 1972; Hailman, 1977b; Hinde, 1974; Schleidt, 1973).

Another approach is to observe the responses of spiders to experimentally manipulated signals such as cardboard models and spiders with altered appearances. Crane (1949b) in particular carried out experiments of this type; but because of her emphasis on searching for releaser effects, few of which she found, her positive findings in these studies seem to receive less notice. Evidently, many signals were found to have non-releaser effects which she labeled variously as cumulative, directive, or excitatory. She also pointed out the often unpredictable nature of a given test spider in a given experiment. Rigorous demonstration of non-releaser effects tends to require comparatively large sample sizes; unfortunately, sample sizes in Crane's experiments were generally small or unstated. However, her experiments were valuable preliminary studies which should provide hypotheses for future researchers to investigate further.

Adaptive Significance

Courtship

EARLY DEBATE Few questions in biology are more basic than those of function or adaptive significance; but these are also among the most difficult when it comes to obtaining rigorous evidence in support of hypotheses (Hinde, 1975). This has not discouraged heated discussion about the functions of displays in spiders, and this has been useful in bringing to our attention the types of data that are needed. The courtship displays of males have attracted the greatest attention, and these will be discussed first (for brief reviews, see Crane, 1949b; Platnick, 1971).

As the ultimate causation of male courtship (displays and especially badges), Peckham and Peckham (1889, 1890, 1909) advocated what they referred to as sexual selection. However, the greater part of their energies seems to have been devoted to arguing against what they perceived as Wallace's (1889, 1890) denial of communication *per se*. Actually, Wallace's views were more complex than this; but somehow the separate issues of sexual selection as a particular communicatory function of courtship, and whether courtship had any communicatory functions at all became intertwined in a way that hindered discussion of sexual selection as such.

Later Montgomery (1908, 1910), Berland (1914, 1923, 1927), Thomas (1929), and Millot (1949) took views with similarities to Wallace's and more or less denied a communicatory function for salticid courtship. Berland's view, for example, seemed to be essentially that the displays of male spiders are generated by the physiological effects ("excitement") concurrent with preparation to copulate and that no further explanation is needed.

Montgomery pointed out possible proximate causes of the male's displays. He argued that the sexually excited male is at the same time cautious in approaching the female, and as a result he performs courtship. Montgomery stated that these movements may be exaggerated, and he conceded that females may be stimulated by the male's courtship. In effect, he came remarkably close to spelling out the conflict theory of ritualization (see Hinde, 1970) which was later to become of major importance in ethology. He also pretty much expressed the view that the male's behavior is communicatory, despite other remarks to the contrary. What he argued strongly against was the notion that the male performs this behavior "in order to" affect the female. Apparently he did not see that "in order to" could refer to a teleonomic explanation (Mayr, 1974a) with no requirement that males have conscious goals.

Some of Montgomery's comments were more relevant to ultimate causes, however. He argued that males tested the females' receptivity and identified themselves to the females with their courtship. These ideas were not clarified, but we should realize that the Peckhams did not very clearly specify what they meant by sexual selection either. Later discussions of the functions of salticid courtship revolved more around ideas of species recognition, cannibalism reduction, inhibition, and arousal, usually in some combination. I will attempt to outline the essential components of each of these hypotheses but not the subtle differences, often implicit rather than explicit, in the views of different authors.

RECOGNITION, AROUSAL, AND INHIBITION According to the recognition hypothesis, a function of male courtship is to provide information on the basis of which the female determines that the displaying spider is a mature and conspecific male. Although Montgomery emphasized sex-recognition, later authors tended to put more emphasis in species-recognition. According to the arousal hypothesis, a function of courtship is to provide information that stimulates the female, bringing her into a state of readiness to mate. The inhibition hypothesis is that a function of courtship is to provide information that prevents the female from performing behavior that is incompatible with mating, such as predatory behavior (including predation directed toward the male), aggressive behavior, fleeing, or simply continuing something as mundane as walking about. Hypotheses of this sort have often arisen in discussions of animal courtship (e.g., Barlow and Green, 1970; Bastock, 1967), but there is a need to explore carefully just what each of these specifies for salticids. It seems likely that such hypotheses will elucidate the functions of courtship to a degree, but it is questionable whether they will provide a complete explanation. The reason for their incompleteness seems to be not so much that other factors need consideration as supplementary selection pressures, but more that these types of hypotheses seem a step removed from other types, such as those related to sexual selection and reproductive isolation that will be discussed shortly.

The hypothesis of recognition, arousal, and inhibition seem to refer, more or less, to what Smith (1977) proposed as the primary function of courtship, which is to bring about the coordination of activities that is needed in order for the male-female pair to copulate and achieve fertilization. This raises the question of whether these requirements alone are sufficient for explaining the rich diversity of courtship displays found in the Salticidae and the seemingly great time and energy that the spiders devote to this activity. For example, salticids have sometimes been reported courting without interruption for hours at a time, after which mating may or may not follow (Carmichael, 1969; Jackson, 1978b; Richman, 1977b). That coordination alone should require so much time is questionable.

Part of what is meant by "arousal" may be physical changes in the female's genitalia that form a necessary preliminary to copulation (Gerhardt and Kaestner, 1937). How important this type of physical adjustment is in the salticids is unclear. For example, this was suggested by Crane (1949b), but she also noted that males successfully copulated with anesthetized females. Coordination of endocrinological, neurophysiological, or other physiological changes might also be considered (Savory, 1925, 1928). However, whatever the particular requirements entailed in coordination, eventually it would seem necessary to ask if these requirements themselves have adaptive significance. We seem to be facing what Richards (1927) referred to as the question of female "coyness" in insects. We might partially explain the function of male courtship by finding that females require arousal and/or inhibition effects from courtship before they will accept the male's sperm, but this leaves the question of whether this need itself (i.e., coyness) is an adaptation. Otherwise, we apparently conclude that female coyness lacks a function as such, that it is simply an inevitable result of adaptations to other requirements, such as predation, predator escape, etc.

REPRODUCTIVE ISOLATION The recognition hypothesis can be expanded to encompass the hypothesis that courtship has a function as an isolating mechanism. The basic assumption is that wastage of gametes in interspecific matings has been an important selection pressure acting on the salticids; and the ability of courtship to decrease this type of waste has been a major selection pressure shaping the specific courtship behavior of different species.

Few hypotheses have held a more important position in zoology than that of courtship's being an isolating mechanism, yet supportive evidence is surprisingly scarce (e.g., Cullen, 1972). In studies of salticids, as in those of most animal groups, there has been an unfortunate tendency to accept this hypothesis as the conventional wisdom without giving adequate attention to the need for strong evidence. Simply to demonstrate that sympatric species differ in their courtship behavior, for example, is not adequate support. To take a facetious example, a salticid species might differ in its courtship from that of a sympatric species of grasshopper, but it seems hardly likely that the selection pressures responsible for these differences involved wastage of gametes between species. As another example, if two sympatric salticid species were found to differ in their predatory behavior, we would not be tempted to conclude immediately that the selection pressures that produce these differences were related to reducing wastage of gametes in interspecific matings.

Evidence is needed to demonstrate that females discriminate between males of sympatric species on the basis of differences in their signals during courtship (see Uetz and Stratton, this volume). The scarcity of overt responses by females has made this difficult to test in the salticids (Crane, 1949b; Drees, 1952). In addition, there is need for evidence that important selection pressures shaping differences in signals have been related to reproductive isolation. One of the most useful types of evidence related to this problem is character displacement of those features of male courtship that females have been shown to discriminate (Waage, 1979). This type of evidence has not been obtained for the salticids.

CANNIBALISM REDUCTION The inhibition hypothesis can be modified to encompass the cannibalism reduction hypothesis. Since spiders are predominantly predators of arthropods, often including species comparable to their own size, cannibalism (intraspecific predation) is a potentially important aspect of their biology. In relation to courtship, we are interested in a particular type of cannibalism, that of females upon males. The cannibalism reduction hypothesis states that a major part of the adaptive significance of courtship is that it provides the female with information that reduces the danger of this type of cannibalism to the male. It seems implicit that males without courtship or with different courtship are selected out of the population by predatory females that either kill or physically injure them. In a weaker form of the hypothesis, males might be selected out of the population by females treating them as prey, regardless of whether death or injury occurs, because predatory behavior is not compatible with copulation.

Apparently, this hypothesis seems very compelling. In one form or another, it probably occurs more frequently than any other in the general literature on spiders (e.g., Darwin, 1871; Dewsbury, 1978; Manning, 1979; Matthews and Matthews, 1978; Romanes, 1882). Experimental evidence supporting this view was provided by Drees (1952). I wish simply to express concern about whether the seemingly reasonable nature of this hypothesis is not obscuring the need for additional evidence. I will mention a few of the points that call for closer scrutiny.

That some salticids are cannibalistic in the laboratory and in nature is certain (Jackson, 1977c), although quantitative information concerning how frequently it occurs is rare; and probably there are important interspecific differences in the significance of cannibalism (see Crane, 1949b; Hollis and Branson, 1964; Peckham and Peckham, 1889). During male-female interactions of *P. johnsoni*, females killed males in approximately 1% of the interactions (Jackson, 1980d). However, the risk of cannibalism works both ways. Males of *P. johnsoni* killed females during courtship almost as often as females killed males. Yet it seems implicit to the cannibalism reduction hypothesis that females should create greater cannibalism pressures on males than males create on females, since it is primarily the males that take the active part during courtship.

Elaborate courtship routines are undertaken by males of many arthropod species, such as fruit flies, butterflies, grasshoppers, and social spiders, in which the risk of predation by females on males would hardly seem a major selection pressure (Jackson, 1979c). The implication that selection pressures related to protection from cannibalism are not necessary for the evolution of arthropod courtship by no means renders the cannibalism hypothesis irrelevant for spiders. However, it does suggest a need for closer scrutiny of this hypothesis.

SEXUAL SELECTION In a sense, serious study of salticid courtship began with Peckham and Peckham and their interest in using this group to study sexual selection. Yet remarkably little of the literature on salticid behavior has been directly relevant to understanding exactly how important this class of ultimate causes may be in the Salticidae. Recently there has been a resurgence of general interest in sexual selection (e.g., Blum and Blum, 1979; Campbell, 1972); and distinctions have been clarified that were blurred when the Peckhams were writing. I expect that a reexamination of sexual selection in the salticids will be a major step toward understanding the proliferation of display behavior and badges in this family.

Darwin (1871) recognized two classes of sexual selection, which Huxley (1938) labeled intrasexual and intersexual (epigamic). Intrasexual selection most commonly takes the form of competition between males for opportunities to mate with females (intermale selection). Intersexual selection most commonly takes the form of female-choice, the type of sexual selection which most concerned the Peckhams and against which their critics seemed to react most strongly. The Peckhams apparently envisaged female-choice of a rather direct sort, with females monitoring the displays and badges of males and on this basis tending to mate more often with males that had certain characteristics.

The descriptions of salticid courtship that the Peckhams provided imply that females normally choose between simultaneously displaying males, which seems unlikely under natural conditions for many (most?) species. However, this is not a prerequisite for the operation of female-choice.

Since there is no evidence of paternal care of offspring in the Salticidae, the adaptive significance of selection among males by females, if it occurs, is most reasonably sought in a tendency for the females' offspring to share the traits of the males. It would be of interest to investigate whether the males that females might tend to choose are ones that also have characteristics giving them relative advantages over other males in respects additional to the advantage arising from female-choice. For example, their abilities to capture prey, escape from predators, resist dessication, and endure high temperatures might be investigated. In addition, the question arises of whether females choose males on the basis of signals that provide reliable information concerning their proficiency at prey capture, predator evasion, or other such activities. In many species (Crane, 1949b; Forster, 1977c; Jackson, 1977a) copulation is frequently preceded by episodes of the female's decamping and the male's following her and renewing his displays (Figure 6.16). These observations suggest the hypothesis that females test the abilities of males to follow them (see Land and Collet, 1974); and possibly a male that is inferior in his ability to follow the female would also tend to be inferior at stalking prey, and his sperm would be likely to produce offspring that share this inferiority. By decamping as a preliminary to mating, the females may be forcing the courting male to perform behavior that provides especially reliable information concerning his capabilities with respect to activities other than courtship.

Males of *P. johnsoni* may or may not dance during courtship, and females tend to mate with ones that dance more often than with ones that do not dance (Jackson, 1981c). When courtship is observed in relatively simple laboratory environments, which has usually been the procedure in studies of salticid behavior, dancing does not give the impression of being such a demanding task for the male. However, in nature the spider must maneuver in the presence of vegetation and rocks in a complex tridimensional environment during predatory and other non-reproductive activities

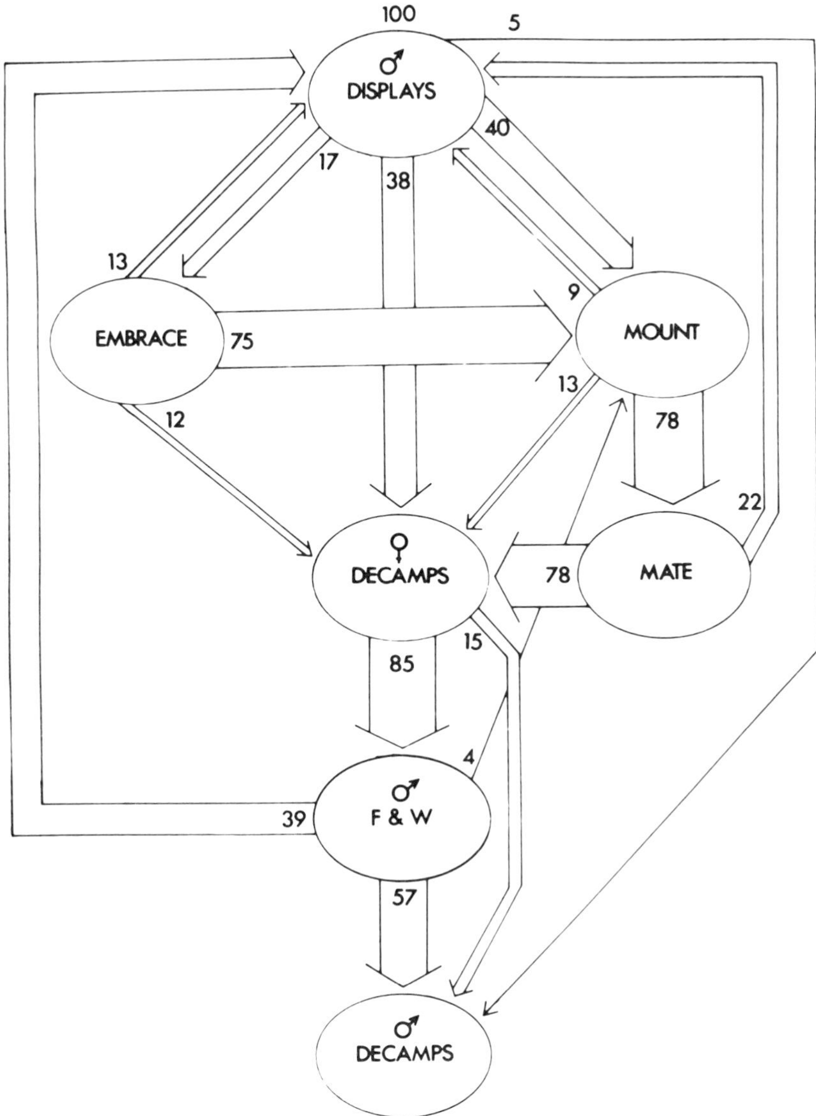

FIGURE 6.16. Summary of 23 interactions between adult males and females of *Phidippus johnsoni*, with both spiders outside nests. Vegetation present. "F and W" refers to "follow and watch" the other spider. Arrows connect preceding with following events. Numbers indicate the percentage of the preceding events followed by the indicated next event. From Jackson (1977b).

as well as during courtship (see Forster, 1979a; Hill, 1979b). This seems challenging not only to the male's motor dexterity but also to his sensory capabilities; and females may increase this challenge by their tendencies to decamp at seemingly unpredictable moments during the male's courtship.

The tendency in *P. johnsoni* for females to mate preferentially with dancing males is more pronounced when the females are non-virgins. Security against the risk of delays or total failure at reproducing may permit more pronounced discriminatory behavior after the female has obtained sperm from the first male (Jackson, 1981c). The sperm of the second male to copulate tends to displace that of the first (Jackson, 1980a). Future investigators of sexual selection in the salticids might profitably pay greater attention to the behavior of non-virgin females.

Other Categories of Display

Compared with the behavior of males during male-female interactions, other categories of salticid communication seem to have attracted less attention. These categories include the display behavior of males during male-male interactions, and the behavior of females and immatures in interactions with other spiders. Displays of these types are probably more common than the literature would suggest, but questions about their functions have not been widely discussed. Crane (1949b), for instance, tended to refer to the displays of females and immatures as "abortive" or "incipient," and Bristowe (1941) attributed the displays that occur in male-male interactions to mistaken identity. The implication of expressions such as these seems to be that these behaviors have no adaptive significance, a view that I would adopt only hesitantly.

MALES Crane (1949b) demonstrated that the males of some salticid species (her hopper group) have distinctly different displays depending on whether they interact with females or with other males, ruling out the hypothesis of mistaken identity for these. She referred to these displays as threat, which seems reasonable since they share characteristics of behavior (e.g., increased apparent size, display of the parts of the anatomy capable of inflicting injury, etc.) referred to as threat in other animals (Eibl-Eibesfeldt, 1970).

Threat is generally viewed as an aspect of aggressive behavior. "Aggression" is notoriously difficult to define, partly because it is not a unitary concept, but the general nature of the phenomenon seems to be behavior that is directed toward causing physical harm to a conspecific individual (Hinde, 1974). It is probably preferable not to attempt more specific definitions.

Given that males are aggressive, the function of displays might be understood as a means by which individuals settle conflicts with minimal risks to themselves by communicating (see Maynard Smith, 1972). However, a complete understanding of the ultimate causation of this behavior requires an acceptable hypothesis explaining why males are aggressive toward each other in the first place. Territoriality and social ranking, which have so much dominated the ethological literature on vertebrates, seem of limited value for understanding salticid aggression (see Crane, 1949b).

The Peckhams hinted at a rather direct relevance of male aggression to intersexual selection, with females watching male-male interactions and this affecting their later choice of males with which to mate. They inferred that males were staging mock battles in order to impress the females (ultimate causation). Although it might be reasonable to explore this hypothesis with species that occur at high densities, supportive evidence is currently lacking.

The Peckhams also hinted at intrasexual selection of a fairly direct type, with males aggressively competing in the presence of females. If the loser decamps and the winner is the only male that remains to court and mate with the female, female-choice need not come into play. In the laboratory, when males of *P. johnsoni* encountered courting or mating male-female pairs, intermale interactions followed. The loser decamped, and the winner usually renewed or initiated courtship with the female and often mated (Jackson, 1980b).

However, intrasexual selection in such a direct sense seems inadequate as an explanation of the functions of intermale aggression because of the uncomfortable fact that males readily interact aggressively in the absence of females. Possibly interactions tend to be more damaging and thresholds of interaction are lower in the presence of females (Crane, 1949b; Jackson, 1980b), but the fact that the female need not be present still needs an explanation.

Males often seem to be defending a personal space around themselves, which they strive to keep free of other males; the adaptive significance of this might be that a male with a larger personal space is less likely to suffer interference from other males when an opportunity to court and mate arises. However, this is still not an entirely satisfactory hypothesis, because simple avoidance behavior would seem to accomplish the same end without the costs (risks of injury, energy and time expenditure, etc.) entailed in aggression. Perhaps there are optimal areas for sexual searching which males are hesitant to vacate; or possibly the males have systematic searching routines, and departure might seriously disrupt these. Factors of these kinds would still be related to intrasexual selection.

FEMALES AND IMMATURES As tempting as it might be to relate all salticid display behavior to sexual selection, we have to deal with the bothersome fact that immature individuals and females have display behavior with similarities to that of matures males; and this may occur in the complete absence of males (see Forster, 1977a; Jackson, 1977a). There may be a tendency for displays of immatures and females to be less intense (lesser durations, higher thresholds, etc.; see Jackson, 1977a), and perhaps there are some generally applicable advantages in aggressively maintaining a personal space (increased predatory efficiency, for example). Display behavior in immatures and females may be related to these factors (see Forster, 1977b), but intrasexual selection acting on the males may intensify the net selection pressures on male display behavior.

Yet another class of salticid displays remains to be considered, those performed by females during male-female interactions. Platnick (1971) expressed concern about the tendency to view courtship as a "one sided activity—a display by the male having certain effects on the female." Although more attention needs to be given to female behavior, the "one-sidedness" many nevertheless reflect an important part of reality. The trend in animal species seems to be for the males to be the more active courters and females to have a larger role in making discriminations. The most important selection pressures seem to be related to the tendency for each reproductive effort of the female to decrease her potential for future reproduction more greatly than is the case for the male. This in turn is related to male-female differences in gamete size (Bateman, 1948; Ghiselin, 1974; Trivers, 1972). To state the trend loosely, males attempt to convince females to mate, not vice versa. In this sense, it seems best not to include female displays in the definition of courtship.

Displays employed by the females may provide information concerning the likelihood that they will attack, decamp, permit the male to mount, etc. The functions of display may be related to preventing, delaying, or facilitating the male's efforts to mate, possibly in relation to intersexual selection. However, unlike the behavior of the males during courtship, the females' displays do not seem so readily interpretable as a sexual pursuit (Jackson, 1978b). It seems more reasonable to interpret the females' behavior as functioning in regulating the male's pursuit and providing an orderly means by which she can make decisions as to whether, with whom, and when she will mate.

Concluding Remarks

Two Basic Questions

I have attempted in this chapter to raise questions about salticid communication. In closing I would like to raise two additional but very basic questions: (1) what is the evidence that communication actually takes place when salticid spiders interact, and (2) what is the evidence that particular characteristics of the spider are indeed signals? Such basic questions have rarely been dealt with explicitly in discussions of the communicatory behavior of animals (e.g., see most of the chapters in Sebeok, 1977), and to be overly insistent on having direct evidence for answers to these questions is potentially stifling to the early stages of research in this field (Beer, 1977). However, the study of salticid communication seems to have reached a stage at which researchers should begin taking a closer look at these questions.

The most widely accepted criterion for objectively demonstrating the occurrence of communication is basically evidence that the behavior of the receiver is different depending on whether or not the signal is present, although this has been expressed in several different ways (e.g., Hailman, 1977b; Klopfer and Hatch, 1968). Using the semiotic vocabulary, we can rephrase this as demonstration that signals have meanings.

The most notable efforts to obtain these types of evidence for salticids have been the experimental manipulations using models and altered spiders by Crane (1949b), Drees (1952), and others, and a transition analysis carried out on *Trite planiceps* (Forster, this volume); these were discussed earlier in relation to meaning. More efforts along these lines would be valuable, but there is also a need for developing quantitative methods and experiments more appropriate for delayed, cumulative, tonic, and other non-releaser effects of signals, since these are an important part of the concept of communication.

Quantitative demonstration of the occurrence of communication associated with particular characteristics of spiders will only partially answer our second question concerning evidence that these traits are indeed signals, since signals are not the only potential source of information to the receiver (Smith, 1977). Signals were defined as those particular characteristics of the sender that are adaptations for information transmission. Evidence is needed showing that traits have been evolutionarily modified (ritualized in the case of displays) by selection pressures related to communication (for discussions of the types of evidence required, see Blest, 1961; Hazlett, 1972). Only a couple of the problems that have to be borne in mind will be pointed out here.

A given characteristic of a spider may have multiple functions. For example, with species in which males dance during courtship, it should be possible to acquire quantitative evidence of a communicatory function for this behavior. However, dancing must surely also have a more mundane locomotory function, bringing the male closer to the female as a necessary preliminary to mounting. Demonstration of any one function should not blind us to other functions.

At some point researchers will have to deal with the question of how much evolutionary modification is required to constitute the criterion for calling a trait of the spider a signal (see Tavolga, 1970). For example, it is quite conceivable that the routine palp-waving behavior of many salticid species conveys information to conspecifics, although this has not been specifically investigated yet. If this kind of palp-waving does prove to be informative, then investigators will need to consider whether ritualization has occurred; if so, how much has occurred and whether it is enough to justify calling this behavior a display.

Research Tactics

The salticids stand apart from other spiders in their heavy reliance on specialized and elaborate visual abilities. Along with certain insects, cephalopods, and vertebrates, this family of spiders constitutes one of the major evolutionary "experiments" related to visual communication. The special advantages and limitations of visual communication compared with communication in other sensory modalities have been discussed by various authors (e.g., Marler, 1968). Visual signals have the advantage of providing very reliable locational information; but for non-luminescent species such as the salticids, ambient light requirements must limit the times and places in which visual communication can occur. Distances over which visual signals may be used are limited, and locational information is available to predators as well as to the intended receivers.

The salticids should provide valuable subjects for comparisons of visual with other forms of communication, not only because of the highly developed visual abilities of these spiders, but also because these comparisons can potentially be carried out using members of the same species or even the same individual. The salticids seem to make important use of non-visual signals, either simultaneously with visual signals or in conjunction with visual signals but segregated into distinct phases; and at least a few species employ separate mating tactics in which visual signals need play no role.

Since the species investigated so far cover only a fraction of the salticid genera, researchers in the future should attempt to document the signals employed by a wider range of genera. At the same time, there is a need for more comprehensive studies of individual species (Savory, 1961). This is partly because communication is interrelated with other aspects of the animal's biology, including its behavior in non-communicatory contexts, its physiology, morphology, habitat specificities, life history, and its relationships with prey, predators, and competitors (Smith, 1977). For example, characteristics of nest-associated courtship seem interrelated with other aspects of nest-use, such as predator avoidance, molting, and oviposition (Jackson, 1980b); and specific aspects of visual communication are related to intraspecific variation in population structure in *P. johnsoni* (Jackson, 1980c).

A complete understanding of salticid communication requires an understanding of the species' particular *Umwelt* (von Uexküll, 1909) and answers to questions about dynamic control, ontogenetic origins, adaptive significance, and phylogenetic factors (Hailman, 1976). Our present understanding of each of these varies considerably from species to species; ontogenetic questions have been hardly investigated at all with salticids. Answers to any one of these questions are likely to assist in answering the others. This interrelationship can be most fully exploited in comprehensive studies of individual species; such studies can best be accomplished by the coordinated efforts of many researchers with expertise ranging from sensory physiology to developmental and evolutionary biology.

Acknowledgments

The author acknowledges gratefully the critical reading of this chapter by Lyn Forster and her suggestions for improvement. Mary Catherine Vick assisted with the preparation of the figures and provided useful comments on the manuscript.

Chapter 7 CHEMICAL COMMUNICATION IN LYCOSIDS AND OTHER SPIDERS

William J. Tietjen
and Jerome S. Rovner
Department of Zoology
Ohio University
Athens, Ohio 45701

Introduction

Most animals communicate with other members of their species at some time during their life-cycle. Among the social animals, communication may be more or less continuous, while solitary animals may communicate only during sexual and agonistic encounters. E. O. Wilson (1975) defined communication as ". . . action on the part of one organism (or cell) that alters the probability pattern in another organism (or cell) in a fashion adaptive to either or both participants." Implicit in this definition is the requirement that at least two animals communicate (the sender and receiver) and that a message having meaning is emitted from the sender in some manner. The action of the sender may be an overt behavior (i.e., a visual or acoustic display) or may be an evolutionarily derived change in the sender's morphology or physiology (i.e., coloration of an appendage, or an odor associated with an excretion).

Of the four channels of communication that have been described in animals—visual, chemical, electrical, and mechanical (Hinde, 1972)—the chemical channel is the most primitive, having probably first served in communication to bring individuals together for the exchange of genetic material (Haldane, 1955). Chemical signals mediating this and other functions were termed "pheromones" by Karlson and Butenandt (1959), who introduced the word to describe chemical "substances secreted by an individual and received by a second individual, in which they elicit a specific reaction." The term has its basis in the Greek roots "pherein" (to transfer) and "hormon" (to excite).

Chemical communication has been demonstrated as the primary mode of communication in many taxa, with the notable exceptions of birds, frogs, lizards, and higher primates (Shorey, 1976). Among the lower invertebrates, for example, pheromones can communicate the injury of a group-member. A sea anemone (*Anthopleusa elgentissima*) produces such an alarm pheromone. The behavioral response of uninjured individuals to the pheromone released by wounded conspecifics is contraction and pulling in of the tentacles (Howe and Sheich, 1975). Among the vertebrates, olfaction is well developed in fish, snakes, and mammals. In mammals, the well-developed olfactory apparatus allows for the analysis and detection of odors used for the recognition of territory, young, mates or social group, and efficient trail-following (Moulton, 1967). Even in humans pheromones may affect behavior at a subliminal level (Comfort, 1971; McClintock, 1971).

Most of the work on chemical communication in animals has centered on the economically important arthropods (Jacobson, 1966; Katzenellenbogen, 1976). In addition to providing information on the sexual behavior of arthropod pests, synthetic or natural pheromones are used in baiting of traps to capture mature males. This method provides efficient sampling for estimating the pest populations in a given area so as to determine the minimum amounts of conventional pesticides needed for control (Marx, 1973). More recently, however, female sex pheromones have been used to disrupt mating in the gypsy moth (*Porthetria dispar*) and pink bollworm moth (*Pectinophora gossypiella*). After broadcast applications of the female sex pheromones throughout the affected area, males were unable chemically to locate conspecific females. Such methods of pest control are often termed "male-confusion techniques" (Cameron et al., 1974; Gaston et al., 1977).

Information Theory and the Physical Properties of Chemical Signals

Any message, be it chemical, electrical, or visual, has some measurable information content. "Information," in this sense, is the capacity to store meaning and not the meaning itself. The level of entropy (degree of randomness) is the measure of this capacity (for review see Gatlin, 1972).

Shannon and Weaver (1949) developed mathematical methods for measuring the information content in communication systems. Their formula

$$H = -\sum_{n=1}^{n_i} p_i \log p_i$$

is used to describe the information content (H) of such systems. If logarithms to the base 2 are used, the unit of information is the *bit*. One bit can switch between two states answering such questions as yes vs. no, on vs. off, positive vs. negative, and so forth. The value p_i is the probability that a specified "question" is answered during the statement of a message.

The above formula implies that an increase in the entropy, or randomness, of a message will increase the information content. This is true to a point, since the message must have higher entropy than the background noise in the environment to be recognizable. If the level of entropy becomes too high, however, the message becomes so complicated that random errors and noise in the system make transmission of the signal unreliable. Messages can be made "self-correcting" in the presence of random errors and noise in the system if some level of redundancy exists through both continuous repetition of the messages and through rules of syntax which make possible error detection and correction.

A communication system requires three parts (Figure 7.1): a transmitter which emits the signal, a channel over which the signal travels, and a receiver which detects and processes the message in some manner. In animal communication, the transmitter and receiver are usually conspecifics. The channel may be air, water, or a solid substratum. Any channel has an inherent level of noise which interferes with the propagation of the signal. If the signal amplitude is not above the noise level, the message will not be received. Channels also may be characterized by their channel capacity: the upper limit at which the channel is considered full. The rate of transmission may therefore be limited by the physical features of the channel. A sand substratum, for example, is a relatively

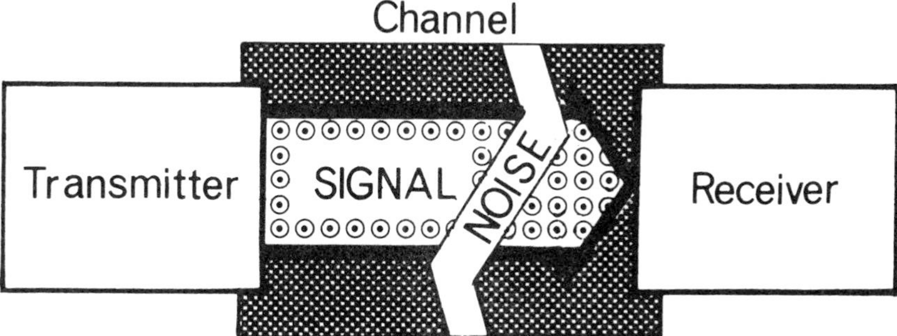

FIGURE 7.1. Conceptual model of a communication system. Noise in the channel interferes with signal propagation between the transmitter and receiver.

good conductor of low-frequency compressional waves, while high frequency vibrations are readily attenuated.

Pheromones are often effective in microgram quantities. The female silkworm moth (*Bombyx mori*), for example, carries less than 1.5 μg of sex pheromone, an amount Wilson (1968) has calculated would be sufficient to activate over one trillion males. Such pheromones are received over an extremely noisy airborne-communication channel. Olfactory noise presented to the male moth includes the pheromones released by hundreds of other species in the area as well as natural and man-made volatile compounds. Information theory predicts that such pheromones would be novel substances in order to be detectable above the background noise.

In a similar vein, Wilson and Bossert (1963) indicated that, as molecular weight increases arithmetically, there is an exponential increase in molecular diversity. They calculated that airborne pheromones would be expected to have a molecular weight of 80-300 and a carbon number of 5-10. Too few compounds can be synthesized with less than 5 carbon atoms; while with carbon numbers greater than 20 molecular diversity would be higher than necessary and the larger compounds would be energetically expensive to build. Furthermore, transport of high molecular weight compounds is restricted in the air-channel since, in general, molecular weight and volatility are inversely proportional (Wilson and Bossert, 1963).

Bossert and Wilson (1963) investigated the physical properties of airborne pheromone transmission. They indicated that, by adjusting the ratio of the number of molecules released (Q) to the response threshold (K; in molecules per cc), the chemical's area of effectiveness (active space) would be affected. A decrease in fade-out time and better localization by the receiver can be effected by decreasing K, increasing Q, or both.

In general, pheromones carry little information content as a message. Sex pheromones usually answer only the question "Is there a mature conspecific female within range?" Such pheromones are often released continuously, thus exhibiting a high degree of redundancy. The air-channel limits the active space of a variety of pheromones through both slow diffusion and disruption from air currents. Thus, concentration gradients are not usually reliable for directional cues over distances greater than a centimeter or so (Wilson, 1968). As we shall see, however, terrestrial trails may provide considerably more information than aerial trails.

Bonner (1963) demonstrated that aggregating slime molds (*Dictyostelium discoideum*) produce a chemical signal in pulses, which provides directional information to isolates and induces movement toward the center of the group. Wilson (1968) suggested that patterned transmission using amplitude or frequency modulation may be possible over short distances in still air. He calculated that information of 100 bits/sec could be expected under such circumstances.

Spiders produce silk, a secretion not common among other taxa. Although silk is used for eggsac and sperm web construction in all species and as a prey-capture mechanism in some species, it also is used as a means of communication in many spiders. Often the silk serves as a channel for vibrational messages, while in other cases a pheromone may be added to the silk. The spiders' use of an external construction for communication is unusual but not unique among animal species.

Evolutionary Considerations

Origin of Chemical Communication

Wynne-Edwards (1962) suggested that mammalian pheromones were derived from "leaking" metabolites that originally had another function. Among the Arthropoda, cuticular wax glands may have acquired a pheromone-producing function (Shorey, 1976); however, in some insects the pheromone may have been derived from metabolites related to the diet (Hendry, 1976). Kittredge and Takahashi (1972) found that the sex pheromone of the shore crab (*Pachygrapsus crassipes*) was the molting hormone crustecdysone. They suggested that the communicatory function of crustecdysone developed from "accidental" leakage of the molting hormone into the environment. Tests to determine if the molting hormone, ecdysterone, was the sex pheromone of the wolf spider *Pardosa ramulosa* indicated it was not; however, ether extracts of the female spider's integument suggested a lipid structure (Sarinana et al., 1971).

Primitive spiders probably were adapted to hiding in holes and crevices. As they rushed out from their retreats to capture prey, deposition of draglines would have resulted in a fringe of silk radiating outward from the mouth of the shelter (Kaston, 1964; Kullmann, 1972a; McCook, 1889). It is easy to see how the silk fringe could take on the function of alerting the inhabitant of the crevice to the presence of a potential prey item. The primary mode of detection would be through vibrations sent along the silk threads into the nest. Later on, males that detected conspecific females through reception of chemical subtances deposited with the silk would be expected to begin courtship behavior earlier than non-detecting males. Under these circumstances cannibalism would be reduced (Platnick, 1971). The deposition of dragline and resulting fringe of silk provided a preadaptation for one type of chemical communication among the primitive spiders.

Reproductive Isolating Mechanisms

Reproductive isolating mechanisms are an important evolutionary consideration in chemical communication. Two species may be reproductively isolated through premating or postmating mechanisms; the former are the more efficient as they avoid gamete wastage. Premating mechanisms may occur at one of three levels: (1) seasonal and habitat isolation, (2) mechanical and physiological isolation, and (3) ethological isolation, the latter including those behaviors associated with chemical communication (Mayr, 1974b). Slight differences between two species in the structure of a pheromone can provide for efficient reproductive isolation. Many arthropod chemoreceptors will exhibit a response (as indicated by neurophysiological recording techniques) to only a few closely related compounds (Kaissling, 1971; Shorey, 1976).

Wilson (1968) suggested that mutation may affect the structure of the pheromone that is produced, thus altering the information content. If, as a class, sex pheromones are the most structurally complicated (Wilson and Bossert, 1963), then mutations may cause speciation in some cases. In addition to changes in pheromone structure, a change in the receptor sites of the receiver could also cause evolution and isolation of a species (Roelofs and Comeau, 1969).

Closely related species often have similar or identical pheromones. Shorey et al. (1965) demonstrated, through the absence of behavioral specificity and by gas-liquid chromatographic analysis, that the sex pheromones of the cabbage looper moth (*Trichoplusia ni*) and the alfalfa looper moth (*Autographa californica*) were similar, if not identical. Comparable data were reported for the analysis comparing the tobacco budworm (*Heliothis virescens*) and the cotton bollworm (*Heliothis zea*). These results suggest that additional premating mechanisms, such as differing host specificity or the timing of mating, prevent hybridization in the field.

While many male spiders respond only to the sex pheromone of a conspecific female (Engelhardt, 1964; Jackson, 1978a; Kaston, 1936; Tietjen, 1977), there may be a lack of specificity in the contact pheromones on the webs of the female black widow spiders *Latrodectus mactans* and *L. hesperus*. Males of one species court when placed in the unoccupied webs of females of the other species (Ross and Smith, 1979), a finding which the authors attribute to the allopatry (and consequent lack of selection favoring divergence) of these two species. Likewise, Hegdekar and Dondale (1969) indicated that male *Pardosa moesta* exhibit courtship behavior when exposed to female *Schizocosa crassipalpis* silk. Although *P. moesta* and *S. crassipalpis* are not isolated through seasonal or habitat features, mechanical isolation (genitalic differences) and the differing courtship behavior of these two species make hybridization unlikely in the field. A similar situation exists in *Schizocosa rovneri* and *S. ocreata*, with males of one species courting in response to the female's silk (and pheromone) of the other (Uetz and Denterlein, 1979). However, at least partial species-specificity was indicated by significant differences in certain aspects of the response to conspecific vs. heterospecific silk.

In all such laboratory studies, the prolonged stimulus deprivation of the test animals could alter response thresholds, which may result in courtship being triggered by subnormal stimuli. This factor was considered by Engelhardt (1964), who indeed found that males freshly captured in the field courted only conspecific females when tested soon afterward in the lab, whereas lab-reared males courted heterospecific females. Platnick (1971) indicated that differential reliance on chemical, mechanical, and visual cues during courtship between species could be incorporated into traditional systematics research to separate spiders at the higher taxonomic levels.

Methods of Study

Chemical communication may be studied through several levels of analysis: morphology and physiology of chemoreceptors, chemistry of the pheromone, and the ecology and behavior of the organisms. All modes of study have been used among the economically important arthropods (Jacobson, 1966; Shorey, 1976); however, among the Arachnida, only the ticks have had analyses performed at all levels (Meinwald et al., 1978).

Chemical analyses present special problems, as researchers must work with μg or mg quantities of complex and novel substances. After a long series of chemical preparations, for example, only 12.2 μg of the sex attractant was obtained from 10,000 female American cockroaches "milked" over a 9-month period. The isolated pheromone was behaviorally active at dilutions down to 10^{-14} μg (Jacobson et al., 1963). There have been no attempts to isolate the sex pheromones of spiders. It is known that the pheromones of various spiders are soluble in or are quickly inactivated by common inorganic and organic solvents (Gwinner-Hanke, 1970; Hegdekar and Dondale, 1969; Kaston, 1936; Sarinana et al., 1971). For example, by using benzene as a solvent, Collatz succeeded in transferring the sex pheromone of *Tegenaria atrica* (Agelenidae) from one web to another that had previously failed to elicit courtship in males (Collatz in Weygoldt, 1977). Research on arachnid pheromones so far has been largely at the behavioral level; however, a few researchers have attempted to apply ecological considerations to their behavioral work (Dondale and Hegdekar, 1973; Tietjen, 1977, 1979a,b; Tietjen and Rovner, 1980).

Pheromone Reception

Morphological and Behavioral Aspects

Most research on reception of pheromones has been on the insect olfactory sense, where, in animals such as the male silkworm moth (*Bombyx mori*), the morphology of the antennae allows these structures to act as "molecular sieves." The sensitivity of the chemoreceptors is such that they may function as "molecule counting" devices (Kaissling, 1971; Schneider, 1964).

A sensillum is a specialized area of an arthropod's cuticle with a minimum of three cells, all of epidermal origin. Two cells are formative, secreting the cuticular components, while the third is a sensory neuron, the dendrites of which often penetrate a cuticular hair (= seta). Several hair types and a variety of plate organs are found in the Insecta. Most insect hair sensilla have one to several receptor cells, while in bees the olfactory plate organs may have from 10 to 30 neurons (Payne, 1974; Schneider, 1969).

Spiders react to olfactory cues provided by the strong odors of essential oils such as cedar and lavender (see the reviews of Bristowe and Locket, 1926; Kaston, 1936); however, these experiments have been criticized because sensitivity to high concentrations of such irritating odors does not necessarily indicate a true olfactory sense (Kaston, 1936; Crane, 1949b). Witt (1975) coated flies with bitter, non-volatile solution (quinine) and, by utilizing highspeed cinematography, indicated that the contact chemoreception sense was concentrated at the distal part of the spider's appendages (see also Holden, 1977). These data are in agreement with the results of previous mutilation experiments (Bristowe and Locket, 1926; Crane, 1949b; Kaston, 1936).

For many years there has been speculation as to the identity of chemoreceptors in spiders (see Foelix, 1970). Legendre (1958) suggested that a patch of sensory cells located near the base of each palp functions in olfaction, these cells being larger in males. The tarsal organ (Figure 7.2), one of which is found on the tarsus of each leg and each palp, probably is a primary olfactory receptor in spiders (Blumenthal, 1935; Dumpert, 1978; Foelix and Chu-Wang, 1973b).

Based on a similarity in morphology to known insect chemoreceptors, Foelix (1970) suggested that the contact chemosensitive sensilla of spiders are the curved, blunt-tipped, steeply inserted hairs found concentrated on the distal portion of the appendages (see also Foelix and Chu-Wang, 1973b). These hairs differ from the proposed mechanoreceptors (Figure 7.3), which are straight, sharp-pointed, less steeply inserted, and found more evenly distributed over the body (Foelix and Chu-Wang, 1973a). The proposed chemo-

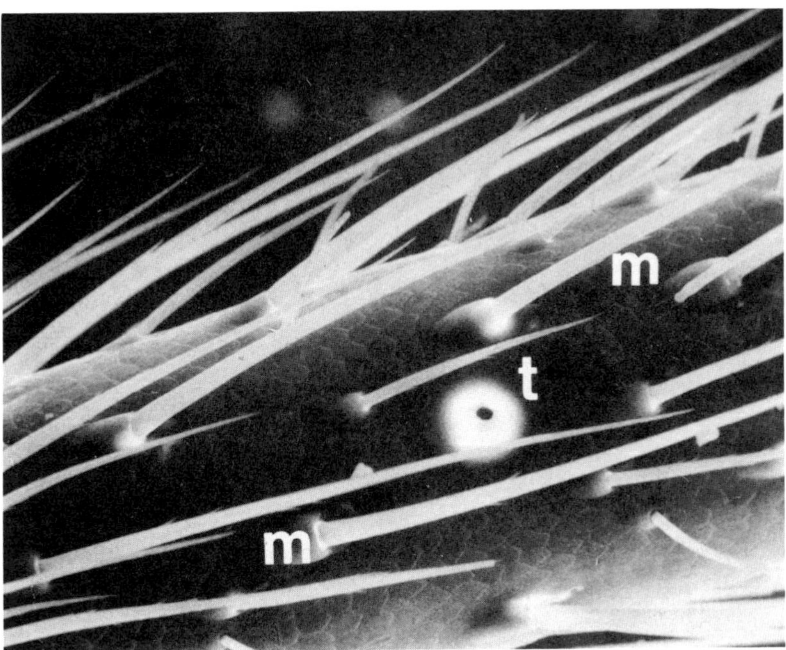

FIGURE 7.2. Tarsal organ of *Araneus diadematus*. The scanning electron micrograph shows a pit-like tarsal organ (t) surrounded by mechanoreceptive hairs (m). (Photograph by R. Foelix and I.-W. Chu-Wang, N. C. Division of Mental Health Research.)

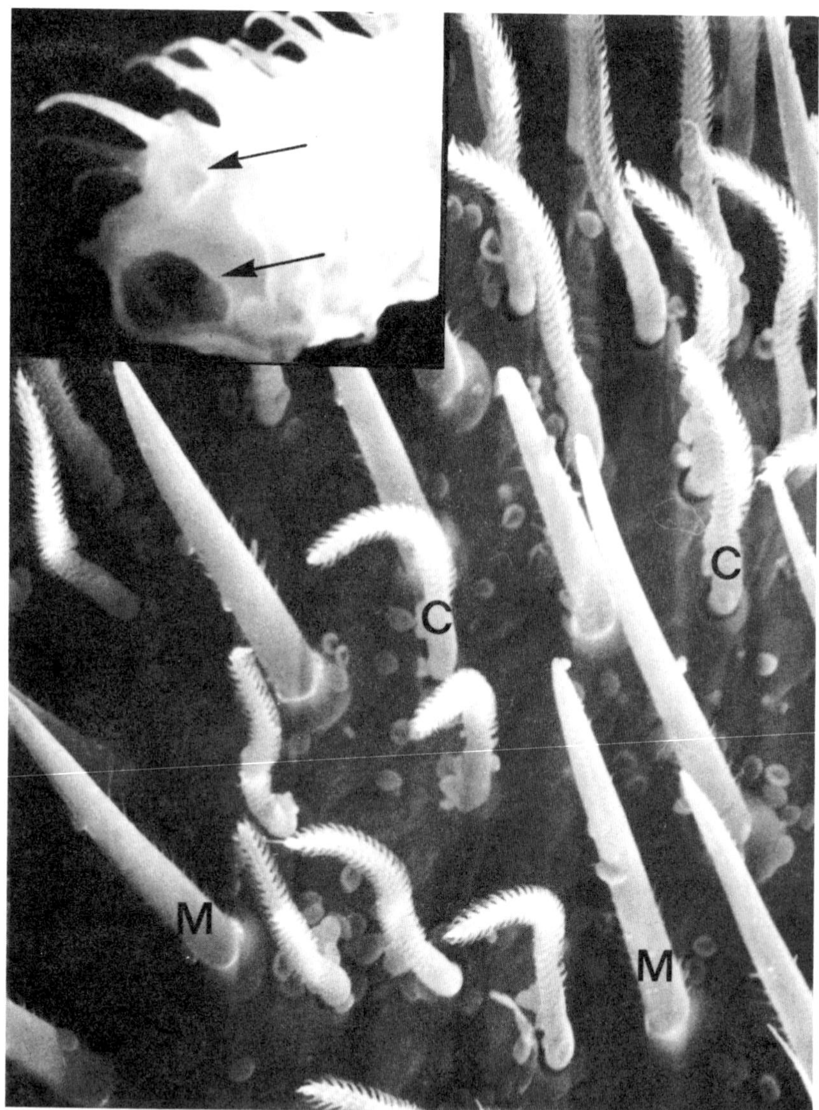

FIGURE 7.3. Sensilla on the adult male palp of *Lycosa punctulata*. The curved hairs (C) are contact chemoreceptors, while the straight hairs (M) are probably mechanoreceptors. (The circular objects on the palpal cuticle are probably spores.) Inset: two lumina (arrows) are revealed within a severed hair.

sensitive setae have more neurons associated with each receptor than do most insect sensilla. In spiders, there are 19 dendrites running the length of the hair shaft and two mechanoreceptive dendrites found in the socket of the sensillum. While the structural properties of the hairs indicate that they are probably contact chemoreceptors, an involvement in olfaction (detection of airborne chemicals) cannot be ruled out. Sensilla with a similar structure have been described in a variety of spider species, including the social spiders *Agelena consociata* and *Stegodyphus sarasinorum* (Krafft, 1975b; Kullmann, 1972a), in theridiids of the genus *Latrodectus* (Ross and Smith, 1979) and in various lycosid spiders (Kronestedt, 1979; Tietjen and Rovner, 1980).

Electrophysiological Studies

There are two main methods for recording the neuronal activity of sensilla among the insects: (1) electroantennogram recording and (2) recordings from single sensilla (Payne, 1974; Schneider, 1957). Among the spiders, recordings are of two general types based on the second of the methods used in insect research.

In the simplest approach, a recording electrode is positioned over the distal end of the sensillum while the indifferent electrode is placed in the hemolymph (Figure 7.4A). Recently, this method has yielded data which support Foelix's suggestion that the curved, blunt-tipped hairs function as chemoreceptors (Drews and Bernard, 1976; Harris and Mill, 1977b). The spiders' chemoreceptors responded to a variety of halide salts and acids; however, because the distal pore was occluded by the recording method, neither contact nor olfactory pheromone stimulations were possible. In addition, mechanoreceptor properties were also hypothesized to be included in the chemosensitive hairs, as is the case in blowflies (Dethier, 1966).

A second, more sensitive method depends on positioning the recording electrode directly within a sensory neuron cell body innervating the sensillum (Figure 7.4B). Dumpert (1978) used this method to record potentials in individual tarsal organs of male *Cupiennius salei*. The tarsal organ was excited by a variety of odors, including various acids and tobacco smoke as well as the odor of female (but not male) conspecifics. The tarsal organ of females did not respond when males were the test subjects. Within a single sensillum four cell types were identified, each with a different sensitivity to the test odors.

The changes in sensillum electrical potential, as determined

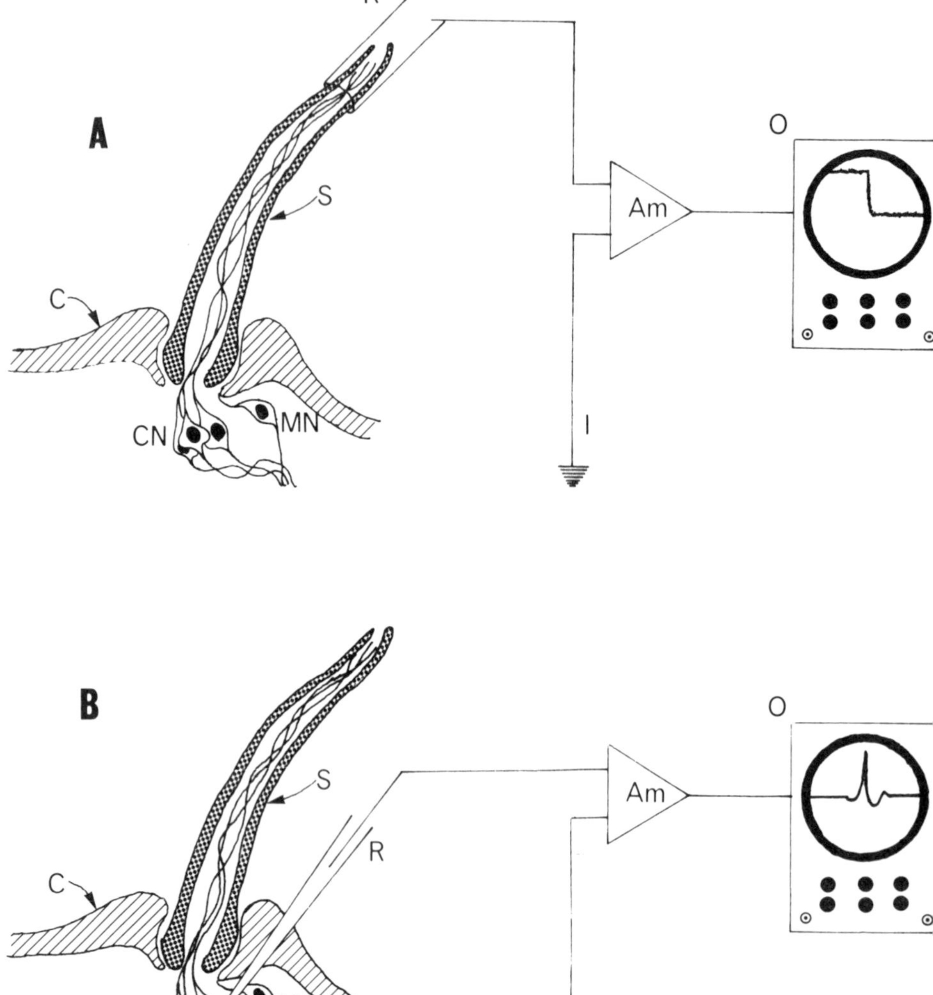

FIGURE 7.4. Electrophysiological recording techniques. Three chemoreceptive neurons (CN) extend their dendrites into the lumen of the chemosensitive sensillum (S). The hair is in a socket formed by the cuticle (C) and provides input to mechanoreceptive neurons (MN). In both recording techniques an indifferent electrode (I) is inserted into the spider's hemolymph. An amplifier (Am) provides gain for the oscilloscope (O). In Figure 7.4A the reference electrode (R) is positioned over the tip of the sensillum. Since many chemosensitive neurons are stimulated, the oscilloscope shows a generalized depolarization. In Figure 7.4B the reference electrode is inserted into the soma of a single chemosensitive neuron. The oscilloscope shows the response of the single cell to the stimulus.

through the above methods, are thought to be mediated through a receptor molecule found on the surface of the pheromone-sensitive neuron. In this model the pheromone binds with the receptor molecule in a manner similar to the lock-and-key theory proposed for enzyme activity (Schiffman, 1974). The interaction between the pheromone and the receptor molecule is thought to change the permeability of the membrane, thus effecting a change in the dendritic electrical potential.

Callahan (1975) has proposed an alternative method for pheromone reception in insects and spiders. He maintains that insect sensilla are functionally similar to man-made antennae and "designed" to receive electromagnetic wavelengths. Pheromones were demonstrated to emit narrow wavelengths of coherent infrared radiation. The sculpturing on the surface of the sensillum (see Figure 7.3) was hypothesized to sensitize the antennae to a specific pheromone wavelength, much as the elements of a television antenna tune the antenna to FM wavelengths. The morphology of the antenna was also hypothesized to function in focusing the received message at a detector located near the distal end of the sensillum.

Behavioral Effects of Pheromones

Pheromones may be classified as to their effects on the receiving animal. Wilson (1968) considered two classes of chemical signals: releaser and primer pheromones. Releaser pheromones stimulate an immediate behavioral effect and may be further classified as to their supposed function (i.e., sex, alarm, or social pheromones). Primer pheromones do not induce a behavioral response, but result in a physiological change in the receiver. The queen bee substance is an example of a primer pheromone which inhibits ovarian development in worker bees (Wilson, 1971). To this date, no primer pheromones have been described in the Araneae.

Releaser pheromones may affect behavior by stimulating or inhibiting responses, or through a change in the receiver's orientation and/or rate of locomotion. An orthokinesis is characterized by a change in the rate of locomotion; for aggregating pheromones, an animal would exhibit a rate of locomotion that is inversely proportional to the concentration of pheromone. A taxis reaction may depend on the receiver orienting to a pheromone concentration gradient or to another stimulus, e.g., airflow or an object in the environment (Farkas and Shorey, 1974; Wilson, 1968). Examples of aggregating pheromones among the Insecta include the colony odors of social insects and the sex pheromones of noctuid moths (Blum and Brand, 1972; Shorey, 1976; Wilson, 1971).

As to the possible role of aggregating chemicals in social spiders, Krafft (1969), using radioactively labeled flies, indicated that an exchange of digestive fluids occurs among feeding *Agelena consociata* and suggested that a social pheromone may be exchanged among individuals during feeding. Kullmann (1972a) stated that in *Stegodyphus sarasinorum* contact chemoreceptors may function in the reception of a colony pheromone.

Later work by Krafft (1975b) demonstrated that a pheromone present on or in the integument of the members of a colony of *A. consociata* is responsible for the inhibition of mutual biting. In a series of trials, he touched vibrating pith balls to the web of a colony, the experimental balls having been rubbed against conspecifics or coated with various portions of conspecifics—crushed prosoma, crushed abdomen, hemolymph, or integument turned inside out. The last of these was most effective in inhibiting biting, while all yielded some degree of protection compared to a clean pith ball (control). Krafft hypothesized that an integumental pheromone that mediates tolerance is produced by cells of the hypodermis.

Distance Chemoreception

Among the orb-weaving spiders (Araneidae), airborne chemical signals aid male orientation to a female on her web over a distance at least as great as 1 m (Blanke, 1973, 1975a,b; Enders, 1975c). In jumping spiders (Salticidae), distance chemoreception is a secondary releaser of courtship behavior, while vision is the primary releaser (Crane, 1949b). As to the wolf spiders (Lycosidae), some researchers (Bristowe and Locket, 1926; Dijkstra, 1976; Kolosvary, 1932; Vlijm et al., 1963) have suggested that various species also employ olfactory orientation to locate a distant female, but they provided no experimental evidence.

Experimental Evidence in Lycosid Spiders

Tietjen (1979a) used bioassay methods to investigate the possibility of an airborne sex pheromone in four species of lycosid spiders: *Lycosa rabida*, *L. punctulata*, *Schizocosa saltatrix*, and *S. ocreata*. Two olfactometers were constructed: one to test the response of a test male in an open field, the other to provide the test male with a choice among three air streams. In the first case, stimuli (male, female, or control odors) were presented singly, while in the second series, combinations of odors were present simultaneously. A variety of behaviors were scored, including locomotor behavior, sound production, courtship, and threat display, to increase the sensitivity of the analysis.

The test males were assayed for "sexual tone" (Crane, 1949b) by introducing them into a container which had recently been vacated by a conspecific adult female. Tests of sexual tone followed each olfactometer test immediately. Only males of high sexual tone (those that exhibited courtship behavior within 5 min) were used in this analysis. Sample size per test for the two olfactometers ranged from 60 to 180, depending on the species. Male *S. ocreata* were not run in the 3-choice experiment, as they were continuously active and their behavior could not be accurately recorded.

In the open-field olfactometer, males of the genus *Lycosa* did not show a behavioral response to a hidden female which could be attributed to an airborne chemical communication. On the other hand, males of both *Schizocosa* species exhibited a change in behavior that was attributed to a female-produced olfactory pheromone. Male *S. saltatrix*, for example, showed a decrease in locomotory behavior when in the presence of a hidden female by covering in a given time period less than two-thirds of the distance traversed when in the presence of a hidden male or a control. The distance traveled by test males did not significantly differ between tests with hidden males or controls.

In nature, male *S. saltatrix* wander over the surface of the leaf litter more than do females, which remain stationary under the litter (Tietjen, unpubl. data). A "cloud" of pheromone could be expected to build up under leaves occupied by adult females. Active males that encounter the cloud would be expected to exhibit a decrease in the speed of locomotion, thus increasing the chance that the male will find the stationary female through chemical or other means (vibration or vision, for example).

On the other hand, male *S. ocreata* did not exhibit an orthokinetic reaction to hidden females but did exhibit a taxis response by approaching hidden females. Although test males also approached hidden males, the approach paths differed qualitatively from those taken in response to a hidden female (Figure 7.5). The latter trails were sinuous, with the test male often stopping to initiate chemoexploratory behavior (rubbing of the dorsal surface of the palp against the substratum). The paths in response to hidden males appeared to be visually mediated (in response to the stimulus apparatus), since they were straight-line and no chemoexploratory behavior was observed.

Male *S. ocreata* also exhibited an increase in directed exploratory

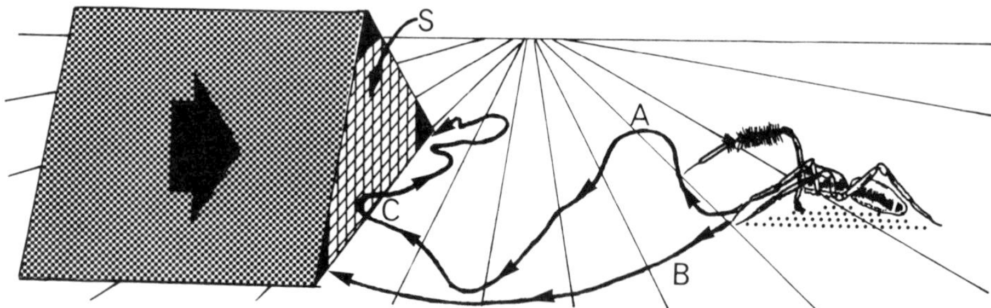

FIGURE 7.5. Responses of male *Schizocosa ocreata* to hidden stimuli. A control conspecific male or female is restrained within the stimulus cage (S). Air flows over the restrained animal (large arrow) and into the arena. The test male's approach path in response to a hidden female is sinuous (A) and includes chemoexploratory behavior (C). Test males less than 10 cm from the stimulus cage took up to 2 min to traverse such paths. The response to a hidden male, however, was a relatively straight-line path which was traversed in less than 5 sec (B).

behavior in response to a hidden female, when compared to a hidden male or control. Directed exploratory behavior was characterized by the test male approaching, touching, and initiating chemoexploratory behavior in response to an isolation chamber containing the hidden stimulus. Directed exploratory behavior was observed in 35% of the tests involving a hidden female, 5% of the tests involving hidden males, and was not observed in response to a control.

In nature, the behavior of male and female *S. ocreata* is similar to that observed in *S. saltatrix* (Cady, 1978). A pheromone cloud would be expected to build up near a resting female, thus directing the wandering male to her in a manner analogous to that discussed for *S. saltatrix*. As will be shown later, the differential reliance on olfactory chemical communication between the *Lycosa* and *Schizocosa* genera was related to a difference in microhabitat preference.

Recently, Dumpert (1978), using electrophysiological methods, identified the odor receptor as the tarsal organ in a lycosid relative, the ctenid, *Cupiennius salei*. The tarsal organ was sensitive to the odor of conspecific females, thus providing additional experimental evidence for airborne chemical communication in lycosids.

ROLE OF APPENDAGE MOVEMENTS

Many insects, including ants, moths, and bees, exhibit vibratory movements of the antennae when in the presence of an olfactory pheromone; an elaborate system of musculature has evolved to mediate these movements (Schneider, 1964). Recently, Schmitt and Ache (1979) utilized electrophysiological recording techniques to study the effect of antennal vibration in the spiny lobster *Panulirus argus*. Behavioral observations indicated that antennal flicking reaches its maximal rate at the onset of chemical stimulation. They suggested that flicking increased the lobster's temporal resolution of its chemical environment by facilitating reception at multiple points on the antennal surface. In addition, antennal vibrations facilitated movement of molecules past barriers imposed by antennal morphology and onto receptor surfaces.

Leg-waving behavior has been observed in several spiders including members of the Lycosidae (Engelhardt, 1964; Kaston, 1936; Rovner, 1968b) and the Salticidae (Crane, 1949b). Koomans et al. (1974) indicated that leg-waving behavior occurs in subadults and adults of the genus *Pardosa* and that chemical stimulation provided by the female's silk increased the frequency of leg-waving in adult male conspecifics. They proposed that leg-waving behavior was an agonistic display ritualized from exploratory movements.

Both *S. saltatrix* and *S. ocreata* incorporate leg-wave behavior into their normal locomotory behavior, with leg-waving being most vigorous in *S. ocreata*. Such movements probably are functionally related to the antennal movements studied in the spiny lobster, serving to increase the likelihood of detection of chemicals by the receptors (tarsal organs) located on the legs. Leg-wave behavior associated with normal locomotion is not seen as often in the *Lycosa* species, although leg-waving is observed in response to chemical stimulation (Rovner, 1968b). The relatively high level of leg-waving in males of the *Schizocosa* species compared to the *Lycosa* species suggests that the former species are more dependent on airborne chemical cues than on contact chemoreception when searching for the females. In addition, the more vigorous leg-wave of *S. ocreata* as compared to *S. saltatrix* may be related to the former's need for increased spatial resolution in a taxis response to odors.

Trail-Following Behavior

A more direct means of orientation to a distant pheromone source than orthokinesis or movement against the concentration gradient of a gaseous chemical is trail-following. The trails employed by certain eusocial insects and by tent caterpillar larvae are received by the follower as a very narrow and much-elongated pheromone cloud above the substratum (Fitzgerald, 1976; Wilson,

1971). Thus, insects use olfactory receptors to detect airborne molecules. On the other hand, the use of contact chemoreception for following a trail has been demonstrated in honeybees (Martin, 1965).

Species Differences in Trail-Following

In some species of lycosid spiders, males perform trail-following when they trace a female's silk dragline (Bristowe and Locket, 1926; Engelhardt, 1964). Tietjen (1977) investigated trail-following behavior in two species of lycosid spiders, *Lycosa rabida* and *L. punctulata*. Females of both species were put on a leash and induced to lay silk trails either directly on the substratum (= ground lines) or suspended above the substratum (= aerial lines). In addition, male-produced silk or imitation lines (nylon thread or human hair) were laid contiguously or discontiguously with the end of a female's trail.

Male *L. rabida* followed aerial trails laid by females but did not follow ground lines. Furthermore, male *L. rabida* followed imitation or male-produced lines if such trails were laid contiguously with a female's silk and the males were already in the process of following the female's dragline. Males would occasionally examine isolated imitation or male-produced trails, but did not exhibit courtship or trail-following behavior in response to such lines. Male *L. punctulata*, on the other hand, followed both ground and aerial trails laid by conspecific females. In contrast to *L. rabida*, male *L. punctulata* did not follow imitation or male-produced draglines under any conditions.

On the basis of these behavioral observations, Tietjen (1977) suggested that a pheromone is necessary to put male *L. rabida* in a following mode of behavior. Once following has begun, tactile cues become important. This mechanism was hypothesized because male *L. rabida* did not follow ground lines, which provide relatively little in the way of tactile cues compared to aerial trails. In addition, the occasional following of male and imitation lines (up to 8.5 cm) indicated that mechanical cues alone would suffice if the male was already following a female's silk. In contrast to *L. rabida*, male *L. punctulata* were described as being more sensitive to chemical cues because they followed ground lines and did not follow imitation or male-produced trails. Male *L. punctulata* therefore depend on chemical cues both to initiate and maintain trail-following behavior.

Fine Structure of Trail-Following

Tietjen and Rovner (1980) further compared the trail-following behavior of these two species by utilizing highspeed cinematography and morphological methods. The trail-following behavior of unrestrained male *L. rabida* and *L. punctulata* was filmed from top, bottom, side, and front views at normal speed (18 frames/sec) and highspeed (36-180 frames/sec). The resulting trail-following sequences were later analyzed frame-by-frame.

The film analysis indicated that the two species exhibit similarities in their trail-following behavior. After encountering and "tasting" the female's dragline, the male straddles the silk thread so that it passes medially beneath him and between the palps (Figure 7.6). The palps, rather than the first legs or other appendages, play the primary role in trail-following, with only the medial surface of the palp being used to contact the silk during most of the process. Both species exhibit a "palpal sliding" mode of trail-following as well as a "ballistic" component. Palpal sliding is shown when the male maintains trail contact with one or both palps while moving forward. The "ballistic" component is characterized by a continued forward movement without the palps making contact with the trail, i.e., without direct mechanical guidance from the silk.

Some modes of trail-following were characteristic of only one species. "Palpal rotation," for example, was seen only in *L. rabida* and was characterized by intermittent contact with the trail. The palps rotated forward while maintaining contact with the line and then looped outward and away from the trail. The palps were used in alternation; however, only one palp tended to make contact with the trail. The silk was often displaced more than 1 mm from its original position when contacted by the palp. The movements seen during a palpal rotation sequence were similar, although not as stereotyped as the behaviors seen during courtship display (Kaston, 1936). This suggests that the visual courtship display of male *L. rabida* represents a ritualized trail-following behavior. Along similar lines, Bristowe and Locket (1926) and Crane (1949b) suggested that the courtship displays of lycosid and salticid spiders have their origins in ritualized chemoexploratory movements.

The palpal movements of male *L. punctulata* rarely resulted in displacement of the female's silk. The most common mode of trail-following in this species, the "forward loop," was similar in general form to the palpal rotation described in *L. rabida* except that the palp maintained contact with the dragline. In contrast, male *L. ra-*

FIGURE 7.6. Trail-following by a male *Lycosa rabida*. (A) The male encounters the dragline of a female conspecific and stops wandering. (B) The palps are used to "taste" the silk. (C) The dragline is straddled, one palp on each side of the thread, and the male begins to follow the trail.

bida usually moved the palp more than one palp width from the trail during palpal rotations. Movements of the palps of male *L. punctulata* could be adequately described with two axes while those of male *L. rabida* required three dimensions.

Male *L. punctulata* exhibited another mode of trail-following which was dissimilar to anything observed in *L. rabida*: "palpal placement." This was characterized by the male's placing the tip of its palp on the line and keeping it fixed at this single point while he continued to move his body forward. The palp was later lifted from the silk and placed forward at another point. These movements resulted in the palps being "walked" along the trail.

Earlier behavioral tests (Tietjen, 1977) have suggested that male *L. rabida* use mechanical cues and depend less on chemical cues than male *L. punctulata* during trail-following. The highspeed film analyses support this view of the use by male *L. rabida* of mechanical cues, since the female's trail was usually displaced by the palps during following, while in *L. punctulata* it was not displaced.

Distribution of Chemosensitive Sensilla

In both species, the medial palpal surface contacted the dragline during all modes of trail-following. This suggests that the distribution of chemosensitive sensilla may relate to the use of the palp during trail-following. In addition, since neither adult female, nor penultimate (instar before adult) male spiders followed silk trails, a difference was expected in the number and/or distribution of sensilla when compared between sexes or age classes. Sexual dimorphism of antennae has been described in a variety of insects, including the silkworm moth (*Bombyx mori*) and the honeybee (*Apis mellifera*) (for review see Kaissling, 1971).

In the study of lycosid spiders, maps of chemosensory hair distribution on the palps of the adults of both species and sexes, and on the penultimate males and females (*L. rabida* only) were drawn with the aid of a stereomicroscope equipped with a camera lucida. Adult male palps of both species were also recorded in scanning electron micrographs.

It was found that the distribution of chemosensitive sensilla was related to the use of the palps during trail-following behavior. The highest number of these hairs was observed on the medial surface, followed by the dorsal, lateral, and ventral surface for males of both species. The distribution was not as dramatically affected by position on the palp for adult female and penultimate male *L. rabida* (Figures 7.7, 7.8). Adult males of both species had approximately three times the number of chemosensitive sensilla as conspecific

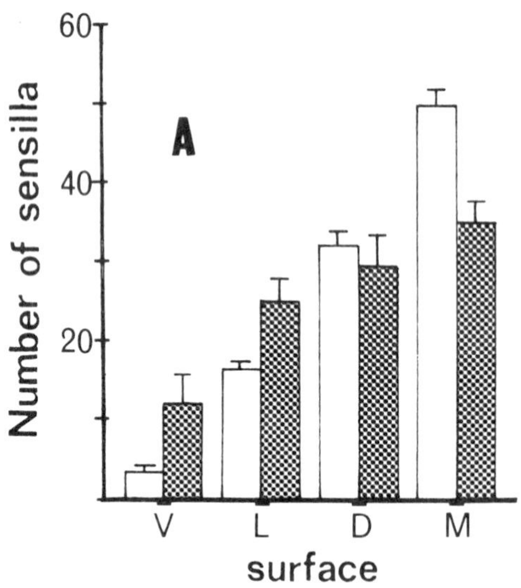

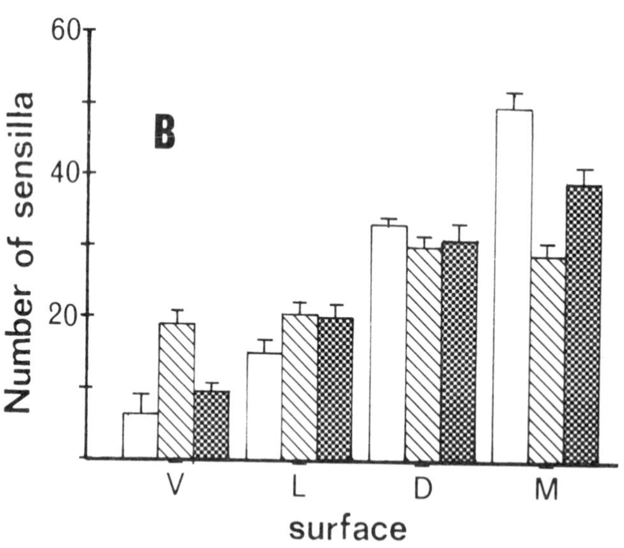

FIGURE 7.7. Percentage distributions of chemosensitive sensilla on the four surfaces of a spider palp: ventral (V), lateral (L), dorsal (D) and medial (M). The stippled bars represent percentage distributions among these surfaces for adult females, the striped bars for penultimate spiders (both male and female), and the open bars for adult males. The indicated variation is the standard deviation. (A) Male and female *Lycosa punctulata*. (B) Male, female, and penultimate *Lycosa rabida*.

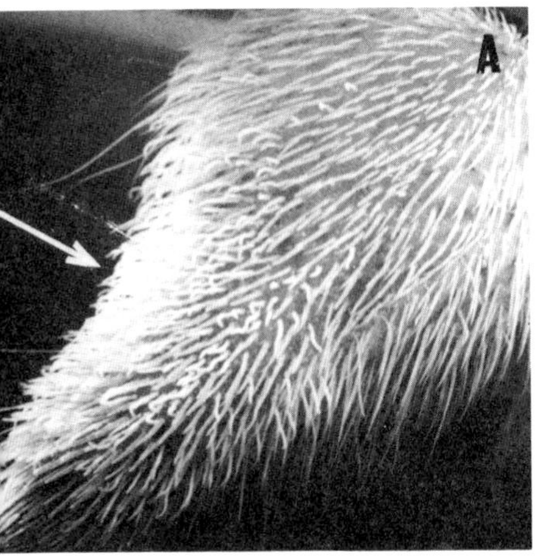

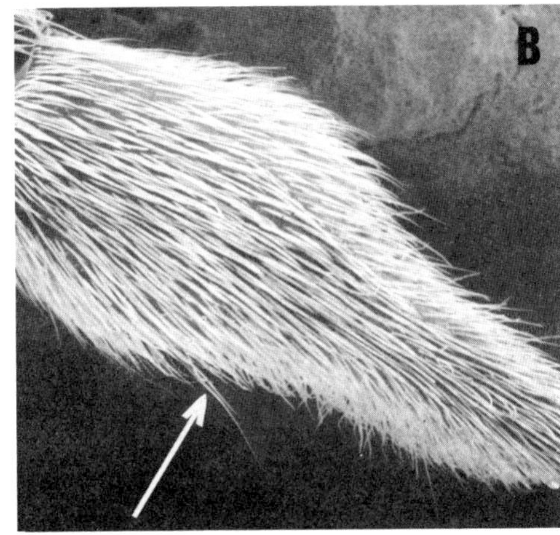

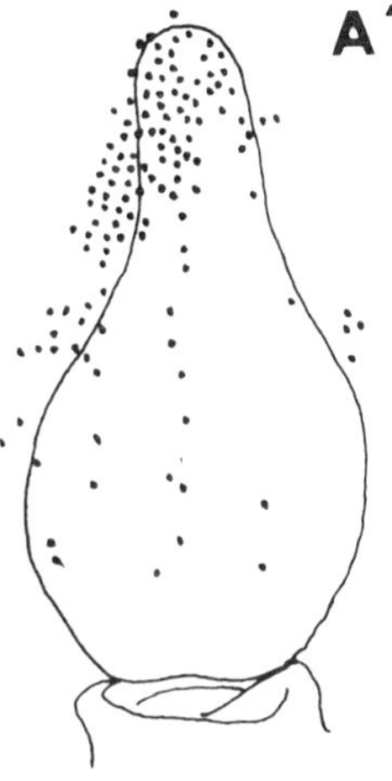

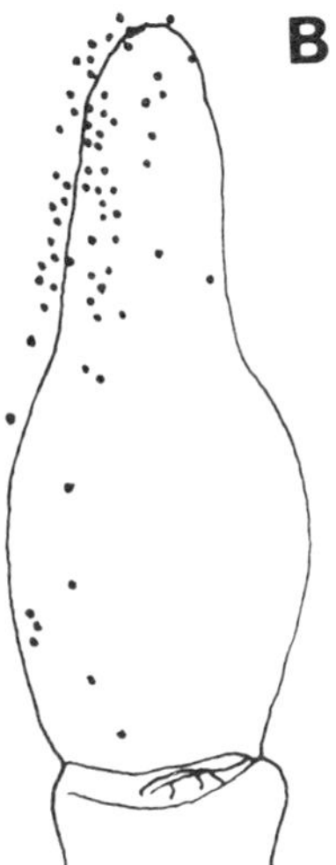

FIGURE 7.8. Scanning electron micrographs of the left palps of wolf spiders, as seen in dorsal views. The partially visible medial surface of each palp is indicated by arrows. Chemosensitive sensilla are concentrated on the medial surface. The diagrams are distributions for the chemosensitive hairs on the dorsal surface of the palps, each spot representing the tip of each hair. (Palps were examined under a dissecting microscope for this purpose.) Note the higher number of chemosensitive hairs on the palp of *L. punctulata* (A′) as compared to *L. rabida* (B′).

females. For *L. rabida*, there were no differences in the number of chemosensitive hairs among penultimate males, penultimate females, and adult females. Comparisons between species indicated that, while the adult females had similar numbers of chemosensitive hairs, adult male *L. punctulata* had approximately 54% more chemosensitive sensilla than did adult *L. rabida*.

The results described above suggest that the distribution of chemosensitive sensilla in males of these two species is related to trail-following behavior. (1) The medial surface of the male's palp has the highest concentration of chemosensitive hairs and is the region exclusively used during trail-following. (There is no such concentration of chemoreceptors on the palps of females or immatures.) (2) The sexual dimorphisms (and age-class effect in *L. rabida*) suggest that most of the chemosensitive sensilla on the adult male's palp are involved in pheromone reception. (3) The greater number of hairs on the palps of male *L. punctulata* compared to *L. rabida* is consistent with the hypothesis that the former species relies more strongly on chemical cues than does the latter. (Such between-species differences were not observed in adult females.)

Possibility of Directional Cues

Wilson (1962) indicated that the information content in the terrestrial trails of fire ants (*Solenopsis saevissima*) was similar to that present in the waggle dance of the honey bee (*Apis mellifera*). Unlike the trail of a spider which utilizes a contact sex pheromone, the trail of the fire ant is in the form of an "odor tunnel" through which the ant moves.

The spider's addition of a mechanical component to the trail may allow more information to be transmitted in the message than is possible in the ant odor trail. For example, the terrestrial trail of the fire ant contains no inherent directional information (Wilson, 1962). This is probably not the case in the two *Lycosa* species. Tietjen (1977) indicated that males of both species tend to follow more often in the direction the trail is deposited than in the incorrect direction. He suggested that the males are not using a pheromone concentration gradient, since silk lines over a month old still induce trail-following and/or courtship. (If a concentration gradient were employed, then the fadeout time of the pheromone would be on the order of a few seconds to minutes, according to Bossert and Wilson, 1963.)

It is possible that the mechanical properties of the silk line provide directional cues to trail-following males (Tietjen, 1977). When depositing a trail, spiders fix the line to the substratum with tangles of silk called "attachment disks." As the dragline enters the attachment disk it is tightly wound. On leaving the attachment disk, the silk consists of separate threads which later join to form the single dragline. Tietjen speculates that if the silk were pressed down at any point between two attachment disks and plucked on either side of this point of immobilization, the side with the higher resonant frequency would be the correct direction for the male to follow the line. This is because the loose threads leaving an attachment disk would dampen vibrations more readily than the single thread entering a disk. When male spiders encountered a female's trail, they usually explored the trail with alternating movements of the palps. Such behavior may represent both sampling for pheromone and testing of mechanical properties of the line.

Sex Pheromones

Pheromones that stimulate sexual behavior have been found in all major taxa (Frings and Frings, 1968; Shorey, 1976). Contact sex pheromones have been described in various families of spiders. Bristowe and Locket (1926), Kaston (1936), and Platnick (1971) provide reviews of early research on this subject. In a variety of spiders, a contact sex pheromone associated with the female's silk and/or integument will induce courtship behavior in conspecific males (Blanke, 1975b; Dondale and Hegdekar, 1973; Dumais et al., 1973; Farley and Shear, 1973; Hegdekar and Dondale, 1969; Jackson, 1978a; Rovner, 1968b; Tietjen, 1977, 1979b; Witt, 1975). Male *Pardosa amentata* respond to the female's dragline with both courtship and searching movements (Dijkstra, 1976), while male *P. ramulosa* and *Schizocosa ocreata* exhibit mainly searching movements in response to the line (Sarinana et al., 1971; Tietjen, unpubl. data; Uetz, personal communication).

Among the social spiders, sex pheromones may be relatively unimportant. Jackson (1978a) indicated that in *Mallos gregalis* a female's silk did not release courtship in conspecific males. In this social species, males incorporate a sexual advertising routine into their daily activity budget. These results were contrasted to those of two non-social, but related species, *Mallos trivittatus* and *Dictyna calcarata*, in which a female's silk did release courtship behavior in conspecific males (*ibid*).

Passive Pheromone Deposition

Deposition of dragline represents active marking of the substratum. In some arthropods, passive marking occurs through the nonspecific liberation of pheromone from the integument. For example, honeybees (*Apis mellifera*) and the colonial wasp (*Vespa vulgaris*) deposit a "footprint substance" from the tarsal integument that serves to attract colony members to the nest entrance (Butler et al., 1969). Bristowe and Locket (1926), Richter (1971), and Richter et al. (1971), after sealing the spinnerets of female spiders and having them walk over substrata, indicated that male *Tarantula barbipes* and *Pardosa amentata* exhibit searching behavior in response to passively deposited pheromones. However, passive marking has not been found in other members of the Lycosidae (Dondale and Hegdekar, 1973; Kaston, 1936; Tietjen, 1979b).

Male-Produced Pheromones

Among various insects, e.g., bark beetles, male-produced pheromones often serve to attract and stimulate conspecific females for mating. Lepidopteran males typically immobilize a conspecific female before copulation with a flight-arresting pheromone (for review see Shorey, 1973). Likewise, there is a possibility that certain male wandering spiders produce pheromones. Engelhardt (1964) suggested that female *Trochosa* "recognize" conspecific males through an olfactory pheromone. Future research may reveal that the predatory behavior of some female spiders is inhibited by a male-produced pheromone.

Some male moths deposit an antiaphrodisiac on females after mating, rendering them unattractive to other males. Such male-male inhibiting pheromones yield increased reproductive efficiency in the competition for a single female (Gilbert, 1976; Hirai et al., 1978). Similarly, certain male spiders may be using chemical cues as part of a mating interference strategy. Studies of various web-weavers (e.g., Robinson and Robinson, 1973) reveal the performance of silk-deposition by males on the webs or even on the bodies of their mating partners. As to the wolf spiders, Richter and van der Kraan (1970) indicated that male *Pardosa amentata* produce increased amounts of silk in response to the female sex pheromone and during courtship. Male *L. rabida* and *L. punctulata* similarly deposit greater amounts of silk during trail-following than during other, non-sexual activities (Tietjen, unpubl. data). Also, male *L.*

rabida exhibit a decrease in directed exploratory behavior when in the presence of a hidden male, compared to the same situation with a hidden female or a control (Tietjen, 1979a). Finally, the possible presence of a male pheromone may have caused a decrease in courtship behavior that was noted in experiments designed to test whether *Lycosa* spp. have a pheromone that is passively deposited (Tietjen, 1979b). The likelihood that chemical signals may mediate certain male-male interactions merits further investigation.

Ontogeny of Chemical Communication

Juveniles of most animal species do not respond to the adult sex pheromone. In some insects, the lack of response is correlated with a lower number or a different morphological type of chemoreceptor in the juvenile as compared to the adult (Kaissling, 1971; Shorey, 1976). In a similar vein, Tietjen and Rovner (1980) demonstrated a difference in the distribution and number of chemosensitive sensilla between penultimate and adult male *Lycosa rabida*, as did Kronestedt (1979) for various species of Palearctic lycosids.

In most species of arthropods investigated, juveniles do not produce a sex pheromone (Shorey, 1976). However, Crane (1949b) found evidence for the onset of sex pheromone production in late penultimate female salticids. The occupancy by males of molting chambers constructed by penultimate females of various species also suggests onset of sex pheromone release prior to adulthood in those species (Bristowe, 1958; Jackson, 1977b).

Among orb-weavers, e.g., *Araneus diadematus* (Meyer, 1928) and *Cyrtophora citricola* (Blanke, 1972), only mature females deposit a male-stimulating substance on the web, females of the latter species not releasing pheromone until 1-2 weeks post-molt. As to the lycosids, sex pheromone output by female *Lycosa rabida* begins about 2 weeks after the final molt (Rovner, 1968b). Likewise, female *Schizocosa crassipalpis*, *S. avida*, *Pardosa moesta*, and *P. saxatilis* deposit a pheromone only as adults; furthermore, the previous mating history of the adult females does not affect the response of males to the female's silk (Hegdekar and Dondale, 1969).

As to the cessation of sex pheromone output, female lycosid spiders that construct and carry egg sacs can no longer deposit draglines (Vlijm et al., 1970), which suggests that active pheromone deposition is effectively stopped at that time. In the araneid *Cyrtophora cicatrosa*, females cease emitting an airborne pheromone soon after copulation. In this spider, the pheromone is not secreted until 3-4 days after the female's final molt, after which the levels of pheromone emission continue to increase until 10-20 days post-molt. This is followed by a decrease in the concentration of the chemical until it no longer stimulates males (Blanke, 1975b). A similar phenomenon has been demonstrated in various salticids (Crane, 1949b).

Environmental or physiological conditions unrelated to adult age may affect the daily levels of pheromone released by female lycosid spiders, just as was presumed for the variations in the courtship tendency of salticid spiders (Crane, 1949b). During a series of experiments, Tietjen (unpubl. data) found that two female *L. punctulata* which were previously producing draglines that were stimulating to males stopped pheromone production. In each case, 5 males did not court in response to the above female's silk, but would court in response to other female's silk. Within 3 days pheromone production in the two females returned to levels that would induce courtship behavior in males. Similar variations in pheromone production have been reported in the boll weevil (*Anthonomus grandis*). A 100-fold difference between high and low daily pheromone production was found for the male-produced pheromone. Highest release levels occurred between 0700-1300 hr, increased with age, and became arrhythmic in constant light, suggesting that pheromone release was cued by photoperiod (Gueldner and Wiygul, 1978). The nature of the factors affecting temporal variation in the level of pheromone secretion in spiders merits investigation.

Behavioral Ecology of Chemical Communication

Arthropods often display behaviors appropriate for pheromone release that are related to physical features of the habitat or to other environmental factors. Female moths for example, climb to an elevated perch during pheromone release; this results in a trail that is undisturbed by structural features at ground level. At very low wind speeds they may vibrate their wings to disperse their sex attractant; above a certain wind speed, pheromone emission ceases (Shorey, 1964, 1976).

Among the Araneae, few studies have attempted to relate the ecology of the animals to behavior during chemical communcation. Hegdekar and Dondale (1969) indicated that the contact sex pheromone of various lycosid spiders remained active for a period of at least 4 weeks in the laboratory. They suggested that the effects of rain or dew would inactivate the pheromone on the female's dragline, preventing a pheromone buildup under natural conditions (Dondale and Hegdekar, 1973; see also Tietjen, 1977).

In studies concerned with chemical communication in wolf spiders, Tietjen (1977, 1979a) hypothesized that microhabitat preferences probably affected the mode of chemical communication observed in two genera. Two forest litter-dwelling species, *S. saltatrix* and *S. ocreata*, responded weakly to an airborne pheromone but did not exhibit trail-following behavior. Since moisture quickly inactivates the pheromone associated with the silk (Dondale and Hegdekar, 1973; Hegdekar and Dondale, 1969), it was hypothesized that a moist-litter habitat precludes the possibility of contact chemical communication involving trails in such species.

Two species, *L. rabida* and *L. punctulata*, inhabiting the relatively drier herbaceous stratum of fields exhibited trail-following behavior but not olfactory orientation to distant chemical sources. Moisture from morning dew and rain is probably important in these species to inactivate the pheromone found on trails laid the previous day. This increases the efficiency of the system, since males would not follow "old" lines, i.e., those which do not lead to the receptive females. Apparently, the inherent accuracy of trail-following makes olfactory orientation unnecessary in the dry herbaceous stratum.

Tietjen also suggested that a differential reliance on chemical vs. mechanical cues during trail-following may have been shaped by the different stratum preferences of the two *Lycosa* species. *L. punctulata* is found in the lower levels of the herbaceous stratum, levels characterized by a relatively large population of heterospecific spiders as compared to the upper levels (Whitcomb et al., 1963). The greater number and diversity of spiders at lower levels would be expected to result in an increased heterospecific silk density, making accurate trail-following within a species important. Furthermore, it was suggested that the greater foliage density in lower levels of the herbaceous stratum results in a greater number of attachment points along the length of the trail when compared to the upper stratum. A greater reliance on chemical rather than mechanical cues would therefore increase the accuracy of trail-following in the lower stratum.

L. rabida, on the other hand, is found in the upper stratum, an area characterized by both a lower interspecific silk density and a decreased number of attachment points. In this species, continuous sampling for pheromone during trail-following may not be as important, compared to *L. punctulata.*

Interspecific Chemical Communication

Most pheromones are species-specific chemicals, a property that serves to promote the reproductive isolation of sympatric populations (Mayr, 1974b; Shorey, 1976). However, instances of interspecific communication have been reported in several arthropods. Among the social insects, for example, trail-sharing between species occurs in many tropical ants and termites (Wilson, 1971). These associations range from parasitism to commensal relationships.

Tretzel (1959) experimentally demonstrated that various species of spiders "recognize," probably through chemical stimuli, the webs of *Coelotes terrestris* with whom they share the same habitat. The heterospecific spiders exhibit slow-walking behavior after contacting the silk of *Coelotes* and usually are not captured by the web owner after foreleg contact. In contrast, spiders from another habitat typically are captured by *Coelotes.*

Remarkable examples of interspecific chemical communication occur in various non-web-building spiders, e.g., *Celaenia excavata* and the bolas spiders *Mastophora* spp., both of which release chemicals that mimic the sex attractant pheromones of certain noctuid moths. This idea is supported by two types of evidence: (1) only male moths are caught by the spiders (McKeown, 1952; Eberhard, 1977); (2) prey approach from downwind with a behavior like that used for aerial trail-following when orienting toward female moths (Eberhard, 1977). It was suggested that the energy expended by bolas spiders during prey-capture by means of chemical deception was similar to that of a like-sized, web-building spider. The possibility that such prey attraction by chemical mimicry occurs even in web-building araneids was suggested by the behavior of male saturniid moths (*Hemileuca lucina*) in the vicinity of the webs of *Argiope* and *Araneus* species (Horton, 1979).

Outlook

At the time of this writing, unlike the situation in insect pheromones, there has yet to be an attempt to identify the molecular structure or sites of production and release of any spider pheromone. Only recently have arachnologists begun to utilize methodologies other than behavioral assays in the investigation of chemical communication among the spiders. Advances in electrophysiological recording techniques and in morphological analysis promise a greater understanding of chemical communication in this taxon, eventually comparable, we hope, with the level attained in some species of the economically important arthropods. Communication theory has not yet been applied to chemical communication in spiders. An informational analysis of trail-following vs. olfactory communication, for example, would allow students of arachnology to compare modes of communication both within the Araneae and between the spiders and other taxa.

Acknowledgments

Many of the data described in this chapter were collected under grants from the National Science Foundation, BMS 7101589 and BNS 76-15009, to J. S. Rovner (Ohio University). During the final revision of the chapter, W.J.T. received support from National Science Foundation grant BNS 75-09915 to P. N. Witt (N. C. Division of Mental Health Research). All communications should be sent to J.S.R.

The following manuscript was prepared after completion of this chapter:

Roland, C. and J. S. Rovner. Chemical and vibratory communication in the aquatic pisaurid spider *Dolomedes triton*. J. Arachnol. (in press).

Chapter 8

SPIDER INTERACTION STRATEGIES: COMMUNICATION VS. COERCION

Susan E. Riechert
Department of Zoology
University of Tennessee
Knoxville, Tennessee 37916

Introduction

"The remarkable thing . . . isn't that they [stink bugs] put their tails up in the air—the really incredibly remarkable thing is that we find it remarkable. We can only use ourselves as yardsticks." John Steinbeck, *Cannery Row.*

Communication can be defined as the conveyance of information from one organism to others. The recipients may utilize this information in determining the behavioral state of the communicator and in choosing an appropriate behavior for themselves. General discussions of the mechanics of communication and its properties are available from Burghardt (1970), Smith (1977), and Marler (1977).

Here I am most interested, however, in the use of communication by animals involved in conflict situations—specifically in competition for limited resources. In such cases the inverse of communication is coercion, wherein "force" replaces "encouragement." By virtue of its indirect operation, communication in settling disputes has long been considered altruistic in nature; by utilizing communication rather than physical contact against an opponent, an antagonist is reducing its opponent's potential loss of fitness from injury and thereby benefiting the population of which both are members (Marler, 1977). Communication therefore should be best developed in cases where the genetic relatedness of individuals is high, since relative losses to individual fitness will have little effect on the composition of the population's gene pool. The development of communication in other groups would require the operation of group selection.

In a 1973 paper, Maynard Smith and Price provided an explanation for the use of communication in settling disputes which does not require genetic relatedness or group selection. Gaming involves the principle of optimization—an individual should attempt to minimize its energy expenditures (loss of fitness) while maximizing its energy gains (increase in fitness). The greatest benefit to fitness can thus be derived from a dispute by expending only that energy necessary to win. Since warlike behavior or coercion involving physical contact has associated with it large fitness costs both in terms of absolute energy expenditure and in its potential for resultant injury, natural selection should favor the use of communication (conventional behavior patterns) rather than coercion in situations where disputes are of frequent occurrence and/or where a loss does not mean total loss in fitness to the antagonist (e.g., fatality or total loss of reproductive opportunity). Both or all antagonists involved in a dispute can thus receive fitness benefits from utilizing communication rather than coercion. This theory concerning the development of communication does not require genetic relatedness among the individuals engaging in it nor does it require the operation of group selection; the process of individual selection suffices.

Major theoretical contributions to our understanding of the behavior of animals in conflict situations have been made by Parker (1974, 1978), Maynard Smith and Price (1973), Maynard Smith (1974, 1976), and Maynard Smith and Parker (1976). Recent work by Hines (1977), Bishop and Cannings (1978), Caryl (1979), and Hammerstein (1981) has provided further clarification. It is not known, however, to what extent the principles of optimization explain the behavior of animals in conflict situations. Many investigators can offer qualitative observations which suggest that various vertebrates may be utilizing game strategies. However, the quantitative data necessary to test the predictions of game theory can best be obtained through study of the behavior of invertebrates in conflict situations. The important empirical contributions are coming largely from studies with arthropods, probably reflecting the new interest in arthropods generated by a better understanding of the complexity of the insect nervous system (Huber, 1978).

Detailed studies of agonistic behavior in the arthropods have been made by Crane (1957), Reese (1962), Hazlett (1966, 1968, 1972), Dingle and Caldwell (1969), Baker (1972), and Brockmann (1979). These studies deal with competitive behavior in insect and crustacean species (e.g., dragonflies, butterflies, wasps, hermit crabs, and stomatopods). Recent work in the area of spider social behavior indicates that despite their solitary tendencies, spiders frequently do engage in disputes with conspecifics over mates, for space (territories), and for rank (social hierarchies). These studies have demonstrated that there is recognition of conspecifics as competitors (as opposed to potential prey or predators) and that communication rather than fighting is often used in encounters with conspecifics.

Competition for Mates

Of all the types of intraspecific interactions in spiders, courtship is undoubtedly the best studied. This reflects the complexity of a behavior which has been developed to limit cannibalism while providing information concerning the degree of receptivity of each participant toward the other. Associated with mating is male-male competition for females. Though widely studied in other animal groups (Trivers, 1972; Barash, 1977), little is known about the competition for mates in spiders. Many papers on spider courtship, however, do make reference to male-male "threats" in the presence of a female of the species: the jumping spiders (Salticidae: Bristowe, 1929; Crane, 1949b); the hackled band weavers of the family Dictynidae (Billaudelle, 1957); the wolf spiders (Lycosidae: Bristowe, 1929; Rovner, 1967); the comb-footed spiders (Theridiidae: Gwinner-Hanke, 1970); the sheet-line weavers (Linyphiidae: Rovner, 1968a); and the orb-weavers of the family Araneidae (Christenson and Goist, 1979). In none of these cases was actual contact (coercion) observed at the onset of an encounter, and in most cases one of the competitors retreated in response to a threat display offered by the opponent.

One would expect male-male agonistic behavior to be most pronounced where the probability of encounter is highest (i.e., characteristic of those species in which males guard females during the period between maturation and receptivity). Rovner (1968a) has observed guarding in the sheet-line weaving spider *Linyphia triangularis*. Males of the bottle-brush spider *Nephila clavipes* are also known to guard the webs of potential future mates (Christenson and Goist, 1979), while guarding the female retreats has been observed in both jumping spiders (Jackson, 1976b) and orb weavers (*Metepeira spinnipes*: Uetz, personal communication). Most of the detailed information concerning male-male competition for mates is available from studies of web guarding. Rovner, as well as Christenson and Goist, consider this behavior to be territorial in nature since a web of definite area is defended. In both species, intrusion is met by specific vibratory displays often exhibited by the male owner as it approaches the intruder. This is followed by orientation toward the opponent and subsequent "threat" displays by one or both of the individuals. The smaller of the individuals may withdraw at any time during the interaction. If withdrawal does not occur following "threat," contact through pushing, grappling, and biting follows. Injury may thus occur in a limited number of cases. Rovner (1968a) states that although injury does occasionally occur in agonistic interactions among male linyphiids, only rarely is a losing opponent eaten by the victor. The use of communication before coercion is thus observed in this competitive context.

Competition for Space: Territoriality

Numerous definitions of territoriality exist. Most require that there be occupation of exclusive areas which are maintained through behavioral interactions. Territorial behavior is a common phenomenon among vertebrates, providing a variety of benefits. It is not as well known among the invertebrates, and in the cases observed, usually serves a mating function—by holding a territory, the male increases its probability of a successful mating (dragonflies: Johnson, 1964; Bick and Bick, 1965; Campanella and Wolf, 1974; wasps: Lin, 1963; spiders: Crane, 1949b; Rovner, 1968a; Christenson and Goist, 1979). Because males actively defend specific sites in association with females during courtship periods, territorial behavior serving a mating function in invertebrates is readily observed. In this respect it is similar to the territorial behavior of most vertebrates which utilize elaborate visual and vocal displays in settling territorial disputes.

However, territorial behavior need not be this obvious. In their reviews of territoriality, Brown (1975) and Wilson (1975) both state that its operation in many animal groups may be overlooked because of a low frequency of contact between conspecifics. Territoriality serving some function other than the guarding of a potential future mate has been observed in a number of spiders (Lycosidae: Vogel, personal communication; Linyphiidae: Ross, 1977; Agelenidae: Riechert, 1978a; Araneidae: Uetz, personal communication; and various social spider species representing several families: Kullmann, 1968a; Blanke, 1972; Buskirk, 1975; Jackson, 1977d; Brach, 1975). In all of these cases, a regular spacing of spiders results from the withdrawal of individuals following encounters with conspecifics. Territoriality appears to be even more widespread among spiders than the limited number of studies indicates. For instance, I have completed pattern studies on nine species of spiders representing five families, and all exhibit the regular spacing King (1973) attributes to territorial systems.

Territory size seems to range from the area occupied by the web itself to as much as a meter radius around the web (Riechert, 1978a). Defense of the web is commonly observed in territorial systems serving a mating function, as mentioned earlier for the orb-weaver *Nephila* (Christenson and Goist, 1979) and the sheet-line weaver *Linyphia* (Rovner, 1968a). In the colonial species, territorial behavior determines the location of a specific spider's web within the colony and also serves to space the webs out. The best competitors then assume locations where food is in most abundant supply (Buskirk, 1975; Rypstra, 1979). There is evidence to suggest that the spacing of a large number of spider species, including the within-colony spacing of the colonial spiders, is energy-based. For instance, *Metepeira spinnipes*, a colonial orb-weaving spider, exhibits geographic variations in colony size and spacing patterns which are inversely related to habitat severity and/or prey availability (Uetz, personal communication). An energy-based system has also been described for the solitary species *Agelenopsis aperta* (Riechert, 1978a) in which the area maintained in addition to the web is dependent on the number of prey available relative to that needed for growth and reproduction. Thus, within the same species, spiders are more tolerant of conspecifics at closer distances in those habitats where prey are more abundant than in more rigorous environments. That this system may be operating in a large number of spider species is suggested by the median areas of occupation of the spider populations shown in Figure 8.1. Note that individuals of the populations sampled all exhibited exclusive areas of occupation considerably greater than those required for their webs or burrows.

It appears as if interactions over territories serving energy needs may be frequent in many spider species, especially since this behavior is exhibited by members of both sexes through much of the life cycle (i.e., until onset of maturity for males and for the entire life of the females). As in disputes over mates, these interactions are settled primarily by visual and vibratory signaling (Buskirk, 1975; Vogel, 1972; Ross, 1977; Riechert, 1978b, 1979). There is also a definite size bias favoring the larger individuals in many cases (Blanke, 1972; Ross, 1977; Riechert, 1978b).

Competition for Rank Position

Social hierarchy is similar to territoriality in that it provides the successful competitor with preferential access to some limited resource such as mates or food. It lacks the space component, how-

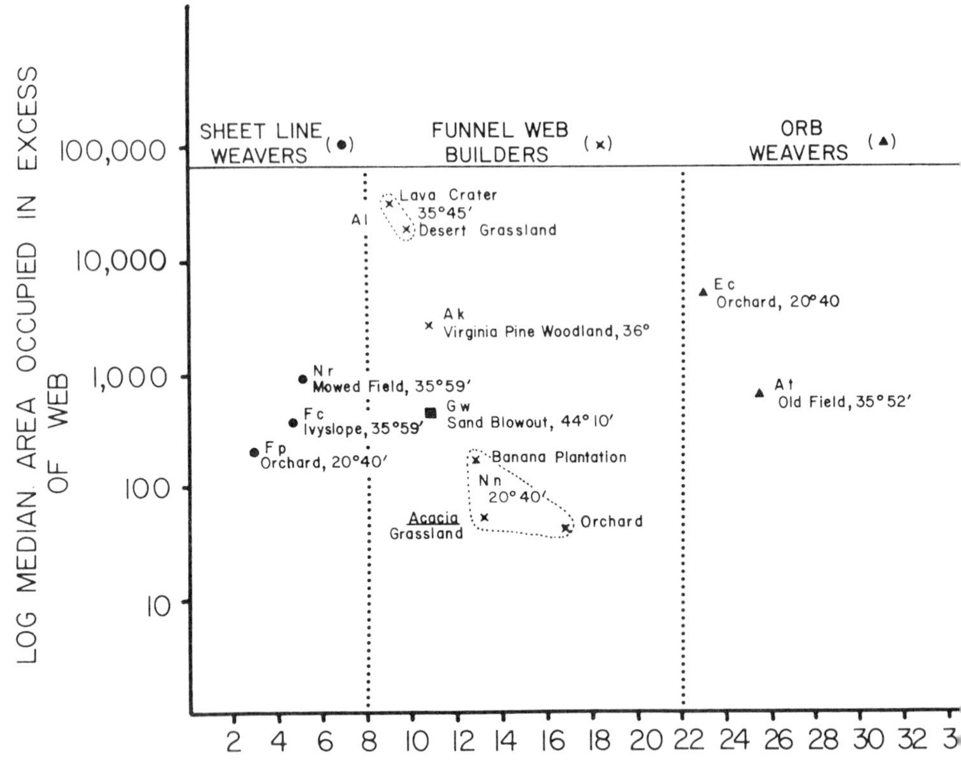

FIGURE 8.1. Plot of median area occupied by adult females of various spider populations (i.e., web plus surrounding habitat). Based on results of block size analysis of variance. Encircled areas include populations representing same species. Represented are: Agelenidae, Ak = *Agelenopsis kastoni*, Al = *Agelenopsis longistylus*, and Nn = *Novalena nsp*; Araneidae, At = *Argiope trifasciata*, Ec = *Edricus crassicauda*; Hahniidae, Nr = *Neoantistea riparia*; Linyphiidae, Fl = *Florinda coccinea*, Fp = *Frontinella pyramitela*; and Lycosidae, Gw = *Geolycosa wrightii*. Guild representation of each species indicated by symbol shown at top. ■ = burrowing wolf spider. Latitude and habitat type noted for each population. All data collected by author and field assistants.

ever, and merely involves the establishment of a pecking order, or dominance hierarchy, commonly known for many farm animals. Dominance hierarchies have been observed both in the more social spider species (Krafft, 1971a; Lubin, 1974) and in some wandering spiders (Salticidae: Crane, 1949b; Lycosidae: Aspey, 1976, 1977a,b; Dijkstra, 1978). The species exhibiting this social structure characteristically exhibit high population densities and associated frequent interactions among conspecifics. Theoretically, hierarchies are the favored social structure in cases where it would be energetically infeasible to maintain territories due to the frequent need for their defense (Carpenter and MacMillan, 1976; Middendorf, 1979). The establishment of a dominance hierarchy thus eliminates the energy and time investments required in frequent disputes over limited resources. The stability exhibited in rank position of *Schizocosa ocreata* (= *crassipes*) studied by Aspey (1977a) seems to verify this function of the hierarchical system. As in other contexts, rank within the system is determined largely by visual display (Aspey, 1976, 1977a; Dijkstra, 1969), and a size bias is present in at least some cases (Lubin, 1974; Dijkstra, 1978).

Games Spiders Play: A Case Study

Though the study of spider agonistic behavior is only in its infancy when compared with the wealth of material available on spider taxonomy, web-building prowess, and sensory perception, the literature just reviewed indicates that spiders, at least as adults, are not the cannibalistic creatures they are frequently characterized as being. In fact, it appears that their solitary existence reflects a social structure which is maintained through communication among conspecifics.

The significance of intraspecific competition and the mechanisms by which it operates are best known for the funnel-web building spider *Agelenopsis aperta* (Riechert 1978a,b, 1979). Members of this species engage in disputes over web sites and associated territories. The behavior exhibited in these agonistic encounters appears to follow the principles of optimization as they are delineated in game theory. The following is an outline of the system—its operation and significance.

Agelenopsis aperta

Agelenopsis aperta is a member of the funnel-web spider family, Agelenidae. Representatives of the genus *Agelenopsis* are prominent in grassland habitats throughout North America. This particular species is a western representative, with populations continuously distributed between northern Wyoming and southern Mexico. *A. aperta* also occupies a wide variety of habitats: desert scrub, desert grassland, lava bed, pine encinal, desert riparian, and orchards and banana plantations, to mention a few.

The spider's horizontal web is non-sticky, with an attached funnel extending into some feature of the surrounding habitat. The funnel permits escape from unfavorable thermal conditions; and individuals in desert habitats spend most of their time in this structure. However, when external temperatures are within a favorable range (between 21° and 35°C), individual spiders sit at their funnel entrance where prey hitting the web-sheet may be detected and captured (Riechert and Tracy, 1975). The time available for prey capture, then, is dependent on the microenvironments characteristic of specific sites, and *A. aperta* has been shown actively to choose its web sites on the basis of thermal properties as well as local prey availability (Riechert, 1976, 1979, unpublished data).

Web-site discrimination in many habitats is associated with reproductive payoffs or benefits. For instance, those spiders occupying a recent lava bed area in south central New Mexico who select sites offering shade, insect attractants such as flowering plants and fecal material, and protection from thermal radiation through association with litter substrates can achieve thirteen times the reproductive potential of spiders occupying sites lacking these characteristics (Riechert and Tracy, 1975). Excellent quality sites are also in very limited supply in the area, with the population existing at near saturation levels (Riechert, 1979).

Intraspecific Competition

Although competition for available sites is often great, it is further increased by the fact that these spiders tend to be intolerant of neighboring individuals. The distance within which a spider will interact agonistically with a conspecific web owner has been shown to be mediated by energy needs and thus to vary with the severity of the habitat occupied (Riechert, 1978a). Briefly, territories serving bioenergetic functions exhibit the following characteristics: (1) territory size varies with the size of the occupant; (2) territory size varies with the availability of food within specific habitats; and (3) the energy received from the territory places the holder at a competitive advantage over individuals lacking territories. All three of the characteristics are observed in *Agelenopsis* territoriality, the most important to our discussion here being the question of competitive advantage (3). Table 8.1 shows the benefit accrued to an individual through occupation of a territory for two habitats. It also shows the loss in fitness associated with sharing a territory with a conspecific web owner in these habitats. It is apparent from these data that territorial behavior permits *Agelenopsis* to optimize its food intake.

TABLE 8.1. Relationship between energy need, energy available, and energy utilized for two populations of *Agelenopsis aperta*. (Data from Riechert and Tracy, 1975 and Riechert, 1978a.)

	Habitat	
	Desert Grassland	*Desert Riparian*
Energy consumed given unlimited food supply (mg dry wt./day)	20.58	22.20
Median territory size	1.44 m^2	0.56 m^2
Energy available (mg dry wt./day)	60.03	73.04
Average capture rate	60%	47%
Energy captured (mg dry wt./day)	36.02	34.33
Ingestion efficiency	66%	66%
Energy consumed (mg dry wt./day)	23.77	22.66
Energy available to each spider if territory is shared by 2 individuals (mg dry wt./day)	14.26	13.60

Encounters, then, fall into one of two categories: (1) that between a roving individual and a web owner; and (2) that between two neighboring web owners (Figure 8.2). Although it is difficult to estimate the frequency of territorial contests, a low estimate can be obtained from turnover rates in territory ownership. The desert grassland population under discussion here in May-June of 1978 thus exhibited an average daily turnover rate in territory ownership of 9.86% ± 0.79% ($N = 700$). This turnover estimate must be viewed as a minimum estimate of encounter frequency in that a roving individual's probability of winning a territory is extremely low, with the exception of those cases in which it has a distinct weight advantage (> 30% larger; Riechert, 1978b). Thus the turnover rate reflects only the number of disputes per unit time involving larger intruders and smaller territory owners.

A fairly good estimate of the size of the floating portion of the population, however, can be obtained through removal experiments. In June of 1978, for instance, 36 territory owners were removed from a 900 m^2 area in the desert grassland study area. After ten days, 19 new individuals occupied territories within this area, 18 of which were unmarked, indicating that they had not been identified as territory owners in other parts of the study area. We can thus conclude that about 35% of this population in the spring of 1978 had been excluded from territory ownership and were "floating."

The size of the floating population and thus the frequency of territorial encounters should vary with season and year, reflecting

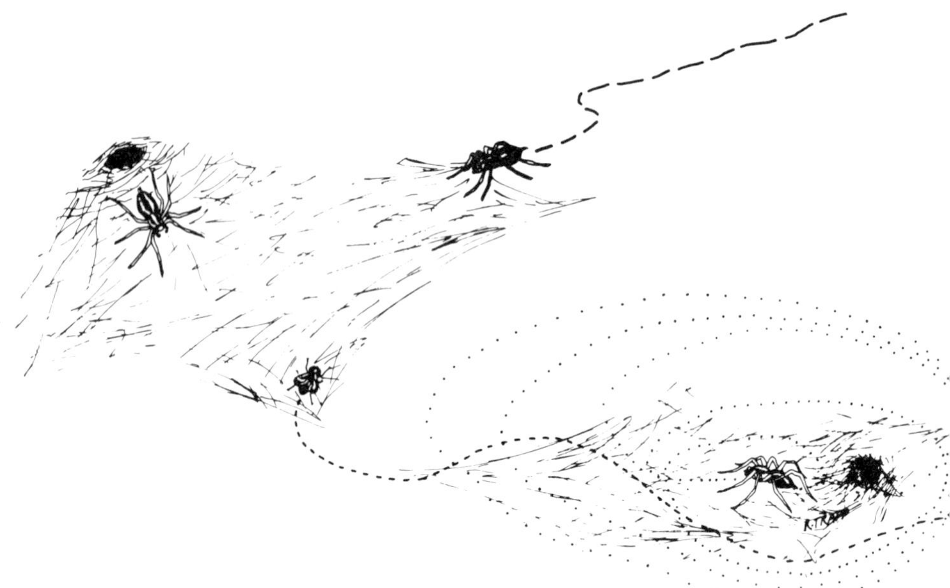

FIGURE 8.2. Relative probability of territorial dispute resulting from encroachment by a floater versus a neighboring web owner for *Agelenopsis* occupying desert grassland study area in south-central New Mexico.

proximate and ultimate environmental effects. In a proximate sense competition will be relatively low during a cool-wet year or season as compared to drier times, since individuals may realize adequate prey abundances and favorable thermal environments at any site within the habitat. Figure 8.3 demonstrates the predicted levels of fitness achieved at various quality sites for individuals living under different environmental conditions. (Quality herein refers to spider preference for particular microhabitat types: excellent = high preference exhibited, average = acceptable, poor = unacceptable except under unusual conditions. For the desert grassland study area under consideration here, preference is synonymous with reproductive success.) Note that I expect all individuals to contribute offspring during favorable times, while only those individuals occupying excellent quality territories will do so during drought years.

In an ultimate sense, competition for territories should be highest in the generation following good years. The population is at a high at these times, in part because of the fact that *A. aperta* exhibits a reproductive numerical response toward changes in prey

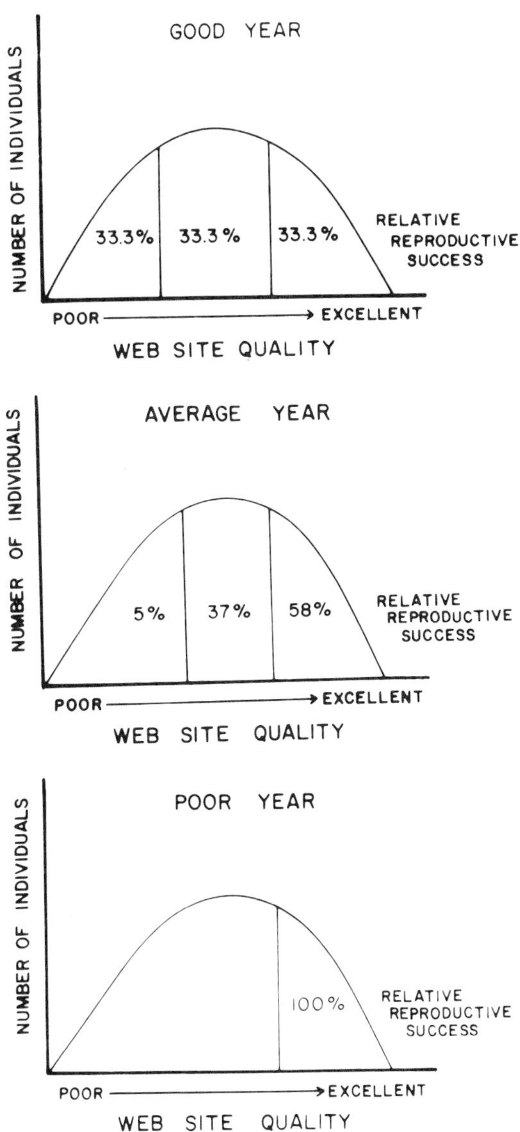

FIGURE 8.3. Relationships between web-site association and reproductive success under different environments for *Agelenopsis aperta*, desert grassland habitat. Numbers under curve represent contribution of offspring for that quality site relative to others in the habitat. Values for average year represent actual data. Values for other years are predictions only.

abundances (the more prey a spider receives, the more offspring it produces; Riechert and Tracy, 1975) and in part due to the fact that a greater proportion of the population will be reaching maturity and producing eggs prior to the onset of the summer rains. The summer rains are an important limiting factor to the desert grassland population in that most of the adults are drowned in the early part of July (Riechert, 1974a). Egg production must therefore be completed prior to the onset of the rains. (The egg sacs are impermeable to water.) Data supporting this hypothesis are available from 1977-1978 and 1971-1972. The 35% floater estimate obtained for 1978 represented a relatively high period of competition pressure following a mild 1977, whereas an estimate of 5% floaters obtained for the same population in 1972 reflects a drought experienced in 1971. From Figure 8.3 we can surmise that the small proportion of floaters observed during this period reflects a lower population size, which in turn resulted from the fact that only a limited number of individuals from the previous generation contributed offspring (those occupying excellent sites).

The Nature of the Interaction: Game Playing

Analysis of the pathways by which interactions over territories progress indicates that *A. aperta* appears to be following many of the predictions of game theory. We can define a game as a means of reaching a decision in a conflict situation in which a win by one individual means a loss to its opponent or opponents. The theory was first developed by Von Neumann and Morgenstern in 1944 to explain human behavior in conflict situations. It has since frequently been used by companies in their bargaining practices and also by the military. In the 1960's evolutionary biologists first applied game theory to biology, and more recently several workers have developed the theory as it might apply to the behavior of animals in competitive situations. Most notable among these workers are John Maynard Smith and G. A. Parker, who have provided the impetus for empirical studies and further theoretical contributions.

Before developing my arguments on spider agonistic behavior as an example of optimization strategy, it should be noted that the findings presented here were obtained from experimentally induced encounters over actual territories in two study areas: desert grassland, south-central New Mexico; and desert riparian, southeastern Arizona. The behavior exhibited in these experimental encounters, however, is not significantly different from that recorded for 33 natural interactions over territories ($P < 0.70$): the overall similarity in behavior between the two data sets is 90% (from the Coefficient of Similarity; Curtis, 1959).

Assumptions

Gaming is primarily a problem in optimization, with individuals attempting to minimize costs while maximizing gains. It has associated with it four basic assumptions: (1) the objective is known to all players; (2) the rules and procedures are set; (3) a terminal point in play exists; and (4) each player receives a payoff at the end, + or −. Each of these assumptions will be considered separately in the following discussion.

CONTEST OBJECTIVE The first assumption concerns the contest objective—specifically, that both players are cognizant of what the contest is over. The contest objective of conspecific *A. aperta* is to obtain or maintain possession of a given territory. Two observations made in connection with these encounters indicate that *A. aperta* are mutually aware of the objective of their dispute. First, intruders released at the edge of an occupied web did not necessarily encroach upon that web. (They did so 86% of the time.) Second, the winners of the disputes always remained on the webs, exhibited control of the funnel, and immediately engaged in activities associated with ownership such as adding silk to the web and prey capture. This activity often occurred in the presence of the losing individual before its final retreat from the vicinity, and seemed to function in signaling ownership to the losing spider (Riechert, 1978b).

RULES The procedural assumption essentially maintains that communication is occurring and that the game assumes one of several pathways following set rules. We can assume that the rules are determined by natural selection and that the pathways, then, are governed by such contexts as the relative size of the two opponents, the value of the disputed resource, the thermal conditions present on the sheet at the time of the interaction, etc. The territorial disputes under discussion here follow one of two major pathways, with withdrawal from the contest possible at any point.

Figure 8.4 is a kinematic representation of transition probabilities between the major functional categories of behavior exhibited in the "Graduated Risk" game of Maynard Smith and Parker (1976) involving spiders of equal size in disputes over excellent territories.

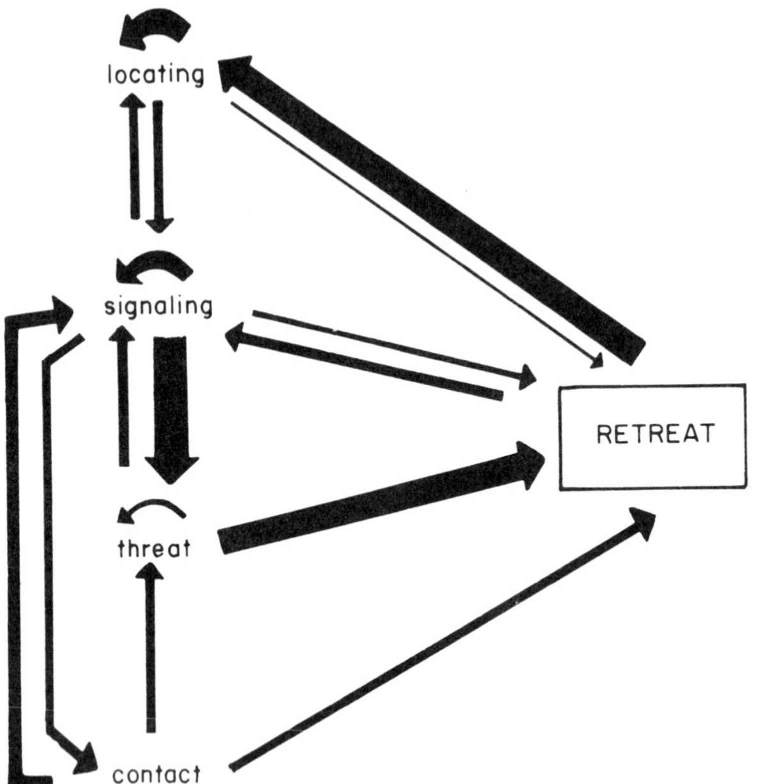

FIGURE 8.4. Transition probabilities between successively higher cost behavior patterns (locating → contact) for territorial disputes between evenly matched *Agelenopsis*, desert grassland habitat. Width of arrow represents probability of transition indicated.

All territorial disputes between *A. aperta* are initiated with subtle movements made by one or both contestants at opposite ends of the web-sheet. This group of behavior patterns has been termed "locating" behavior since such acts as "adjust position," "turn," "palpate web," and "spread legs" orient the opponents toward one another as well as serve an assessment function to be discussed later in this chapter (Riechert, 1978b). It thus is largely an information-gathering stage in the contest.

In the Graduated Risk Contest, if one of the individuals does not withdraw from the territory following locating, there is usually a slow escalation of the contest through "signaling," "threat," and finally "contact" behavior. Following an approach by one or more of the opponents, the contestants engage in lengthy exchanges of vibratory and visual displays. This behavior is termed "signaling," indicating its possible role in communication. Numerous views exist as to the purpose of signaling. For instance, it may be a a bluffing phase in which individuals try to increase their apparent size through such behavior patterns as "rear-up," "wave legs," "flex," and "stilt." It also may communicate the intended degree to which each individual will escalate to warlike behavior or persist in a contest. Maynard Smith (1974) feels such displays should not provide cues as to the degree to which contestants intend to persist, since it is difficult to see the selective advantage of conveying such information. It follows, then, that there should be no differences in the intensity of signaling between eventual winning and losing spiders. Caryl (1979) tested this prediction using data obtained from the literature. He found that information pertaining to the probability of attack was not transferred during agonistic displays but that information concerning potential withdrawal might be exchanged.

My findings with signaling in *A. aperta* appear to contradict theory to a greater extent. For instance, there are distinct differences in the patterns of signaling behavior exhibited by eventual winning and losing *A. aperta*. Winning spiders exhibit significantly lower stereotypy (greater variability) in their displays toward opponents than do losers (Riechert, 1978b). I originally suggested that the use of this unpredictable behavior is analogous to the "protean displays" exhibited by prey in the avoidance of predators (Humphries and Driver, 1967, 1970). By using unsystematic behavior chains, the contestant might confuse its opponent, putting it simultaneously into conflicting states. In a conservative system, such confusion should lead to the withdrawal of the opponent. The classical explanation for less stereotypy, however, follows the reasoning that the better communicator is the winner: long chains of diverse behavior patterns reflect the accurate transmittal of information by an individual concerning its ability to win an escalated contest. This is the Type II Redundancy of information theory (Gatlin, 1972). Dawkins and Krebs have recently (1978) contributed a new function to communication—manipulation or persuasion. This view is based on the observation that redundancy and repetition are highly effective advertising techniques in our world where the goal is to persuade the consumer to run out and purchase the given product or service. The data are not available to test among these alternatives at this time, though the evidence definitely favors the view that communication is occurring during signaling in these territorial disputes. Its use for bluff, persuasion, manipulation, conveying intention or motivational state, or as tactic to create confusion may, in fact, vary with the specific role of the contestant as intruder or owner, smaller contestant or larger, and so on. Note that, despite the fact that Caryl concerned himself with escalated contests, Maynard Smith restricted his arguments concerning the lack of information transfer to contests in which escalation is impossible. The high potential for injury in the disputes discussed here may favor accurate information transfer and the consequent withdrawal of the "less motivated" spider. At any rate, if neither of the contestants does withdraw following signaling, escalation to higher cost behavior patterns (i.e., "threat" and "contact" behavior) usually occurs.

An alternate pathway entails the use of threat and contact behavior immediately after information is exchanged through locating behavior (the "Hawk-Dove" game of Maynard Smith, 1974). Thus the lengthy signaling is bypassed and escalation to potentially damaging behavior is immediate (Figure 8.5). This pathway is often observed in contests involving inequality in the weights of the two contestants, where injury to the larger individual is highly improbable and activity on the web sheet in a stressful thermal environment is energetically costly.

We can conclude from this information that the pathways exhibited in these disputes are non-random and can be assigned probability functions for given contexts. Natural selection appears to have favored the development of some set rules for use in these territorial contests, and thus Assumption 2 of game theory is met.

TERMINATION OF PLAY Game theory also assumes that a terminal point in play exists. In these territorial disputes play is ended when one of the contestants leaves the territory. This may occur within a few seconds or may take as long as 20 hours after the initiation of the bout.

Contest length for *A. aperta* is to some extent dependent on context. Maynard Smith (1974) predicts that it will follow a negative binomial distribution based on the function:

$$p(x) = \frac{1}{V}e^{-x/v}$$

where V is the utility or value of the contested resource, x is the cost of displaying, and $p\ (x)$ is persistence time. Figure 8.6 repre-

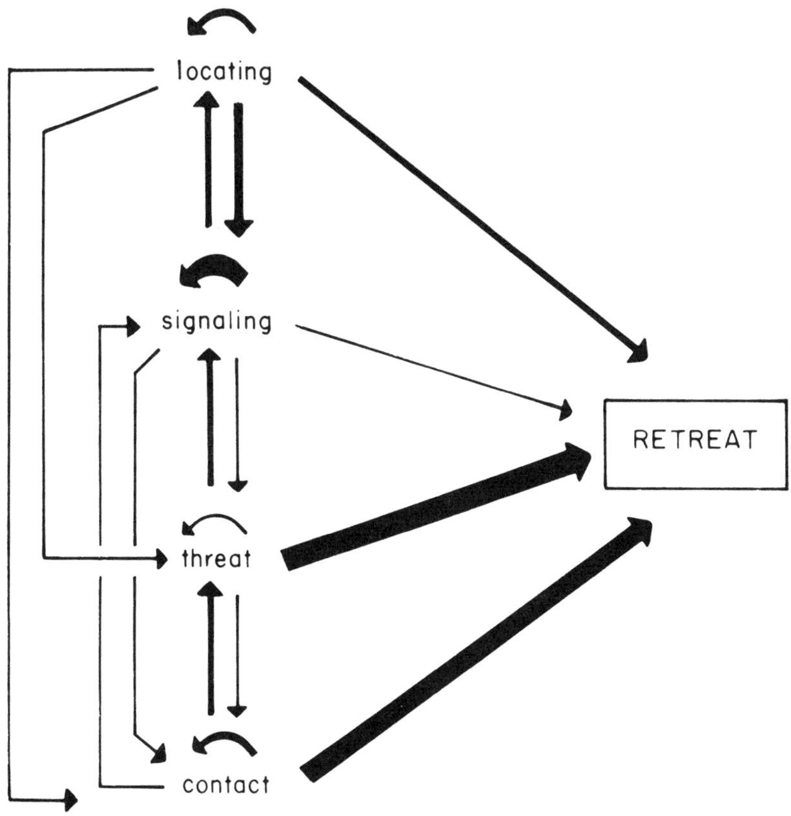

FIGURE 8.5. Transition probabilities for territorial disputes of *Agelenopsis* involving a weight asymmetry. Details in Figure 8.4.

sents the frequency distribution of contest times in the territorial disputes of *A. aperta*. Three groupings exist, the latter two of which appear to follow the negative binomial distribution. The first group represents an extremely short time interval between encroachment by a much smaller intruder and its retreat. Intermediate times reflect contests between more equal-sized spiders over territories of varying quality with persistence presumably increasing with quality of the site. The final category represents contests over excellent quality sites in which the owner is smaller than the intruder. It reflects the high degree to which owners will persist in a contest

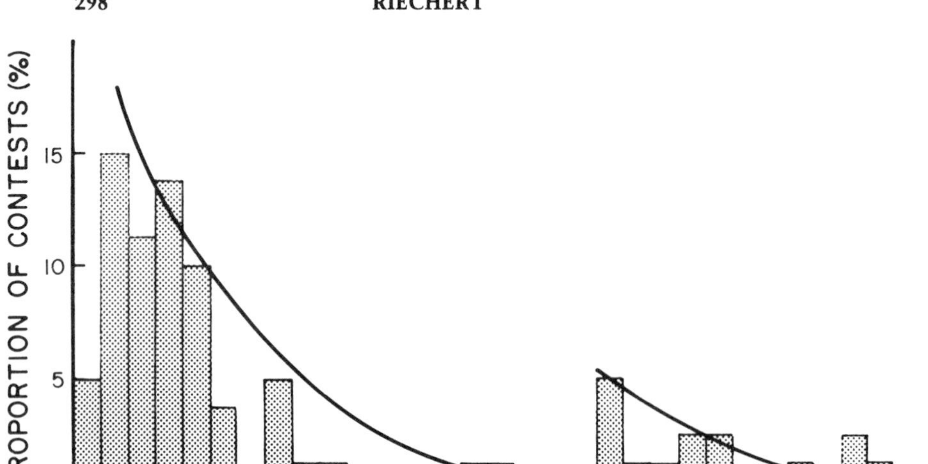

FIGURE 8.6. Negative binomial distribution fitted to the frequency distribution of persistence times for territorial disputes of *Agelenopsis aperta* in different contexts in desert grassland habitat. Pause time not included.

over a valuable site despite the fact that their probability of winning might be relatively low.

The median duration of a dispute occurring over a territory in the desert grassland habitat is 1,560 sec (26 minutes). The major portion of the dispute time, however, is spent motionless (98.44 ± 0.18%; Figure 8.6), with chains of action patterns occurring at irregular intervals. One of my students once remarked, after spending many hours in the desert heat waiting for two spiders to terminate an interaction, "Spiders are tremendously patient creatures!" This phenomenon does not appear to be unique to spider agonistic interactions but may be exceptionally prominent here. Perhaps spiders perform the "freeze" behavior for much the same reason that other groups exhibit displacement behavior (i.e., due to conflicting drives). It may also, however, represent a strategy in which each individual is waiting for the other to provide some additional information as to its intentions, in the sense that a defensive position might provide more options than an offensive one. If this is the case, we would expect pauses in activity to be greater in the more closely matched contests fitting the Graduated Risk category than in the Hawk-Dove contests involving a definite size bias. In the latter case, the role of the two individuals is more predictable, and thus the intention of an individual is clear to its opponent. The frequency of pauses of greater than 20 seconds is significantly higher for the Graduated Risk contests ($\overline{X}$ = 10.9 ± 2.4) than for Hawk-Dove contests ($\overline{X}$ = 2.95 ± 0.69; $P < 0.001$: Mann-Whitney Test), supporting the view that "freezing behavior" might represent a strategic ploy.

Within a dispute there may be a series of "bouts," each being initiated by an approach and ending with a retreat by one of the individuals to at least the web edge but usually off the web. As many as thirteen bouts have been observed in one contest, and bout number lends itself to prediction just as does contest pathway, if certain factors such as the relative size of the contestants and value of the resource are known.

PAYOFFS At the end of a contest each player receives a payoff in fitness. For the winner of a territorial dispute, the payoff will be the value of the resource minus the cost of the contest to the individual. Resource value might be proximate (an immediate payoff) or ultimate (a future payoff). For *A. aperta* the proximate value of winning a territory will be the rate of prey it receives per day through occupation of the site. Its ultimate payoff then is the reproductive potential it achieves as a result of the level of prey it has consumed over its time of occupation of the site. Note that optimization does not predict that individuals will engage in a dispute over a given territory if the benefit to be gained from a win is less than the cost necessary to achieve it, and also does not predict that the probability of withdrawal from a contest increases the closer cost comes to benefit (Parker, 1974).

The loser receives a negative payoff to fitness. This includes the weight loss associated with time spent without a regular food supply, energy expended in the search for a new territory, and the cost of this and possibly additional territorial disputes and/or the construction of a web at an unoccupied site. The ultimate negative payoff would be either mortality suffered as a result of injury incurred in the dispute itself or during the period following the dispute and prior to successful occupation of some other territory. During this time individuals are undoubtedly placed under food and predation pressures.

Average payoffs can be expressed in terms of daily weight gains and losses to our contestants (Figure 8.7). Thus, the winning spider in a contest over a territory in the desert grassland habitat in New Mexico gains an average of 3.32 ± 0.56 mg/day while the losing individual receives a larger negative payoff averaging −8.57 ± 1.76 mg/day during the period it lacks a territory (largely water loss).

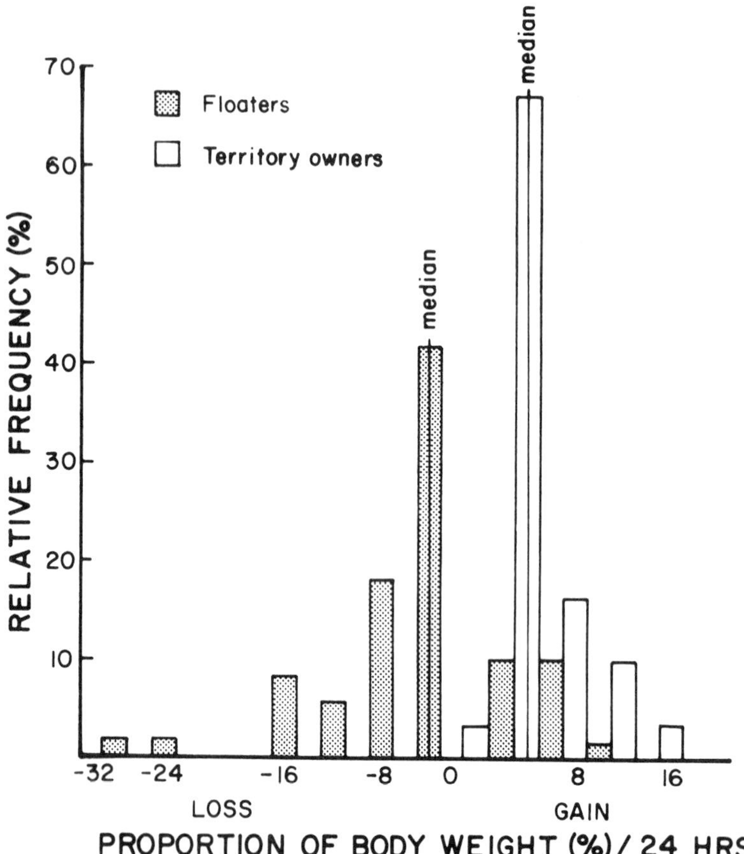

FIGURE 8.7. Frequency distribution of rates of weight change associated with floating versus territory ownership in desert grassland habitat of *Agelenopsis*. Data collected from multiple captures of marked individuals during 1978 and 1979 field seasons.

Few individuals actually suffer mortality as a direct result of injuries received during the disputes (less than 1%). Mortality resulting from indirect causes such as weight loss and predation might be higher, but it is difficult to estimate. Perhaps the most important aspect of fitness here is loss in reproductive potential—a parameter which is directly related to the amount of food an individual consumes and thus to its body weight (Riechert and Tracy, 1975). Weight may be equally important to male spiders in terms of the energy needed in searching for mates. It is also correlated with male mating success in that female *A. aperta* are less receptive to smaller males than to larger ones (Riechert, unpubl. data).

Climate imposes an additional constraint on the reproductive success of members of the desert grassland population. This study area is located in the Chihuahuan desert and as such receives heavy summer rains which begin in early July. Much of the grassland adjacent to the lava bed is covered by floods following these storms, and the adult spider population is virtually eliminated at this time (Riechert, 1974a). Those individuals maximizing their food intake through occupation of excellent territories mature and lay their eggs prior to the onset of the rains. If flooding does occur within a specific area, the egg sacs, being impermeable to water, survive the flooding; and only offspring contributed by these exceptionally fit parents are represented in the subsequent generation. Whereas in many spider populations a lower rate of weight gain results merely in slower maturation and delayed egg production, in this population it may result in no egg production at all.

Context: Assessment Strategies

Basic to optimization is the idea that individuals will select a behavioral strategy appropriate to a specific context. While Maynard Smith has emphasized the evolutionary stability of specific strategies in his theoretical work, Parker (1974) contributed the idea of contest asymmetries and the need for individuals to make assessments of them. Included in the possible asymmetries are differences in the probability of winning and in the value of the disputed resource. Each has associated with it a number of behavioral alternatives consisting of various persistence and escalation decisions—specific local strategies being favored for certain contexts. Maynard Smith has since termed individuals which adjust their agonistic behavior to specific roles as "Assessor Players" (Hammerstein, 1981). He uses *Agelenopsis* as exemplary of this type of strategy.

RHP ASSESSMENT One important asymmetry occurs when the contestants are equally matched. In these cases the contest outcome is biased toward the better competitor, which is frequently the largest opponent. The probability associated with winning against a specific opponent is called an individual's Relative Holding Power (RHP). If RHP disparity does exist within a contest, the preferred local strategy should differ for each contestant.

The outcome of *Agelenopsis* disputes over territories is significantly biased by the relative weights of the two contestants, with the larger spider winning 91.2% of the time ($P < 0.001$). It follows that if behavior is energy-limited, competing *A. aperta* should be making strategy decisions based on RHP assessment. The "locating" behavior exhibited at the initiation of a dispute appears to serve in RHP assessment in that such behavior patterns as "turn" or "adjust position" can provide weight information through tension changes in the web. These movements, for instance, led 30% of the time to the immediate and final retreat of much smaller intruding spiders (RHP difference > 30%) (Riechert, 1978b). That weight is being assessed through these behavior patterns was tested by introducing artificially weighted spiders to occupied territories. The weights of the intruders were effectively doubled while adding little to their apparent visual size by gluing a flattened lead shot to the dorsum of the abdomen. Table 8.2 shows that territory owners responded to the smaller spiders as if they had much higher RHP's.

TABLE 8.2. Probability of initial territory owner retreat for natural vs. artificially weighted intruders in *Agelenopsis aperta*.

		Natural Weight	*Artificially Weighted*
Intruder Weight			
Advantage	$\geq$ 10%	.81	.83
Difference	$\leq$ 10%	.22	—
Disadvantage	$\geq$ 10%	.13	—

As predicted, spiders utilize this assessment in determining the local strategy that each exhibits for the remainder of the bout (Table 8.3). Maynard Smith and Price (1973) categorized contestants as belonging to one of five groups: Hawk, Mouse, Bully, Retaliator, and Prober-Retaliator. An individual that exhibits the "Hawk" pattern is one that escalates to fighting behavior immediately following RHP assessment and continues the warlike behavior throughout the remainder of the contest. A "Mouse," on the other hand, is an individual who tends to retreat or withdraw regardless of the behavior exhibited by its opponent. The "Bully" category is assigned to those individuals who escalate to contact behavior initially but drop down to more conventional behavior if the opponent does not withdraw. "Retaliator" and "Prober-Retaliator" are similar local strategies, though the Prober tends to initiate escalation in a bout whereas the Retaliator selects its behavior largely in response to that exhibited by its opponent.

TABLE 8.3. Relative frequencies of contest strategies associated with size and status inequalities between conspecific *Agelenopsis aperta* from the desert grassland population.

	Retaliator	*Prober-Retaliator*	*Hawk*	*Mouse*
Close Size				
Owner	100%			
Intruder	87%	13%		
Owner Larger				
Owner	48%		52%	
Intruder	69%			31%
Intruder Larger				
Owner	100%			
Intruder	62%		38%	

The predominant local strategy exhibited by *A. aperta* in these disputes appears to be that of a Retaliator (Table 8.3), though RHP differences and status as territory owner or intruder do lead to some variability. Intruders, for instance, tend to initiate escalation in equal RHP contests and hence may exhibit the Prober-Retaliator pattern which is primarily offensive in nature. On the other hand, the Hawk and Mouse patterns are observed in contests involving weight inequalities. For instance, the larger contestants will often engage in warlike behavior immediately following RHP assessment. The Hawk appears to be a favored strategy under these circumstances since it greatly shortens interaction time in an often hot and stressful environment on the web-sheet, while the size bias is such that injury to the larger contestant is highly improbable. The favored local strategy in response to a Hawk is that of a Mouse (i.e., retreat or de-escalate) since the probability of winning is low and the potential for injury high. A much smaller territory intruder usually exhibits the Mouse in response to a play of Hawk by its opponent, whereas a smaller territory owner responds with the Retaliator strategy (Table 8.3). The apparent contradiction of economics in the case of the smaller territory owner is explained by two observations: (1) all of the data presented in Table 8.3 represent contests over excellent sites; and (2) by exhibiting the Retaliator pattern, territory owners were able to sway bouts in their favor 90% of the time when the weight disadvantage was less than 10% (Riechert, 1978b). We can hypothesize, then, that the high value of this resource warrants the risk and energy expenditure associated with a local strategy that has a greater probability of success than a less energy-consuming losing one.

RESOURCE ASSESSMENT Game theory predicts that by utilizing RHP assessment and the local strategy appropriate to the result of the assessment, the contest cost and potential for injury will be minimized. Actual escalation to contact and potentially damaging behavior thus will rarely occur. In fact, Parker (1974) feels that warlike behavior will only occur in cases where either RHP assessment is poor or where the value of the contested resource is extremely high. While competing for high quality territories, however, *Agelenopsis* frequently escalates to potentially damaging behavior. In part this deviation from theory reflects the favored use of the Hawk under the circumstances discussed earlier. But even closely matched contestants sometimes engage in warlike behavior (Figure 8.4). Since RHP assessment appears to be quite accurate, one must consider the possibility that *Agelenopsis* might be making resource value assessments in which clues related to short-term gains in fitness are monitored and adjustments in contest behavior made accordingly. This hypothesis was tested through comparisons of contest cost with site value, wherein "value" consisted of the absolute quality of the territory in terms of proximate and ultimate benefits (described earlier).

Territory quality. If we compare contest cost and resource quality in contests involving equally matched competitors from the desert grassland population in New Mexico, we find that a logarithmic relationship exists with proportionately larger energy expenditure required for a small increase in site quality at the high end of the scale (Table 8.4). This relationship is significant for total cost ($P < 0.0005$) and for the individual cost to the territory owner ($P < 0.005$) (Cox and Stuart trend test).

We can conclude that a resource assessment is being made and utilized, but by whom? From these results it is difficult to determine whether each individual had made its own assessment or whether the territory owner had made the assessment and the intruder was merely cuing in on its opponent's behavior. A group of induced encounters involving two intruders and no original site owner was used to distinguish between the two alternatives. The relationship between web-site quality and contest cost in this group of samples ($N = 22$) was not found to be significant ($P < 0.50$). Thus the intruding spider must be relying to some extent on the information provided by the owner in making its escalation and persistence decisions.

TABLE 8.4. Relationship between contest cost and territory quality for desert grassland population of *Agelenopsis aperta* (Riechert, 1979).

	Contest Cost	Benefits of Ownership: Proximate	Benefits of Ownership: Ultimate
Site Quality	(Mean and Standard Error in Joules)	(Predicted Dry Weight of prey available/day in Joules)	(Predicted Reproductive Potential in mg eggs produced)
Poor	$6.0 \times 10^{-6} \pm 1.6 \times 10^{-6}$	11.2	70.4
Average	$4.3 \times 10^{-5} \pm 5.5 \times 10^{-6}$	55.6	417.0
Excellent	$1.4 \times 10^{-4} \pm 2.7 \times 10^{-5}$	84.5	653.0

The fact that the territory owner is responsible for making the quality assessment poses a problem relative to the termination of a contest. In contests where an owner bias exists, it is commonly assumed that the intruder or losing spider determines the length of time a contest will continue. Does the relationship existing between contest cost and site quality imply that the owner, as the Assessor, determines when to end the bout? The findings are in actuality compatible with theory in that to a large extent contest cost is related to the degree of escalation achieved in a contest rather than to persistence. An explanation is required.

As I discussed earlier, the intruder tends to be offensive in its behavior. As such it initiates escalation during the signaling phase of the contest. The owner can determine the subsequent level of interaction by responding to this initial escalation in one of three ways: (1) if the quality of the territory is high, it should retaliate by meeting threat behavior with threat and contact; (2) if the territory is of average quality, the owner might respond to a threat with signaling, thereby bringing the contest back down to a conventional level; or (3) if the site is of very poor quality, it might even respond with a retreat and avoid any chance of injury. If this is indeed the mechanism by which contest cost is tuned to site quality, the losing spider, regardless of its status as owner or intruder, can still determine contest length.

The actual cues used by the owner in assessing the extent to which it is willing to escalate are as yet undetermined, though hunger levels have been shown to be utilized by hermit crabs in their persistence decisions in similar contests (Hazlett and Estabrook, 1974). Since spiders are frequently subjected to uncertain prey availability and often exist under food limitation (Anderson, 1974),

I do not expect that hunger is the cue that *A. aperta* is using to judge the quality of the territory it occupies. It may, however, use cues similar to those used in locating the site initially (i.e., the presence of appropriate vibratory stimuli from flying prey, the availability of structural features necessary for web construction, and the presence of a favorable thermal environment) (Riechert, 1976).

It is interesting that silk investment in the webtrap itself does not appear to be included in the escalation decision, despite the fact that there may be greater payoffs associated with a larger and stronger web-sheet. Since *Agelenopsis* adds silk daily to its web, and web size and thickness affect prey capture success, there should be selection pressure for the inclusion of web-quality information in contest decisions. The data presented in Figure 8.8, however, show that there is an inverse relationship between web size and territory quality. This appears to be related to the fact that hungry spiders that have enough biomass to expend energy in web constructions will have the largest webs. Satiated spiders do not need to increase their capture rate and, by the Principle of Stringency (Wilson, 1975), will not make such wasted energy and time investments. In the experiments completed for Figure 8.8, 10% of the satiated spiders even stopped building the sheet portion of their webs and merely accepted prey offered to them with a forceps at their funnel entrances. At the other end of the continuum are the spiders at poor sites, who apparently cannot afford the energy expenditure requisite to increasing the size of their catching area.

There is another reason why web quality might not be used in persistence decisions. Since the sheet can be readily destroyed by wind, rain, and animals, its significance relative to the actual potential for production of that site is negligible.

Territory availability. Game theory predicts that the level of escalation reached in a contest over a disputed resource will reflect the importance of the resource to both contestants. In addition to site quality, therefore, we must consider site availability, since ultimately a resource is less valuable if it is in ready supply than if it is in limited supply.

The availability of territories to the desert grassland population in New Mexico was compared to that occupying a desert riparian habitat in the Chiricahua Mountains of southeastern Arizona (Riechert, 1979). The spiders in the riparian habitat are afforded shade by a tree canopy and also enjoy about twice the prey availability realized by the desert grassland population (Riechert, 1978a). Prey are abundant here in part because of the ameliorated thermal

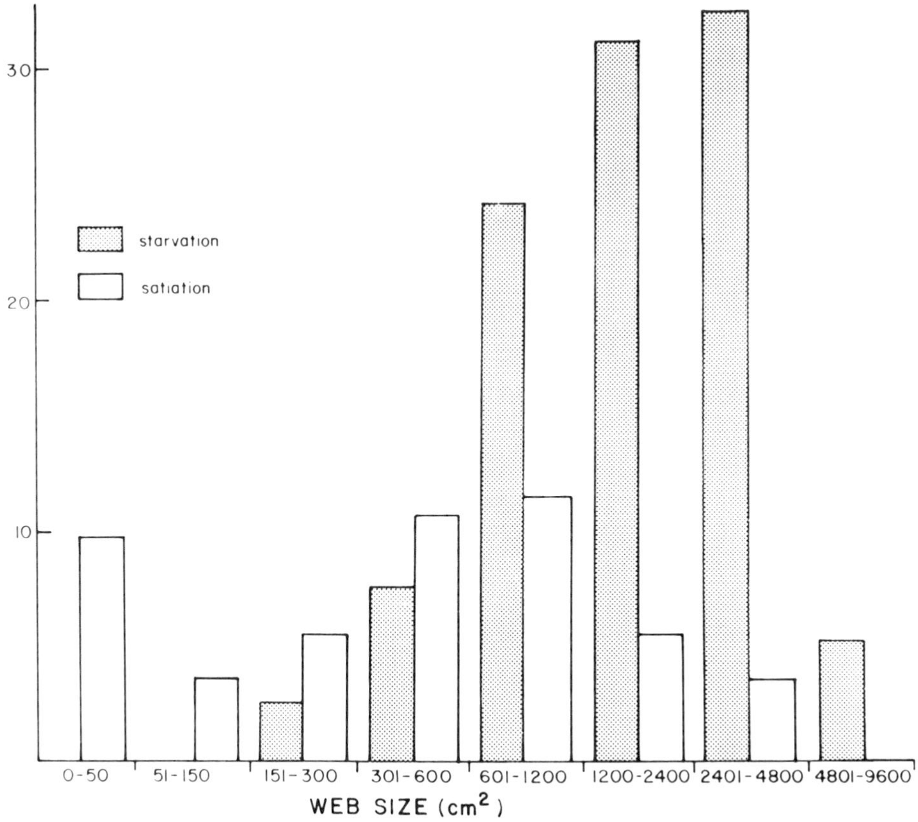

FIGURE 8.8. Frequency distribution of web sizes for *Agelenopsis aperta* encountering high versus low numbers of prey.

environment created by the trees and in part because of the presence of a permanent stream which traverses the habitat. The relative saturation of the desert grassland and desert riparian study areas in 1978 is presented in Table 8.5. Though the riparian habitat supports a larger density of spiders (110 adults/1000 m^2)than does the desert grassland habitat (42 adults/1000 m^2), spiders are far closer to saturation or carrying capacity in terms of site availability in the desert grassland because there is less suitable habitat available.

The behavior of spiders in disputes over sites in the two habitats supports the principle of optimization in that the average total cost of a contest completed by members of the desert grassland population is over two times greater than that of a contest in the desert

TABLE 8.5. Between-population comparison of web-site availability to *Agelenopsis*, based on cover estimates of various microhabitat types and associated censuses of spider occupation (Riechert, 1979).

	Site Availability (%)		*Degree of Saturation (%)**	
	Grassland	*Riparian*	*Grassland*	*Riparian*
Site Quality				
Excellent	3.0	7.8	100.0	48.0
Average	8.4	83.1	86.3	11.8
Poor	88.6	9.1	2.1	6.1

* All values adjusted for median territory size of the population (Riechert, 1978a).

riparian habitat (Table 8.6). Since the means for the total number of acts and the number of different kinds of acts were not significantly different between the two populations, the different costs obtained appear to reflect the use of the higher cost contact behavior by the desert grassland population as opposed to greater variety and longer interactions. The variances of these two parameters do differ, however, with the desert grassland exhibiting longer tails in the distribution. This indicates that the range of payoffs is greater in this habitat. Estimates of spider reproductive success while occupying different quality sites in the riparian habitat indicate that, unlike the situation in the desert grassland study area (Table 8.4), site quality is not related to reproductive success, with individuals occupying the least favored sites receiving more prey and as a result being capable of expending more energy in egg production than spiders in other locations. (Poor site = 450.6 mg; average = 361.0 mg; excellent = 396.9 mg.) Web-site selection in this highly favorable habitat appears to be dependent simply on the availability of structural features required for web construction rather than on the thermal or prey abundance conditions that largely determine web-site quality in the more rigorous desert grassland study area.

There is little evidence to suggest that population differences in agonistic behavior corresponding to site availability reflect active assessment of site availability by individual spiders. Since the territory owner has been shown to make resource value assessments and maintains a given territory for extended periods, it is unlikely that this individual has access to site availability information. An alternative mechanism is selection pressure for increased aggression in populations under competition for a limited number of sites. Since the population sizes of *A. aperta* remain relatively con-

TABLE 8.6. Between-population comparison of *Agelenopsis* contest behavior.

	Desert Grassland		*Desert Riparian*				
	Mean	*Stand. Err.*	*Mean*	*Stand. Err.*	*Test Used*	*Significance*	*Conclusion*
Total cost (in Joules)	2.9×10^{-5}	7.4×10^{-6}	1.2×10^{-5}	2.8×10^{-6}	Mann Whitney	$P < 0.05$	Means differ
Number of acts	17.3	19.5	10.3	5.4	Siegel Tukey	$P < 0.05$	Variances differ
Number of different action patterns	8.4	5.6	6.6	2.3	Siegel Tukey	$P < 0.05$	Variances differ

stant from year to year as a result of territorial behavior (Riechert, 1978a), the evolution of a specific level of aggression within a population seems to be an appropriate mechanism by which competitive behavior may be adjusted to resource availability.

Evolutionarily Stable Strategies

Thus far I have limited my discussion to the concept of fitness costs and payoffs, which can be nicely equated to the economic utility terms of game theory wherein "utility" expresses in monetary terms the relative attractiveness of alternative choices. There is, however, an important extension of optimality that must be considered—the concept of the Evolutionarily Stable Strategy (ESS).

Maynard Smith is largely responsible for the development of this concept, especially as it pertains to animal behavior (and Price, 1973, 1974; and Parker, 1976; Parker, 1978). This theory overcomes one of the major criticisms of optimization—the fact that there is no single best solution to a problem. The ESS is a best solution as a result of the assumption that the strategy exhibited is dependent on what others in the population are doing (i.e., is frequency dependent). In other words, an ESS is a strategy such that if most members of a population adopt it, there is no other strategy that would give higher reproductive success.

Thus, the strategy (I) is an ESS if the expected payoff in fitness P of I played against itself is greater than that from any other strategy played against I:

$$P_I(I) > P_I(J),$$

where P is the payoff of the strategy in parentheses when played against the strategy in the subscript. There may also be a case wherein the strategy results in an equal payoff both when played against itself and when another strategy is played against it, as in:

$$P_I(I) = P_I(J).$$

Then the additional condition of:

$$P_J(I) > P_J(J)$$

must be met for I to represent an ESS (Maynard Smith, 1974).

Note that we are dealing here with a different sense of the term strategy from what we have used in previous sections. A strategy herein refers to the complete set of local strategies exhibited by an individual in the various roles in which it finds itself. Thus, since assessments are being made of resource quality and potential contest asymmetries, we would expect the ESS for the desert population of *A. aperta* to consist of a number of local strategies, each specific to a given context. As if this is not complicated enough, there is also the potential for a mixed strategy—in a given situation, an individual may play *A* with probability *p* and *B* with probability *q*, etc. Mixed strategies may represent either polymorphism within a population or flexibility within the repertoire of the individual.

Thus far ESS's have been identified for two behavioral systems, digger wasp nesting behavior and mating strategies in toads. Brockmann et al. (1979) found digger wasps to exhibit two strategies associated with procurement of nest—digging and joining. These behaviors persist in a population in stable equilibrium with behavioral flexibility as to which strategy an individual may exhibit at a given time. A mixed ESS has also been observed in male *Bufo bufo* by Davies (personal communication). In this case larger males compete for females in the center of a breeding pond through calling, while the poorer competitors (satellites) may wait at the edge of the pond and sneak matings with females approaching the calling males. In both cases the stability of a given strategy is dependent on what most of the other individuals in the population are doing.

It is apparent from what we have discussed thus far of *A. aperta*'s agonistic behavior that the search for an ESS for the desert grassland population is no simple matter. I have chosen to divide this large evolutionary game into sub- or situation games corresponding to a single contest situation. The subgame concept has been developed mathematically by Hammerstein (1981), and I have applied his mathematical construct to three contest strategies designated as Mouse, Retaliator, and Hawk (defined earlier). Strategy assignment was based on the behavior exhibited by a given individual following the assessment phase of the contest.

Table 8.7 shows the local strategies *A. aperta* should exhibit for each role it might find itself in when competing for optimal sites in the desert grassland habitat. Though the detailed methods by which these equilibria were derived will be discussed elsewhere, the results presented in Table 8.7 were determined from payoff matrices constructed by the following parameters associated with each strategy in a given contest: the probabilities of winning, receiving minor injury, and suffering mortality; the average energetic cost of the disputes; and the average value (utility) of the territories in dispute.

Apparent from the local strategies exhibited is the large number of factors involved in their determination. The set of strategies can

TABLE 8.7. Predicted ESS associated with situation games involving optimal territories for *A. aperta* from the desert grassland habitat in south-central New Mexico.

	Owner	*Intruder*
Equal weight	Retaliator 100%	Retaliator 100%
Unequal weight		
Owner Heavier	Hawk 50% Retaliator 50%	Mouse 57% Retaliator 43%
Intruder Heavier	Retaliator 100%	Hawk 50% Retaliator 50%

perhaps best be understood by remembering that (1) *Agelenopsis* makes an initial assessment of its weight relative to that of its opponent, (2) that territory owners also have some knowledge of the quality of the territory they occupy, and (3) that the strategy exhibited by an individual is influenced by what its opponent does. In other words, strategy choice is non-random.

The role of RHP assessment in influencing contest behavior is seen in the predicted 57% use of the Mouse strategy by smaller intruders who have little expectation of winning the dispute. The predicted use of the Retaliator strategy by equally matched opponents also reflects the role of RHP assessment in strategy determination. The high payoffs associated with the ownership of high quality territories in the desert grassland habitat and the information the territory owner has about site quality should override the RHP decision rule for the smaller owner of such a territory. By exhibiting only the Retaliator strategy as opposed to the Mouse, the owner has a 9-10% probability of retaining possession of its territory. The benefit from ownership of an optimal site thus outweighs the higher cost of displaying.

If an equal-sized or larger-sized opponent finds a Retaliator strategy being used against it, the Hawk strategy would appear to be the appropriate counter, as in the scissors, paper, rock game commonly played by children. Just as scissors cuts paper, Hawk beats Retaliator. In actuality Hawk is not the preferred strategy in this context because the Retaliator frequently will escalate against a Hawk with consequent injury suffered by one or both contestants. Thus the appropriate response to displaying in this RHP context is displaying (Retaliator vs. Retaliator).

Can we determine whether this set of local strategies represents an ESS for this population? Let us limit our question to the behavior of our territorial owners since they appear to manipulate to a large extent the behavior exhibited by their opponents. *Agelenopsis* is definitely optimizing in these contests in that (1) the local strategy exhibited represents only that level of escalation required to win territorial disputes over quality sites, and (2) withdrawal (the Mouse strategy) is predicted if the probability of receiving serious injury is high. The local strategies exhibited by territory owners are also stable to invasion by other strategies, given the continued shortage of available territories to this population. But what about annual fluctuations in desert environments? Given a good year, would not withdrawal be the optimal strategy for the smaller opponent since it can readily find unoccupied sites or possibly even do well as a floater? The question then becomes whether the ESS is behavioral flexibility itself (the ability of an individual to use prior experience in making its strategy decisions) or consists of a series of local strategies each programmed for use in specific contexts.

One can expound at great length on the limits of the invertebrate nervous system, on the fact that other less complex behaviors linked to competition (e.g., territorial behavior) appear to have a strong genetic component, and on the physical adaptations that have evolved to meet the needs of individuals in variable environments. These considerations weigh heavily toward accepting this set of optimal local strategies as an Evolutionarily Stable Strategy. The conservative view, though, is that while the data are consistent with the existence of an ESS, they may be explained in this case by optimality criteria alone. Verification of the ESS for this competitive system, then, requires that a comparison of the contest behavior exhibited under different levels of competition (i.e., during highs and lows in territory and prey availability) be made.

Conclusions

In this chapter I have attempted to demonstrate the extent to which communication might be used to settle conflicts between animals. The prevalence of communication is related to the observation that animals are optimizers, rather than "satisficers." In "satisficing," individuals are concerned merely with wins versus losses (Simon, 1957). Optimizing is a more finely tuned process involving the adjustment of costs to benefits (Parker, 1974). Since *Agelenopsis* adjusts the level of escalation exhibited in these territorial disputes relative to the value of the contested resource and its behavior according to its probability of winning, natural selection must have favored the finer tuning of the optimization process over "satisfication" in its populations.

Spiders occupying a desert grassland habitat in south-central New Mexico exhibit a complex assessment system for adjusting the level of their interactions to their expected payoffs in fitness. I expect this finely tuned system to have evolved in response to a rigorous environment where intraspecific competition is a strong determinant of individual success.

This kind of complexity would be excessive for competitive interactions within populations occupying more favorable environments, where a simple convention which eliminates wasted energy or time expenditure should be exhibited. Thus, what Dawkins (1976) might consider a "paradoxical" asymmetry is frequently used to settle disputes between *A. aperta* inhabiting a desert riparian habitat in southeastern Arizona. The convention here is for the smaller territory owner to retreat out the back of the web funnel following RHP assessment. A domino effect often results with deposed territory owners displacing their neighbors who in turn encroach on other webs. That the intruder-wins convention is favored here is probably related to the abundance and equality of web-sites in this habitat as well as to the ability of floaters to realize weight gains. In other words, the energy and risk of injury required to dispute a territory in the riparian habitat is greater than is warranted by the expected payoff. This same effect was noted by Burgess (1976) in the Mexican spider *Oecobius civitas*, where individuals searching for retreats were observed to displace other individuals who then did the same to others. (No resistance was exhibited by the owners to this intrusion.)

Given this use of communication, it seems inappropriate to attach an altruistic label to it. Communication may have altruistic consequences in a limited number of social contexts, but it has its origins in the behavior of solitary animals and certainly represents an adaptation by which individuals may increase their own inclusive fitness. Why has it evolved? Individuals do not exist in a void, but must constantly interact with biotic and abiotic components of their environment. To do so efficiently requires information gathering and exchange. Communication is not basically cooperative in nature because interactions usually involve inequalities in both costs and payoffs. It must, therefore, be manipulative, persuading other individuals to act to the benefit of the signaler regardless of the consequences to themselves.

It is also time to reassess our classification of the spider as an asocial, highly cannibalistic beast. Although detailed studies of the agonistic behavior of most spiders have yet to be made, sufficient evidence exists in the form of qualitative observations to suggest that they are capable of recognition of conspecifics of both sexes and of using communication to partition resources. It is time to put our observations of cannibalism in the framework of optimality criteria by studying the contexts in which it occurs: (1) overcrowding where retreat following loss of a conventional contest is not possible, (2) where to lose a dispute means total loss in fitness regardless of whether mortality is suffered as a result of the conflict, and (3) where a poor competitor might misjudge its opponent's ability or intention to escalate and thereby exhibit a behavior pattern inappropriate to the context. If we look for these things, we might find that coercion in spiders is the exception rather than the rule.

ACKNOWLEDGMENTS

The author gratefully acknowledges support of the work reported in this chapter from the National Science Foundation, Grants DEB 80-02882 and DEB 74-23817.

Chapter 9 SOCIAL SPACING STRATEGIES IN SPIDERS

J. Wesley Burgess
North Carolina Mental Health Research
Anderson Hall
Dorothea Dix Hospital
Raleigh, North Carolina 27611

George W. Uetz
Department of Biological Sciences
University of Cincinnati
Cincinnati, Ohio 45221

Introduction

Spatial patterns can be contagious. That is, after one becomes aware of interindividual distances, orientation, and movement as group processes, many new faces of group behavior come into view. Not only spiders, in their daily struggles to eat and reproduce, but also a new litter of puppies, a monkey colony at the zoo, or school children in their playground become the source of exciting new examples of group spacing responses all around us. Spiders, puppies, monkeys, and children all have in common the need to secure space to live in; it is the overall species patterns of dividing available space that we will call spacing strategies, which form the subject of this chapter.

The group comprising the spiders, or Araneae, is quite large, containing an estimated 30,000 known species (Kaestner, 1969; Levi, Levi, Zim, 1968). We will have to restrict our examination to those few species whose spacing has been the subject of careful study, paying particular attention to the group-living spiders, in which spatial behaviors are quite striking. Since there has not previously been an overview of spatial strategies in spiders (or almost any other animal, for that matter), we will emphasize three of the most basic questions about spatial patterns, asking: (1) what are their forms, (2) what mechanisms are responsible for them, and (3) what ecological functions do they serve? By form we mean any pattern of a group's arrangement in its environment which can be discerned by quantitative studies of interindividual distances, dispersion, and density. In this approach, a well-chosen behavioral measure is likely to tell far more about the spatial behavior of a group than we can ever learn by subjective observations. What we call mechanisms are the same as short-term or proximal causes of the spatial pattern that is being measured. These may originate in visual, tactile, chemosensory, or vibratory cues produced by conspecifics and transmitted through the environment. Beyond the proximal causes are long-term factors which serve to maintain a spatial behavior in a species for many generations. Thus, most functions we find for a spatial pattern ultimately serve to ensure survival and reproduction of the individuals transmitting those genes which code for it. Spatial patterns may help individuals avoid predation in their environment or they may provide a competitive edge in securing resources like food, web-sites, or available mates. Our approach will be as follows: first we will organize different strategies into a few manageable categories, in order to review what is currently known about spider spacing. Then, techniques used for the quantification of spatial patterns will be examined, and several new studies will provide examples of methodology applied to test hypotheses about spider spacing. Our main intention is to explore the nature of social spacing strategies in this group of animals rather than the specific communicatory mechanisms (i.e., the proximate factors) used in establishing and maintaining the spacing. Further consideration of the proximate mechanisms mediating spacing in spiders—vibratory and chemical signals, agonistic behaviors associated with territorial defense—is provided elsewhere in this volume.

Spatial Patterns

If we were to try to imagine all possible ways animals in a group of any reasonable size might arrange themselves within their environmental space, we would soon give up in despair. Likewise when we observe spiders in a group it is hard to know what kinds of spatial patterns to look for. Some species live in groups of thousands of individuals which crawl about, constantly changing their positions; other species hardly move at all but maintain positions inside a confusing geometric array of tangled, three-dimensional webs. Fortunately for our studies, there are three simple categories of spacing that can be applied to spiders or any other animals and which will aid in our discrimination of spatial patterns.

One pattern seldom, if ever, found in nature is true randomness over large areas. Stars in the sky or pebbles on a beach may approach a random distribution, but living creatures like spiders constantly interact with each other and with their environment in ways which influence their positions in space. Yet our ability to define mathematically many dimensions of a species' spatial pattern and to test for statistical differences from randomness is a powerful tool for analyzing spatial patterns. In these cases we can define random in the context of a Poisson or normal probability distribution and develop from these confidence ranges for testing the distances, density, or orientation between animals in their groups. If a spatial measure shows significant departure from randomness, we can begin to look for evidence of one of two other alternatives: either clumped (also called contagious or underdispersed) or regular (called overdispersed) spacing. In actuality, most animals exhibit a combination of clumped and regular spacing patterns; describing such heterogeneous patterns in ways which remain behaviorally meaningful is still an important challenge.

For ease of organization, spatial strategies of spiders will here be treated in three categories, based on the way spider species characteristically live: solitarily and spaced regularly within their habitat; in web complexes; and in close aggregations. Solitary, regularly spaced species often defend the space around them and do not form groups; instead their individual webs are spread throughout their habitat or they wander alone, without building webs. In web complexes, on the other hand, spiders are found apart on their own individual webs, but these webs are joined together in modules, and may include a communal retreat area. In this sense, web complexes represent a compromise between close aggregations and regular spacing. Close aggregations are found where all animals peacefully share the same space, usually on communal, maternal, or juvenile webs. Close distances and frequent touching are found in groups of these spiders.

Solitary Spiders

Random Spacing

Most spiders are solitary, and exist in close proximity to their conspecifics only during early development and the mating season. Their distribution in space may be random, but more usually reflects the pattern of habitat resources. Cole (1946) found spiders to be randomly distributed under Cryptozoa boards in an Illinois forest. Oddly, spiders were the only group of arthropods exhibiting a random distribution. This finding may be suspect, however, due to the fact that Cole did not identify the spiders below the level of order—while other arthropods were identified down to family, genus, or species. Kuenzler (1958) has shown wolf spider dispersion to be random except in those cases where the environment was not uniform. It is probable that many solitary spiders show distributions that are related to the dispersion pattern of habitat or food resources. The forest-dwelling orb weaver, *Micrathena gracilis*, shows a random dispersion pattern horizontally, but an aggregated pattern vertically, presumably because habitat resources critical to this species are distributed randomly throughout the forest at set heights (Uetz, Derksen, and Biere, unpubl.).

Regularly Spaced Spiders

Regular dispersion is another spatial pattern found in solitary spider species, particularly when resources are limited. Regularity in nature has been measured in solitary female orb weavers (e.g., *Argiope trifasciata, Edricus crassicauda*), sheet-web builders (e.g., *Florinda coccinea, Frontinella pyramitela*), funnel-web spiders (e.g., *Agelenopsis aperta, A. longistylus, A. kastoni*), and burrowing wolf spiders (e.g., *Geolycosa wrighti*) (Riechert, this volume). Regular patterns can even be maintained in the absence of natural environmental stimuli (e.g., *Agelenopsis utahena*, Burgess, 1979a; *A. aperta*, Riechert et al., 1973; Riechert, 1974b, 1978a). In the wolf spiders, silk cues (Tietjen, 1977), agonistic interactions, and visual displays (Aspey, 1977a,b) all provide mechanisms to effect spacing at close quarters.

In more conventionally territorial web-builders, females typically defend their webs from intrusion, as in *Dictyna civica* (Billaudelle, 1957) and *Linyphia triangularis* (Rovner, 1968a), but may permit males to cohabit temporarily. *Agelenopsis aperta* has been shown by Riechert (1976) to defend both its web and the space surrounding it. Their pattern of dispersion is also affected by the fact that they actively select sites which offer protection and high prey availability (Riechert, 1976; Riechert and Tracy, 1975). Such a strategy has an important ecological function: an environmental model of *A. aperta* (Riechert and Tracy, 1975) suggests that spiders with good web sites can derive as much as thirteenfold reproductive increase over individuals living in less desirable sites. Regular spacing probably permits territorial spiders efficiently to partition available prey where they are low in abundance (Riechert, this volume). It may not be clear why the seemingly antisocial strategy of territoriality is included in a study of spider social strategies; yet it is obvious that any successful territorial lifestyle is ultimately dependent on communication. Consider a popular definition of territoriality as "the exclusive use of terrain" resulting from "overt defense or repulsion through advertisement" (Wilson, 1975, p. 262). There is even evidence from other animals that territorial neighbors can provide valuable social stimulation (Fisher, 1954). And in spiders, it is clear that territorial defense of space is not incompatible with group living, since many web-complex builders combine elements of both strategies.

Fortuitous Aggregations

There have been many observations of fortuitously aggregating spiders, which, although normally solitary, may occur in groups as the result of the environment. McCook (1889) has reported that *Metepeira labyrinthea*, a common solitary orb weaver, occasionally joins webs together in areas where microhabitat features permit unusually high densities. Honjo (1977) found *Dictyna follicola* (Dictynidae) occurring in clumped dispersion patterns in man-made structures where fluorescent lights drew large quantities of insect prey. In the same manner, aggregations of *Nuctenea sclopetaria* (Araneidae) occur on docks near water (McCook, 1889) or at Cincinnati's Riverfront Coliseum Sports Arena (Uetz and Allen, unpubl.). Superabundances of insect prey surrounding the lights of the Coliseum cause this nocturnal species to exhibit what has been called the "aggregative response" of spiders to increased prey density (Riechert, 1974b). As the summer season progresses, populations of *N. sclopetaria* build to unnaturally high densities ($100/m^3$) and unusually low interindividual distances. Webs are built daily, and often attached to each other. Over the summer, webs degenerate into what appears as a large communal mass of webbing occupied by hundreds of individuals. At dawn, spiders leave the webs for retreats in cracks and crevices of the building. Retreats are packed together, with spiders < 1 cm apart. Sometimes individuals build webs and abandon them, leaving a large amount of collapsed sticky thread in the tangled mass. Spiders capture prey, exhibit courtship behaviors, and mate on this communal web. Adult females defend a reduced individual space (or the periphery of their orb-web if they build one) on the communal webbing, while males wander about, presumably searching for mates. Spacing is maintained by vigorous plucking of silk lines, web shaking, and grappling with intruders. The behaviors exhibited by these fortuitously aggregating spiders are similar to those exhibited by colonial orb weavers like *Metabus gravidus* (Buskirk, 1975a) or *Metepeira spinipes* (Uetz, unpubl. data).

Colonies with Shared Space

Some spiders appear to have the best of communal and solitary living; they join their individual webs together to form web colonies, and include web areas where conspecifics may sometimes cohabit peacefully. For example, *Dictyna calcarata* and *Mallos trivittatus* live in web complexes containing 1,000-10,000 individual sheet-webs (Jackson, 1977d, 1978a). Females defend their webs against other females but allow males and spiderlings to use the central retreat areas and to feed on captured prey. Spiders are attracted to webs of their own species, apparently as the result of chemical cues (*ibid.*). Orb-weaving spiders also build web complexes. For example, *Uloborus republicanus* (Simon, 1891; see also Wilson, 1971), *U. mundior* (Struthsaker, 1969) and *Metepeira spinipes* (Burgess and Witt, 1976; Uetz and Burgess, 1979) build colonies by attaching orb-webs at their frame threads.

The webs of *Metepeira spinipes* (Figure 9.1) include a three-dimensional space web with a retreat area where spiders sit during the day. Mature males may cohabit with females in these retreats prior to mating. Spiders defend the space of their individual webs against intruders, and prey caught there are not shared. Coloniality in *M. spinipes* varies geographically over a gradient of habitats from severe (desert and high altitude sites) to intermediate (agricultural valleys with seasonal rainfall) to moderate (tropical sites). Social grouping tendencies range from solitary individuals and small groups (4-10) in severe habitats to huge aggregations of hundreds or even thousands of individuals in areas of benign climate and high insect activity. Spacing patterns vary over this gradient as well, with nearest-neighbor distance decreasing as the environment becomes more moderate and prey availability increases (Uetz, Kane, and Stratton, in prep.).

Colonies without Shared Space

Other spider species live in web colonies consisting of numerous individual webs without provision for cohabitation or communal retreat areas. *Metabus gravidus* (Buskirk, 1975a) build colonies of conventional orb-webs, while *Cyrtophora citricola* (Kullmann, 1968a) and *C. moluccensis* (Lubin, 1974) build horizontal orbs surrounded by a buffer zone of space threads. These three species actively defend their individual webs by utilizing agonistic behaviors such as leg-jerks and web-tensioning (Buskirk, 1975a, Lubin, 1974).

Buskirk (1975a) has shown that spacing inside colonies is an intimate part of social interactions. For example, individuals of *M. gravidus* usually maintain a distance of 16-22 cm from their nearest neighbors in orb-webs. However, females in their webs allow males to approach significantly closer than other females. Moreover, intruding spiders release different levels of agonistic responses depending upon their distance from the builder of the web. Despite the agonism which takes place on individual webs during the day, by evening *M. gravidus* take down their orbs and may pass the night close together in a sheltered spot (Buskirk, 1975a).

Another web-complex builder, *Oecobius civitas* (Figure 9.2) (Shear, 1970), employs a strategy that is radically different from web defense. Instead of exhibiting agonistic responses directed at intruders, *O. civitas* leave their webs immediately when disturbed and run away to a vacant web or crevice, or displace one of their neighbors (Burgess, 1976). This peculiar type of spatial non-defense is called a "paradoxical strategy" by Maynard Smith and Dawkins (Dawkins, 1976; Maynard Smith, 1976). Although it is the absolute reverse of conventional territoriality, Dawkins points out that such a strategy can be evolutionarily stable. Its non-combative nature may save energy and injury to individuals (compared with more violent territorial defense), but we suspect its utility may also be enhanced by the surplus of suitable crevices and extra webs which were observed in *O. civitas* web complexes. Clearly this unique example of an alternative strategy deserves further study as well as comparison with conventional spatial defense behavior.

Ecological Functions

Web complexes can protect colony members from predation in at least two ways. First, there is a geometric effect of clustering which can make predators' searching more difficult. In a mathematical model using fish schools as an example, Brock and Riffenburgh (1960) showed a decreasing probability of detection of individual prey by solitary predators as groups grow larger and cluster more tightly together. This effect would presumably hold for colonial spiders as well as other group-living animals.

Spiders may also gain protection from predators by securing a position on the inside of their web complex. Galton (1971), Williams (1964, 1966), and others have noted the protective advantages

of living inside a "selfish herd." These potential benefits were illustrated in a clever mathematical model by Hamilton (1971). For spiders, individuals located on the outside of a colony are most likely to be preyed upon first; thus other colony mates are safer from predatory attacks. Spiders also have a unique advantage not shared by aggregations of other animals: the complexities of tangled layers of silk provide an extra buffer to spiders with inner webs. These hypothetical survival benefits are testable, and it will be interesting to find out how some individuals are left with the more dangerous outer positions. Rypstra (1979) has shown that position within colonies of *Cyrtophora citricola* is important. Spiders

FIGURE 9.1. (Left) Several orb-webs are joined together by common support threads in this web complex of *Metepeira spinipes* spiders found in central Mexico. Above the orbs is a three-dimensional space web with retreat areas. A fine mist of water makes the thin silken threads visible, but distorts the natural geometry of the web. (Below) The web of a *Metepeira* spider contains three structural zones. The individual orb (O) is constructed by a single spider who catches food upon it. Silken threads join the hub of the orb to the retreat (R) where the spider waits for prey. Several round egg sacs are attached above the retreat. The retreat and egg sacs are suspended within the three-dimensional space web (S), which serves to join webs in colonial species. (Original woodcut from McCook, 1889.)

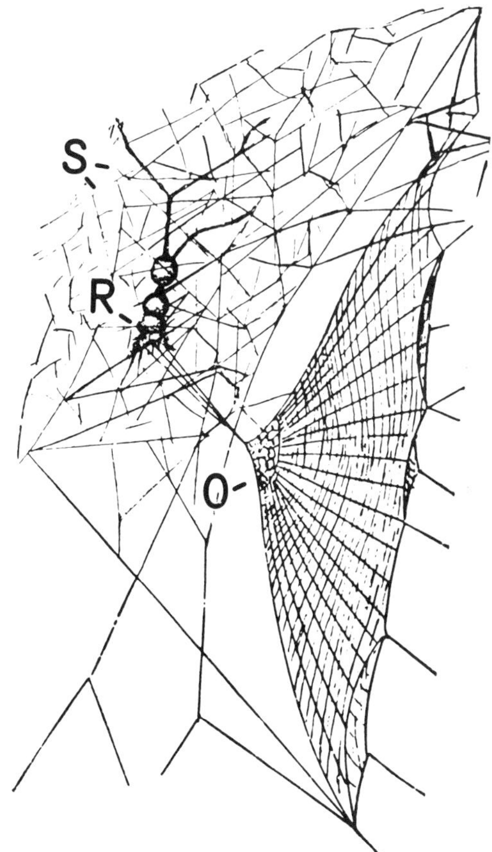

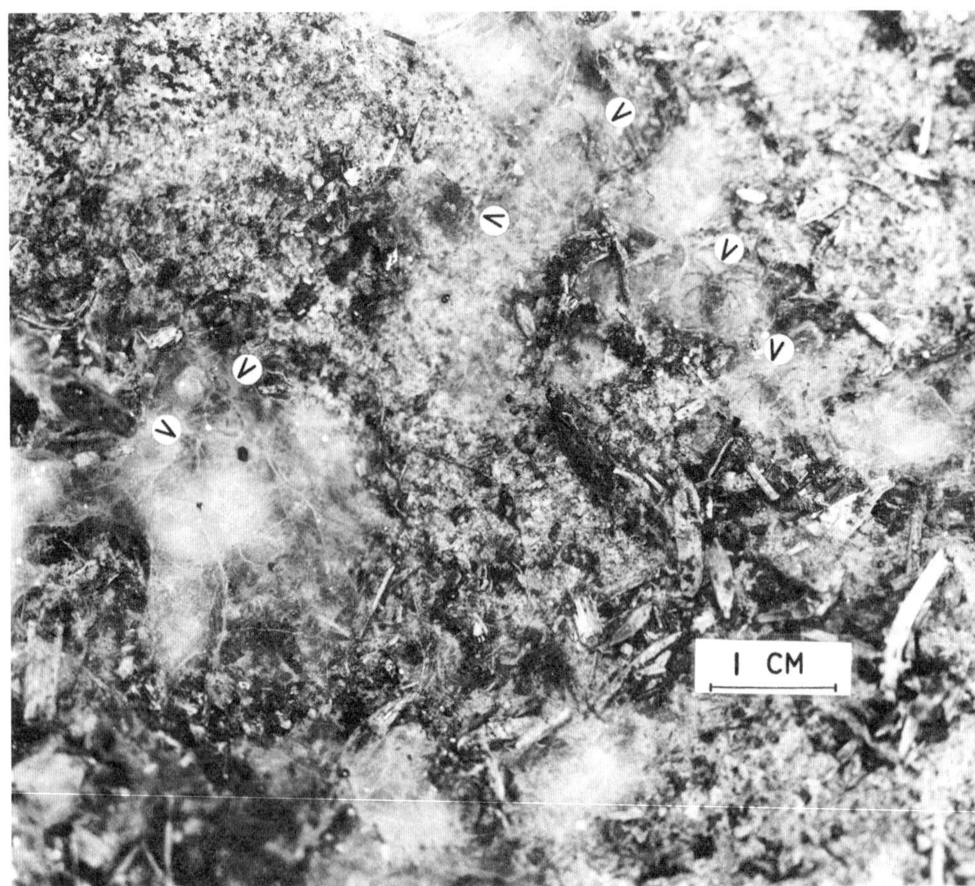

FIGURE 9.2. Web complexes of the tiny spider *Oecobius civitas* are found beneath stones in arid regions of northeastern Mexico. The gauzy white webs are surrounded by a network of silken threads which warn spiders of approaching prey or intruders. Arrows point to several of these cryptic spiders hiding inside their webs.

on the periphery of colonies are vulnerable to predators, while those well inside are protected. However, interior sites receive fewer prey. Intermediate sites (just inside the colony) provide both protection and prey and are thus preferred. A significantly higher number of aggressive disputes over web sites occur in these locations.

Most of the apparent ecological functions of aggregation in web complexes probably involve a competitive advantage over solitary spiders or increased prey-catching efficiency. For example, large web colonies can monopolize habitat areas which are rich in prey resources: e.g., *M. gravidus* builds colonies which span large streams (Buskirk, 1975a); *M. spinipes* fills large cacti (Uetz and Burgess, 1979). In these cases the added support and increased knock-down area provided by many webs surely aid in prey capture (see Buskirk, 1975a). Large colonies are also better able to harvest abundant, clumped food resources (as in the so-called "bonanza strategy" employed by beetles and other invertebrates: Wilson, 1971). Lubin (1974), Rypstra (1979), and Uetz, Kane, and Stratton (in prep.) have noted increased capture efficiency with greater colony size in orb-weaver colonies. Finally, as Jackson (1978a) suggests, the close proximity of males and females living together in web complexes may increase the likelihood of successful mating. All these examples represent ecological advantages over solitary web-building; yet spiders in web complexes do not forego the communicative and architectural advantages of the modular, individual web (see Burgess and Witt, 1976; Witt, 1975).

Spiders Living in Close Aggregations

Communal Spiders

The formation of close aggregations is perhaps nowhere so apparent as in communal cooperative species habitually living together in large colonies which can include thousands of individuals (for review see Burgess, 1978; Krafft, 1970a,b,c; Shear, 1970). Such species include the agelenids *Agelena consociata* (Darchen, 1965b; Krafft, 1970a,b,c) and *A. republicana* (Darchen, 1967) from Africa; the eresids *Stegodyphus sarasinorum* and *S. mimosarum* from the Middle East (Kullmann et al., 1972); and the theridiid *Anelosimus eximius* (Brach, 1975; Simon, 1891) and the dictynid *Mallos gregalis* (Figure 9.3) (Burgess, 1976; Diguet, 1909b) which are found in South and Central America respectively. The group behaviors observed in all these species are remarkably similar in many ways. Group activities take place on a communal web which is constructed jointly by conspecifics and contains both prey-catching surfaces and inner chambers where spiders of all ages and sex classes may be found close together. Predation and feeding are also communal ventures which may involve many spiders utilizing a single prey item at one time.

One mechanism which facilitates aggregation in some species is their mutual attraction to the web colony. Krafft (1970a,b,c) tested

FIGURE 9.3. This close-up shows a "spiders'-eye view" of a communal web colony of *Mallos gregalis* spiders found in northern Mexico. Two large, female spiders can be seen running along the communally constructed web which is spun between tree leaves and twigs. Predation and feeding are other chores shared by colony members.

A. consociata in divided arenas, and found significant clustering around the web, egg sacs, and conspecifics. Such tendencies may be generalized in some species; for example, Jackson (1978a) found that male *M. gregalis* responded equally to their own web and to the web of a closely related, congeneric species. Response to the positions of conspecifics may be sufficient to produce an organized spatial pattern of clumping in small subgroups (Burgess, 1979a).

Specialized communication cues may change aggregation in the groups, as is the case during communal predation (Figure 9.4). In *M. gregalis* the stereotypic buzzing of fly prey sets up resonances

FIGURE 9.4. An aggregation of *Mallos gregalis* spiders feeds on a fly which has landed on the surface of their communal web. The hapless fly is no longer visible: it was quickly covered by a crowd of hungry spiders. The web surface is quite adhesive, but it acquires a thick coating of dust particles in nature. The web threads possess resonant properties which transmit the vibrations of struggling prey across the web to spiders hiding inside. A tunnel leading to the interior of the web is shown at the left (arrow).

which are carried for surprising distances across the sheet-web (Burgess, 1979b). This may be especially important when the sticky web threads become less effective due to windblown dust particles or seasonal rains (Burgess, 1979c). The signal processing properties of the web cause non-prey vibrations to be dampened, including those from conspecifics walking on the web; thus, these communal animals do not mistake conspecifics for prey as may be the case in some solitary web builders. In *A. consociata*, Krafft (1970a,b,c) has also shown that predation is stimulated by web vibration, but the most effective vibrations are rather variable, "noisy" ones more like the movements of the ensnared orthopteran prey on which *A. consociata* feed in nature (Krafft, personal communication).

It is most likely that the pattern of close spacing in communal spiders serves not one but several ecological functions at the same time. The ability to tolerate close crowding inside interior web chambers allows these species to use their webs as protection from predators and extremes of climate (Diguet, 1909b). Living in large groups may also simplify the tasks of finding mates or harvesting large reserves of ensnared prey.

Other advantages may be common to other animals which live in close groups. For example, close, tolerant aggregations may require less energy from individuals than the defense of individual territories. Certainly in the parrot fish *Scarus coricenus* (in which members of the same species can utilize either strategy), communally living individuals suffered fewer attacks and had more time to feed than territorial ones (Robertson et al., 1976). This hypothesis could be tested in spiders by comparing the daily time budgets of closely related species with different lifestyles.

Maternal-Social Spiders

In addition to communal species, some spiders form close aggregations with their young offspring on individual sheet-webs (Burgess, 1978; Kullmann, 1968a). The degree of maternal investment in the young spiderlings varies according to species: some merely share their web space (e.g., Gertsch, 1979), others share prey, regurgitate special nutritious fluids (e.g., Kullmann et al., 1972; Tretzel, 1961), or even signal to offspring that prey is in the web (e.g., Nørgaard, 1956). However, as the young spiderlings begin to mature, they always leave the maternal web and disperse to build their own individual webs. Web-building maternal species include the theridiids *Achaearanea riparia* (= *Theridion saxatile*) (Nørgaard, 1956), *Theridion sisyphium* (Bristowe, 1958; Kaston, 1965), *Anelo-*

simus studiosus (Brach, 1976, 1977), the agelenid *Coelotes terrestris* (Tretzel, 1961) from Europe; the eresids *Stegodyphus pacificus* and *S. lineatus* (Kullmann et al., 1972) from the Middle East; as well as the lycosids *Sosippus floridanus* (Brach, 1976) and *S. janus* (Rovner, personal communication) from North America.

In maternal-social spiders, as in the communal spiders, vibration signals play a part in the mechanisms underlying interactions between the female and her offspring. For example, Nørgaard (1956) has shown that female *A. riparia* can warn spiderlings of danger or alert them to available prey by strumming on the threads of her web. More commonly, maternal females must be able successfully to discriminate their offspring from potential prey striking the web. Female nursery web spiders (Pisauridae) show limited maternal care, standing guard on the web they have constructed around the egg sac until some time after the young have hatched, and protecting them until their dispersal. Interestingly, most wolf spiders (Lycosidae) do not build a web but still exhibit maternal social behavior. Instead of forming their close aggregations on the surface of a maternal web, the spiderlings crawl directly onto the female's abdomen from the egg sac attached to her spinnerets. Young spiderlings are quite persistent: if they are brushed off their mother, they immediately scale up her legs and settle again (Fabre, 1913).

This last-mentioned behavior is apparently mediated by tactile stimuli provided to the offspring by abdominal hairs on the mother (Engelhardt, 1964). In *Lycosa rabida, L. punctulata*, and *Schizocosa avida*, Rovner et al. (1973) investigated the specialized spiny, knobbed, abdominal hairs and found that they provide structures for grasping by the spiderlings. If these hairs are covered with cloth or shaved, young spiders no longer settle on their mother. Other long, smooth mechanoreceptor hairs presumably provide feedback to the female to cease carrying the egg sac and to begin the long fast which accompanies her brood care (Engelhardt, 1964).

In both web-building and webless maternal-social spiders, maternal care probably serves the dual functions of decreasing predation and providing increased competitive advantages for the young spiders. It is likely that predators are less able to detect spiderlings hiding inside the maternal web or huddling together on their mother's back. Moreover, we know from studies of other animals that staying in a group can decrease an individual's chance of being eaten because of predator "saturation." Saturation of predators works like this: because each predator cannot consume unlimited quantities of food at one sitting, it benefits each individual to travel with many other conspecifics. Then if a predator does find the group and feeds until satiation, the remaining individuals will survive. Such strategies are thought also to provide protection for the young of schooling fish (Cushing and Hardin-Jones, 1968; Fishelson et al., 1971). Group-living species which only produce a few offspring at a time, such as colonial nesting birds (Hall, 1970; Orians, 1961; Smith, 1943) and social ungulates (Estes, 1966), can achieve the same "saturation" result by timing reproduction so that all the individuals have offspring at the same time (see Coulson and White, 1956; Darling, 1938).

Competitive advantages probably result from the extra maternal care that spiderlings receive. In species in which there are increased food resources available to spiderlings feeding on regurgitated fluid or prey caught by adults there is likely to be greater survival to older ages. Spiderlings which do not build their own webs or are carried about on their mother's abdomens save energy; this may also be a factor in increasing survival.

Juvenile Aggregations

Many solitary spiders, particularly among the orb-weavers (Araneidae), form close aggregations for several days after leaving their egg sac. This behavior is quite different from the maternal-social pattern, since the mother is never present and the web is spun from a meshwork of draglines left by the spiderlings as they move about (McCook, 1889). Juvenile webs have been the subject of study in several species, including *Argiope trifasciata* and *A. aurantia* (Tolbert, 1977) and *Araneus diadematus* (Burch, 1979).

Both developmental and environmental mechanisms are known to play a role in the spacing of juvenile aggregations. Initially, *A. diadematus* spiderlings stay quite close to their nearest neighbors, but gradually move farther away as they mature (Burch, 1979). Climatic conditions can also affect spacing, however. Tolbert (1977) found that full sunlight causes *A. trifasciata* to cluster close together; the same effect was found for increases in humidity. Since spiderlings eventually spin draglines and float away *en masse* on air currents, their sensitivity to climatic variables may help young spiders select the best moments for aerial dispersal.

Juvenile aggregations probably serve many of the same antipredator functions as maternal-social aggregations, except that no mother is present. Since spiderlings do not feed while on their juvenile web, there is no indication that the experience produces a competitive advantage. On the contrary, Burch (1979) found no survival effects resulting from complete isolation during this period.

Structural and Evolutionary Constraints on Aggregation

Whether species are found to have communal or solitary strategies depends upon which best suits individuals' fitness. Space is somewhat like other resources (e.g., food and shelter) in its ability to be divided up in diverse ways. In our judgment, the spatial strategy which proves most advantageous to a species depends on a combination of individuals' needs, the environment, and the repertoires of physical and behavioral characteristics (called pre-adaptations) which are already available in each species. In most spiders, the major obstacle to group-living is their predatory or territorial tendencies. These obstacles are difficult to overcome, particularly in hunting spiders.

An important pre-adaptation possessed by communal and colonial spiders is the nature of each species' web. Orb-webs, typical of most members of the families Araneidae, Tetragnathidae, and Uloboridae, are peculiarly suited to fill up space with a minimum of materials; long radii serve to funnel vibratory information into a single focal point at the hub where there is room for only a single spider (Burgess and Witt, 1976; Witt, 1975). Moreover, the modular structure of the orb-web does not permit colonial expansion by addition at the periphery—tacking on a little extra web at one corner would destroy the structural integrity of the whole web. These constraints help explain why colonial orb builders join entire individual webs together instead of trying to aggregate on a giant, communal orb. In fact, where communal behaviors are observed, they may take place on a special non-orb structure (e.g., in *U. republicanus*).

In contrast, sheet-web builders must operate under different structural and spatial constraints imposed by their webs' so-called "continuous" design. Continuous structures can be enlarged at their periphery without violating their overall structural integrity, and this is precisely what is done in sheet-webs which span large areas. Sheet-webs also permit the relatively homogeneous vibration transmission that is shown in some species, e.g., *M. gregalis*, which is uninhibited by frame or support threads. The web's ability to transmit vibration over a distance makes possible some of the signal mechanisms of alarm calls, e.g., *A. riparia*, and communal predation, e.g., *M. gregalis* and *Agelena consociata*. Thus, communal and maternal-social species are predominantly found on sheet-webs whose design complements their social requirements.

Spacing may itself be a factor in influencing selection for group-living strategies. It may be possible that aggregations in areas of high prey density result in increased tolerance of conspecifics and in permanent aggregation. Riechert (1978a) has demonstrated a decrease in territory size in *A. aperta* over a gradient of increasing prey availability. This living arrangement may increase fitness due to energy saved (not wasted in territorial encounters), or may have been favored by passive functions such as decreased predator detection and protection. However, it is easy to see that if the members of a non-social species cluster together generation after generation, there may be a tendency to select for more specialized group behaviors which capitalize on benefits to be derived from group living. In spiders these behavioral specializations may include building communal webs, communal predation and prey sharing, alarm signals, tactile stimuli from bodies and webs, web-borne chemical cues, and other adaptations. The recognition that animals may live together because of individual benefit is the starting point in understanding the evolution and function of group living (see Hamilton, 1975; Wilson, 1977).

Techniques for Measurement of Spacing

Up to this point we have discussed spatial patterns without explaining in detail how those patterns can be measured. We will describe some of the most widely used techniques and attempt to clarify the meaning of the measures employed; thus, the potential investigator who would like to measure spatial patterns of his own may be helped.

Unfortunately, no one measure is clearly better than all others; the most appropriate techniques depend on the type of research question being asked. For example, in a field study it is often practical to divide the study area into square sites (called quadrats), in order to record the number of individuals found therein. Such sampling can be used to construct a table of the frequency of occurrence of 1, 2, 3, or more spiders per quadrat. These frequencies can then be directly compared to expected frequencies from a known distribution. Poisson or normal distributions approximate random; binomial distributions tend toward regularity, and negative binomial distributions indicate clumping. For example, Kuenzler (1958) showed that the distribution of three wolf spider species (*Lycosa timuqua, L. carolinensis,* and *L. rabida*) in 0.1-hectare quadrats in old-field approximated a Poisson distribution except where vegetation was clumped.

A problem with such analyses is that the size of the sampling quadrat chosen can influence the results. Also, many species are distributed in three dimensions rather than two. Sometimes, when natural groupings seem to provide suitable sampling units, this difficulty can be reduced. For example, we compared the group sizes of *M. spinipes* occurring in each of several colonies. They were found to approximate a zero-truncated negative binomial and to be significantly different from a Poisson distribution, thus indicating aggregation (Uetz and Burgess, 1979). Farr (1977) used this technique on web distribution in *Nephila clavipes* and found it to be random.

A powerful technique for measuring large-scale dispersion of a species in its environment is block-size pattern analysis (Greig-Smith, 1952, 1961; Pielou, 1977; Thompson, 1954). Block-size analysis can be used to compare the distribution of animals in a variety of quadrat sizes, ranging from small to very large, in order to determine at exactly which quadrat size clumping or regularity occurs. Using these techniques, the regular spacing of the desert funnel-web spider *Agelenopsis aperta* was found to be related to social, reproductive, and habitat functions (Riechert et al., 1973; Riechert, 1974b, 1978a).

A simpler but more limited measure of dispersion is provided by the variance/mean ratio ($S^2/\bar{X}$) (Pielou, 1977), and related measures (see David and Moore, 1954; Lloyd, 1967; Waters, 1959). When animals are sampled in quadrats, their mean density per block and variance among blocks are calculated. High ratios indicate a tendency toward clumping while lower ratios show regularity. This measure can be used in small areas, such as within a spider group (Burgess, 1979a), but users should be aware of the possible interference effects arising from the choice of quadrat size.

A more sensitive way to quantify spatial patterns within a group is by measuring actual distances between individuals. Since animals are presumed to respond to cues from other conspecifics, distance measures have a more obvious application to behavioral study than does quadrat sampling. The mean distance between nearest neighbors is the most often used measure and has, for example, been shown to be important in interactions between colonial spiders (Buskirk, 1975a). Statistical tests are available that suggest the pattern of nearest-neighbor distance (Clark and Evans, 1954); these demonstrate whether nearest neighbors are significantly clumped or regularly spaced. However, Thompson (1956) points out that distances to only the nearest neighbors can tell little about the heterogeneity of spatial patterns in a group. He suggests that distances to many successive neighbors be measured (e.g., mean distance to 1st, 2nd, 3rd . . . nearest neighbors), and presents statistical tests for these measures. Distances to 1st–5th neighbors can also be compared to the simulated results obtained in a mechanical model which takes into account the size of the animals and the shape of their testing environment (Burgess, 1979a). An increased awareness of the applications and drawbacks of different techniques may help to increase the comparability of spacing studies, a quality which has previously often been lacking.

New Spatial Studies

Having examined some of the previous work on spider spacing strategies, we will now present several new observations on spatial patterns in spiders and thereby illustrate the different ways spatial data can be treated. Three specific questions will be addressed: (1) do web complex builders adjust spacing to the constraints of different habitats? (2) is so-called "paradoxical territoriality" as effective in ensuring stable spatial patterns as some more conventional strategies? and (3) are different overall spatial patterns found among individuals or species with different social behavior under the same controlled environmental conditions?

Spatial Patterns in Different Habitats

We have noted that spacing in web-complex builders helps to meet the survival demands imposed upon each species by its environment. But what about species which inhabit several different habitats or microhabitats? Is their spacing adjusted in each different environment, or is spacing relatively static regardless of habitat? Different spatial patterns have been shown for the territorial species *Agelenopsis aperta* in habitats of differing quality (Riechert et al., 1973; Riechert, 1974b, 1978a), but this question has not been addressed in web-complex builders.

Metepeira spinipes, a web-complex builder observed near Tepotzotlan, Mexico, builds on a variety of supports but was found most often either on large maguey plants (*Agave* sp.) or in the midst of willow shrubs (*Salix* sp.) (Figure 9.1) (Uetz and Burgess, 1979). Although both plants are found in the same areas they offer very different environments for spiders. Maguey plants are found on roadsides, field margins, and plantations; they consist of a rosette of smooth, stiff incurving lanceolate leaves which periodically collect water and may be the site of flying insect activity. In contrast, the willows were observed along drainage canals where prey activity was high and colonies were shaded, hung between the willows' leafy, flexible stems.

We compared spacing in 23 maguey colonies and 12 willow colonies by the following method. At midday, color slides were taken of the center of each colony. These photographs purposely had a very shallow depth-of-focus (obtained by using the widest 13.5 aperture of the 5.5 cm Micro-Nikkor lens), so that only a narrow plane through the middle of each colony was in focus. When the slides were later projected at full life-size, the effect was as if a two-dimensional slice were taken from the center of the web complex, thus providing a standardized sample of each colony. Coordinate positions (x,y) of each sharply focused spider were recorded and transcribed onto IBM cards. Distances between every animal and every other animal within each colony were calculated by using the formula $d = \sqrt{[(x - x_1)^2 + (y - y_1)^2]}$ with the help of a computer program. This method is not as accurate as direct measurement, since actual nearest neighbors may be in a different (vertical) plane than the photograph. However, this method does not disturb the spiders or their web structure, and allows a large amount of field data to be collected in a short time. Nearest-neighbor distances estimated from photographs were similar to those measured directly (at a later date at the same site). Mean distances to 1st–4th nearest neighbors were calculated for each colony. Colony means were compared between microhabitats for each nearest-neighbor distance using one way analysis of variance (2-tailed).

Spatial patterns in the two microhabitats were found to be significantly different (Figure 9.5). Spiders in the maguey plants maintained greater distances to their 1st–4th nearest neighbors than spiders in the shrubs, thus showing that habitat can affect spatial patterning in different populations of a single species. We also wished to know if spatial differences would persist in the two groups if they were placed in identical environments. To test this, we collected one colony from each microhabitat and transported individuals in small vials to the laboratory. There, spiders were placed in identical 50 × 50 × 50 cm glass cages kept under controlled conditions of temperature, humidity, and light (see Burgess, 1979a,b,c). Coordinate positions (x,y,z) were measured in three dimensions and distances between 1st–3rd nearest neighbors were calculated using the equation $d = \sqrt{[(x - x_1)^2 + (y - y_1)^2 + (z - z_1)^2]}$. Distances were compared by the Wilcoxon two-sample test (2-tailed) (Figure 9.6).

Although interindividual distances in both colonies increased in

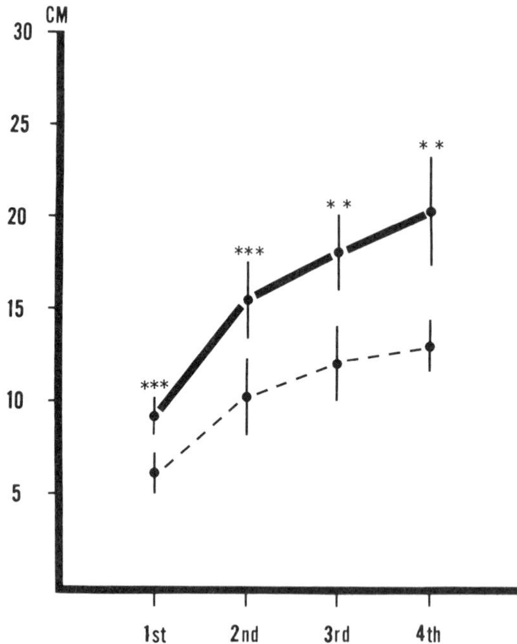

FIGURE 9.5. Spatial differences in colonies of *Metepeira spinipes* from different microhabitats. The ordinate compares mean distances (cm) to 1st, 2nd, 3rd, and 4th nearest neighbors on the abscissa. Greater distances were maintained within colonies built in maguey plants (unbroken line) than in those located in willow shrubs (dashed lines). For 1st–4th distance, F (1,43) = 11.48, 10.04, 8.15, 8.88; $^{**}P < .02$; $^{***}P < .01$.

the laboratory (probably due to growth of spiders), the maguey colony continued to maintain greater distances than the willow colony between 1st–3rd nearest neighbors (Figure 9.6). The behavior of these two colonies should serve as a caution that if differences between spatial patterns occur in different habitats, they may be more complex than a simple response to the physical structures of each microhabitat. Apparently, spatial patterns within populations of a species are not static, but vary according to the type of microhabitat in which they occur.

Further evidence of this tendency in *M. spinipes* comes from recent studies of this species over a wide range of habitats in Mexico (Uetz, Kane, and Stratton, in prep.). *Metepeira spinipes* colonies were studied over a gradient of habitats ranging from arid grassland and high elevation sites (severe habitats) to temperate subhumid agricultural areas (intermediate habitats) to tropical mountain areas with banana and coffee plantations (benign habitats). Nearest-

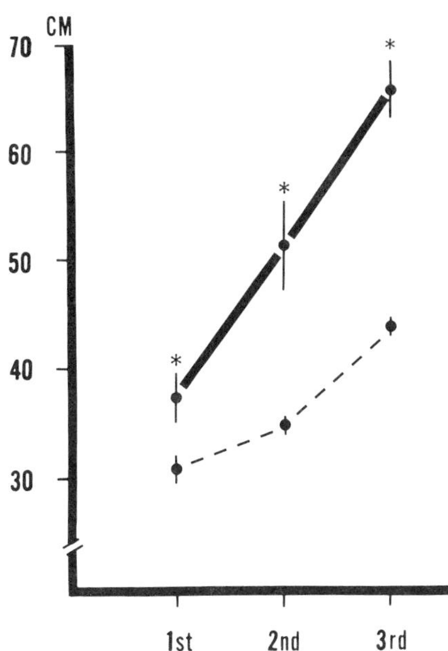

FIGURE 9.6. Spatial differences persist in two *Metepeira spinipes* colonies placed in equivalent, controlled environments. The ordinate compares mean distances (cm) to 1st, 2nd, and 3rd nearest neighbors on the abscissa. The colony originally built in a maguey plant (unbroken line) maintained greater distance than a similar colony taken from a willow shrub (dashed line). For 1st–3rd distances, Wilcoxon (2-tailed) $U_{(4,4)} = 16$; * $P = .05$.

neighbor distance decreased over this habitat gradient and showed a significant negative correlation with prey availability (Table 9.1). Individuals of *M. spinipes* can apparently tolerate conspecifics at closer distances in areas where food is abundant. Riechert (1978a) found a similar reduction in interindividual spacing over a habitat gradient in *Agelenopsis aperta*. In both these species, spacing is mediated by agonistic behavior, despite the fact that one is solitary and regularly spaced and the other is colonial.

Field experiments were conducted to examine further the relationship between food availability and spatial pattern in *M. spinipes* (Uetz, Kane, and Stratton, in prep.). In one site (Toluca, Mexico), colonies were moved from agave plantations in an agricultural valley with moderate climate and high insect activity to a plantation on a higher elevation mountainside with severe climate and low insect activity. After 10 days, 12 out of 20 of the "transplanted" colonies remained and had built webs together. Nearest-neighbor distance had increased significantly to nearly twice what

TABLE 9.1. Nearest-neighbor distances and prey abundance for *Metepeira spinipes* colonies at several sites in central Mexico.

Site Description	*Estimated Prey Abundance (No. /m³/Day)*	*Overall $\bar{X}$ Nearest-Neighbor Distance in cm (± 2 S.E.)*	*Range of Colony Means in cm*	*No. of Colonies Measured*
San Miguel de Allende (El. 2400 m) Arid grassland with *Opuntia* and other cactus species.	49 – 70 ($\bar{X}$ = 59)	17.49 ± 3.17	6.26 – 41.0	24
Toluca (El. 2600 m) Agricultural field with *Agave* border. Large quantity of animal waste.	64 – 116 ($\bar{X}$ = 96)	11.73 ± 1.24	7.50 – 17.00	26
Tepotzotlan (El. 2100 m) Roadside *Agave* bordering agricultural field, with some animal waste.	112 – 160 ($\bar{X}$ = 130)	12.65 ± 1.96	7.30 – 21.70	21
Agave plantation with dumping of feedlot waste.	160 – 300 ($\bar{X}$ = 205)	8.22 ± .70	5.60 – 13.11	37
Lagode Guadalupe (El. 2279 m) East shore of reservoir with large *Opuntia*. Aquatic insects blown ashore by prevailing wind.	296 – 336 ($\bar{X}$ = 316)	6.85 ± .828	3.60 – 14.50	27
Cordoba (El. 929 m) Tropical mountainside with coffee and banana plantations.	600 – 1000 ($\bar{X}$ = 840)	3.78 ± .42	3.00 – 4.47	10

it was in the original habitat. In a second site (Tepotzotlan, Mexico), colonies were moved from an area within an agave plantation where insect activity was high (due to dumping of feedlot waste) to an area in the same plantation where insect abundance was low. The experimental colonies were divided into two treatment groups, one receiving prey supplementation by addition of cow manure to the base of each agave, and the other without prey supplementation. Ten days later, there was no change in group size or nearest-neighbor distance in the treatment group receiving prey supplementation. In the unsupplemented group, there were significantly fewer spiders and a significantly higher nearest-neighbor distance (3 × greater than in the original site). The results of these experiments strongly suggest that *M. spinipes'* spacing patterns vary in response to food availability.

These findings are consistent with foraging studies of a wide range of predatory animals, which show that activity levels and aggression increase with starvation time and decrease with satiation. The rapid changes in spacing in response to prey availability demonstrated by *M. spinipes* are probably due to the fact that these orb weavers renew catching spirals in their web complex on a daily basis. It is during the time of spiral construction that most agonistic encounters and web-site contests occur. A flexible spacing pattern may allow this web-complex builder to survive in habitats where prey availability changes seasonally or annually.

Relative Effectiveness of Spacing Strategies

Most web-complex builders employ agonistic behaviors like leg-jerking and web-shaking to maintain individual space against intruders (Buskirk, 1975a; Rypstra, 1979). The effectiveness of these behaviors is obvious from the above example. In contrast, some strategies clearly involve trade-offs, including so-called "paradoxical territoriality" (Dawkins, 1976). As we mentioned earlier, the only example of this strange strategy that has been found to date is the spider *Oecobius civitas*, which, when disturbed, leaves its web and takes refuge in a crevice or in another occupied web by displacing the occupant. One way to evaluate the effectiveness of this non-combative strategy is to see how consistent are distances within *O. civitas* web complexes (Figure 9.2). If there is little variability in the distances to many successive neighbors, then this would be evidence that paradoxical territoriality is quite effective in maintaining consistent spacing.

Four spider species having a range of social and territorial behaviors were chosen for comparison with *O. civitas*. These included a web-complex species which builds and defends its individual orb-webs (*Metepeira spinipes*); a classically territorial species which defends individual sheet-webs (*Agelenopsis utahena*); a communal-web species in which individuals sometimes defend the zone of "personal space" immediately surrounding their bodies (*Physocyclus dugesi*); and a solitary species for which no interaction or spatial defense is known (*Leucauge regnyi*). We expected spacing would be most consistent in the territorial spiders and most variable in the solitary spiders and that paradoxically territorial *O. civitas* would fall in between. Distance measurements were made in the ways described below.

O. civitas was observed in 1974, south of Guadalajara, Mexico (Figure 9.2) (Burgess, 1976). Webs were found in the protected microhabitats on the undersides of large stones. All 120 spiders' positions were recorded in color photographs of six web complexes. These were projected at full size in the laboratory and positions measured in (x,y) coordinates. Distances were calculated with the help of a computer program.

Observations by J. W. Burgess and G. Uetz on *Metepeira spinipes* were discussed in the previous section. Measurements come from 46 colonies located in fields, along drainage canals, and near roadsides, in both willow shrubs and maguey plants.

Physocyclus dugesi (determined by W. J. Gertsch) were first observed in five large depressions in stone walls near Teotihuacan, Mexico, during 1974. These long-legged spiders build large, connected sheet- and space-webs, which may extend for 10 m (Burgess, 1978). All age-sex classes live on the webs; however, web-shaking and leg-jerking responses are seen, particularly when spiders approach closely. Twenty-six spiders were photographed in five 700 cm^2 samples, which were analyzed in the same way as described for *O. civitas*.

J. W. Burgess observed *Agelenopsis utahena* spiders in Raleigh, North Carolina, during 1978. These spiders build handkerchief-sized sheet-webs in shrubs, crevices, and along buildings. Although webs are sometimes aggregated in favorable sites, individual web space is defended by territorial adults. Distances were measured directly between all spiders in two shrub sites, using a meter tape.

Leucauge regnyi (determined by N. Platnick) were observed in the El Junque rain forest in Puerto Rico. The orbs built by these beautifully colored spiders are often found close together or even connected by their frame threads. Spacing was measured during 1975 in four sites, each 180 × 170 cm, using a triangulation method. We measured between each spider and two of its neighbors with a measuring tape. Later, in the laboratory, these distances were plotted on large (1 m^2) sheets of graph paper. The resulting 33 (x,y) coordinate positions were analyzed with the aid of a computer program.

We can tell little about variability of spatial patterns from the actual distances we measured. For example, *O. civitas* is a tiny spider and also has close distances with low variation. *P. dugesi* live close together on their communal web while the individual orbs of similar-size *L. regnyi* are spread farther apart. Because of the disparity in spider size and spatial pattern, coefficients of variation from each group were used to compare distances between species. Coefficients of variation (C.V.) are used extensively to compare variation between variables of different magnitude; they are derived by dividing each group's standard deviation by the group mean, and are expressed as a percentage (S.D. $\times$ 100/$\bar{X}$ = C.V.; Sokal and Rohlf, 1969). Both 1st and 2nd nearest-neighbors distances were measured, and coefficients of variation were calculated for each colony or group. Mean values for coefficients of variation were then used to compare variability of spatial patterns between species (Figure 9.7): 1st nearest neighbors were presumed to be involved in the defense of space between bordering spiders' webs, while 2nd nearest-neighbors' positions provided a measure of the consistency of spacing beyond web borders. The results indicated that positions in the web complexes of *O. civitas* were quite variable. Differences between species were determined with Dunnett's multiple comparisons test (Steel and Torrie, 1960) (shown in Figure 9.7).

Spiders which actively defend their webs against intrusion (i.e., *M. spinipes* and *A. utahena*) exhibited less variability in the positions of 1st and 2nd nearest neighbor than did *O. civitas* spiders, which retreat from intruders (Figure 9.7). *P. dugesi* spiders, which will defend close personal spaces, were less variable than *O. civitas* only with respect to the spacing of 1st nearest neighbors—both species showed similar high variability in the positions of 2nd nearest neighbors. Spacing in the solitary spiders *L. regnyi* did not differ from *O. civitas*.

These comparisons can help us begin to evaluate the effectiveness of a spacing strategy. Our measures suggest that the non-combative, paradoxical territoriality of *O. civitas* is less effective in preserving consistent intragroup spacing than active spatial defense, and about as effective as no defense of living space whatsoever. Perhaps other advantages compensate for the decreased predictability of conspecifics' positions—for example, the freedom

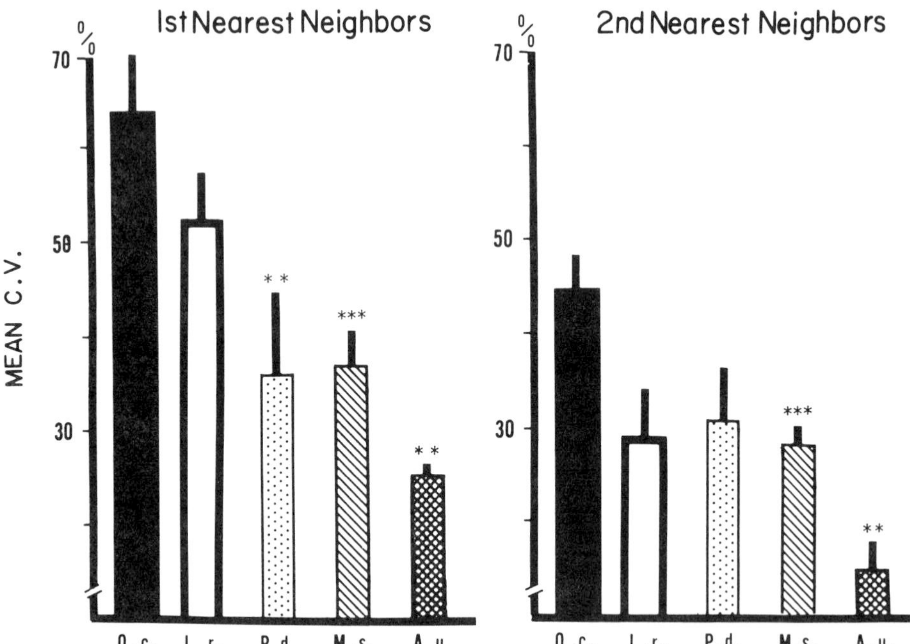

FIGURE 9.7. The consistency of spacing, measured as the mean of coefficients of variation for distances in each group (represented as C.V. on the ordinate), was compared between spiders which show different defensive behaviors. In the web complexes of the non-combative "paradoxically territorial" spider *Oecobius civitas* (O.c.), spacing to 1st nearest neighbors was more variable than in groups of territorial spiders (*Agelenopsis utahena*: A.u.; d = 2.91) or web-complex builders which defend their individual webs (*Metepeira spinipes*: M.s.; d = 3.80) or communal-web spiders which defend close personal spaces (*Physocyclus dugesi*: P.d.; d = 2.81). Spatial variation in solitary spiders *Leucauge regnyi* (L.r.) was about the same as *O. civitas* (d = 1.09). Spacing of *O. civitas* (O.c.) spiders to their 2nd nearest neighbors was more variable than in groups of the two actively defensive spiders (A.u.; d = 2.98 and M.s.; d = 3.12), but was not different from the close-defending spiders (P.d.; d = 1.72) or the solitary spiders (L.r.; d = 1.93). Values of Dunnett's d test for multiple comparisons used d.f. = 58 and 2-tailed probabilities: $^{***}P < 0.01$; $^{**}P < 0.05$ (Steel and Torrie, 1960).

from time and energy expenditures necessary for territorial defense (see Robertson et al., 1976). Further studies of these energetic tradeoffs can tell us more about differences in behavioral specializations.

SPATIAL PATTERNS IN DIFFERENT SPECIES

How can the pattern of spatial behavior in groups of two different species be compared? Such a comparison would necessarily be made under controlled conditions equivalent for both species, since we have seen that the surrounding environment can influence group spacing, and should employ measures which are meaningful for the behavior of the species. In a previous study (Burgess, 1979a), ten spatial measures were used to compare six animal species in homogeneous, open arenas with the spacing of an artificial mechanical model. For all species, the most revealing spatial measures turned out to be the 1st–5th nearest-neighbor distances. It may also be concluded that farther neighbors' distances might reveal additional information about group spacing, and that other comparisons could be used to test spatial patterns in addition to the passive artificial model.

To address these questions, data were taken from two spider species: *Mallos gregalis* and *Agelenopsis utahena*. *M. gregalis* is a group-living spider from Mexico found in close aggregations on a single, communally built web; groups of conspecifics catch prey and feed together. *A. utahena* is a spider of the southeastern United States which builds individual sheet-webs that are sometimes found aggregated in favorable habitats.

The study procedures are detailed in Burgess (1979a). Two 25-member groups of each species were housed in round arenas scaled to each animal's body length: the *M. gregalis* arena = 202 cm^2; the *A. utahena* arena = 962 cm^2 (Figure 9.8). Temperature, humidity, and light regimes were controlled (see Burgess, 1979c). Ten photographs were taken of each of the two groups, each pair of photographs at 24-hour intervals. Later, they were projected full-size for analysis. With the aid of a grid on the floor of each arena (x,y) position coordinates were recorded, transcribed onto IBM cards, and analyzed with the aid of a computer program. Distances to 1st–12th nearest neighbors were calculated for each group.

Two comparisons were made of each species. To test for non-randomness of spatial patterns, distances to successive neighbors were evaluated using the equation

$$\chi^2 = N\,(2\pi m)\, r_n^2\,,$$

which is distributed with 2 N_n degrees of freedom, with N = number of observations, m = density of animals per unit area, and r_n = the mean distance calculated to the nth nearest neighbor (here between 1st–12th nearest neighbors). This computation was derived by Thompson (1956) and provides a remarkably simple test for clumping or regularity in a group. For comparative purposes, the normal approximations to random spacing distances are given by $d = 0.5643 \sqrt{[n/m]}$ (valid for large n's; see Thompson, 1956). In addition to tests for non-randomness, 1st–12th nearest-neighbor dis-

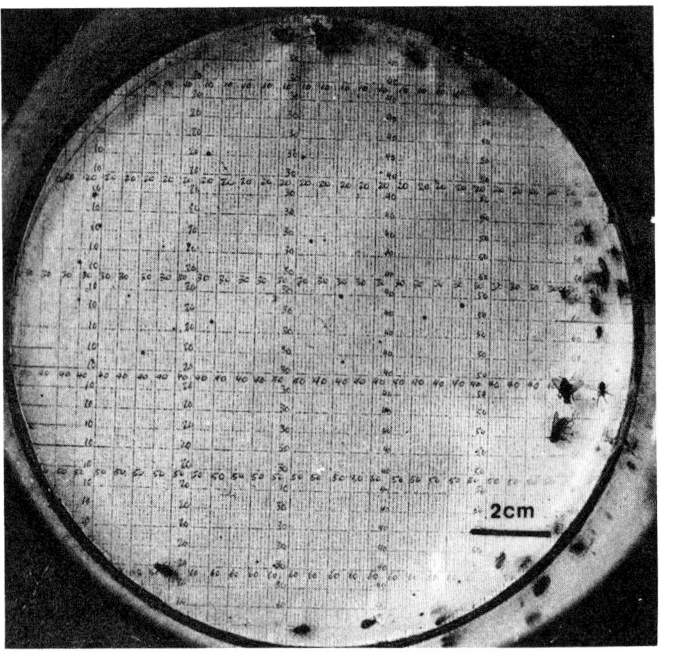
2cm

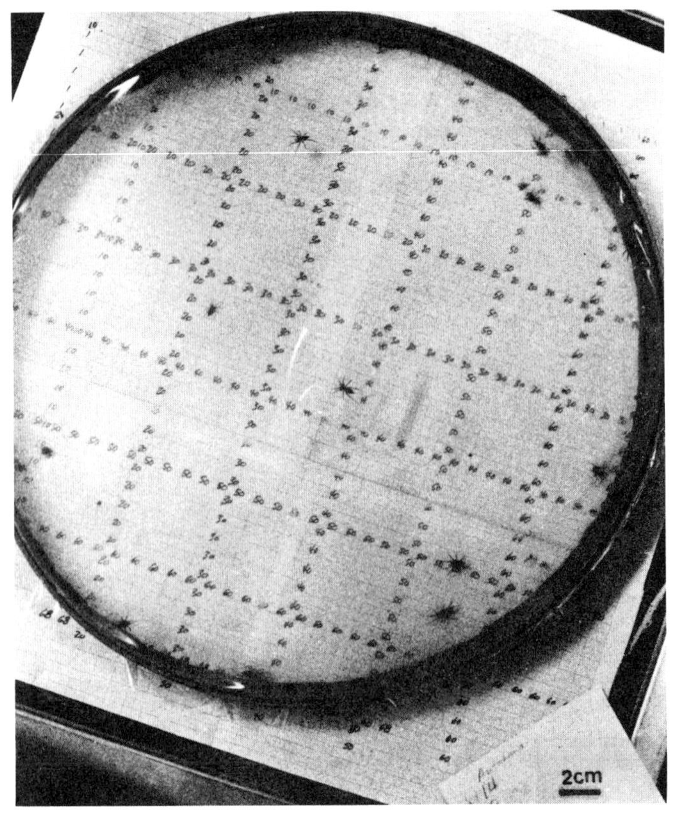
2cm

tances were calculated for the passive mechanical model. This model was made by dropping round marbles into a size-scaled arena and permitting them to roll and collide; this model tests for non-behavioral aggregations which might result from the shape of the arena or the size and passive contact of the animals. The marbles were photographed and analyzed in the same way as were the spiders.

Both species exhibited significantly non-random spatial patterns which differed between the two species (Figures 9.9, 9.10). Comparison with both the random approximation and the mechanical model gave the same results, although the mechanical model showed predicted clumping between 5th–12th nearest-neighbor distances, presumably resulting from mechanical constraints of the arena. The communal spider *M. gregalis* maintained a heterogeneous pattern, with significant aggregation toward 1st–3rd nearest neighbors and significantly regular distances toward 6th–12th nearest neighbors (Figure 9.9). The solitary but aggregative spider *A. utahena* maintained significantly regular spacing between all 1st–12th nearest neighbors (Figure 9.10).

In fact, the spatial pattern exhibited by the communal spider looks much like the solitary regular pattern with the addition of close clumping. Since *M. gregalis* does have close congeneric relatives which exhibit regular spacing, it is attractive to speculate that *M. gregalis* may have initially been a regular-spacing species which added clumping to near neighbors during the course of its evolution toward a communal life style. This hypothesis could be tested by measuring the spatial patterns of other communal spiders and their regularly spacing congeneric relatives.

Conclusions

Spacing strategies, which depend on mechanisms of communication between animals, exist to serve ecological functions for increased survival of individuals. Spacing thus plays an integral role in group social dynamics. Spider spacing strategies can be divided into three general categories: regular spacing, web complexes, and close aggregations.

FIGURE 9.8. Two open-field laboratory arenas were used to measure the spacing behavior of spiders under controlled conditions. (A) Communal spiders *Mallos gregalis* distributed in small spatial subgroups within an arena scaled to their size. (B) In contrast, territorial spiders *Agelenopsis utahena* dispersed across the floor of their arena.

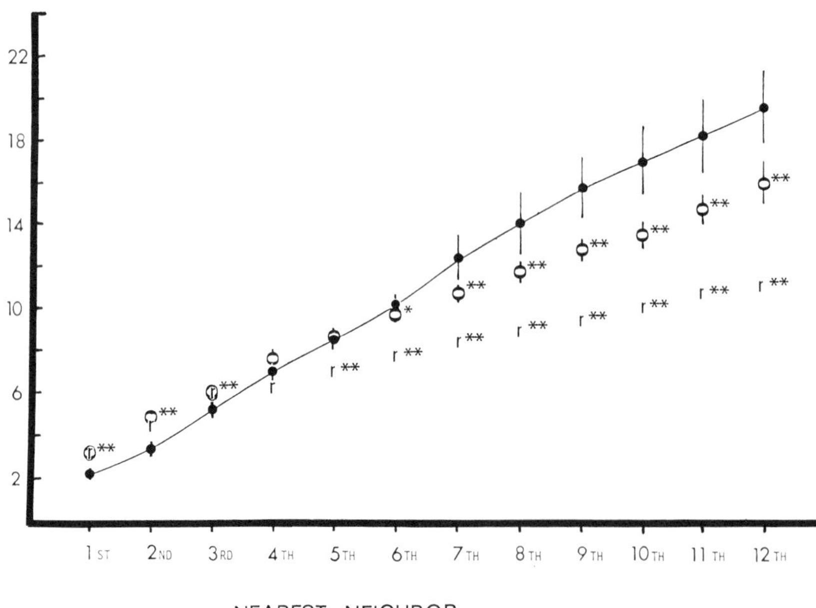

FIGURE 9.9. Spatial pattern of a communal spider *Mallos gregalis* under controlled conditions. Mean distance in body length (cm/0.48) on the ordinate is compared to 1st–12th nearest neighbors on the abscissa. The communal spiders, in both groups of 25 members (filled circles), maintained significantly closer distances toward 1st–3rd nearest neighbors, compared to a mechanical model of passive spacing (open circles) and a random model (r's). Farther neighbors maintained greater distances than either model. F (1, 38) = 4.20-75.52; χ^2 (40-480) = 17.97-1501.05; $^*P < .05$; $^{**}P < .005$.

Regular spacing of solitary spiders depends on communication and defensive behavior between individuals, as well as on selection for protected web sites with high prey availability. This spatial pattern ensures individual spiders access to a certain amount of prey, necessary for growth and reproduction. The size of territory defended by regularly spaced spiders has been shown to vary with accessibility of prey. Regular spacing may also increase the chances of successful mating in dispersed populations.

Web-complex builders combine the benefits of communal and regularly spaced living. Chemical cues and agonistic displays apparently play a part in mating and in defense of space in web colonies. The geometric effects of aggregation provide more protection from searching predators than does territoriality alone; in addition, inner colony members are buffered from attack by their conspecif-

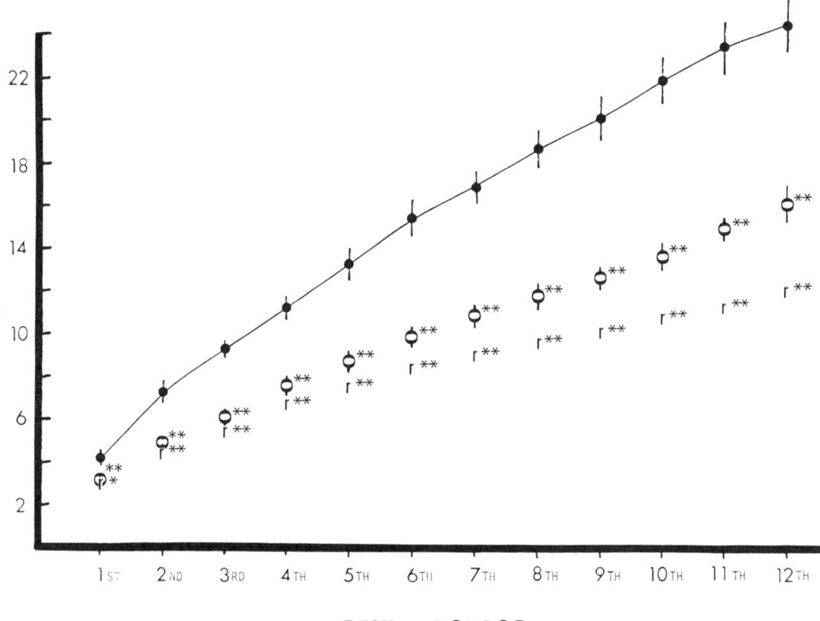

FIGURE 9.10. Spatial pattern of a solitary spider *Agelenopsis utahena* under controlled conditions. Mean distance in body lengths (cm/1) on the ordinate is compared to 1st–12th nearest neighbors on the abscissa. The solitary spiders, in both groups of 25 members (filled circles), maintained significantly greater distances to all 1st–12th nearest neighbors than predicted by either a mechanical model of passive spacing (open circles) or a random model (r's). F (1, 38) = 80.25-250.98; χ^2 (40-480) = 59.20-2033.96; ***$P < .005$.

ics on the outside of the colony. Competitive efficiency is enhanced by the increased ability to penetrate desirable habitat areas; and individual prey-capture efficiency is enhanced by the increased knock-down of prey which results from large webs and sizeable groups. Aggregated spiders are constrained by the communicative and structural properties of different types of webs; these constraints may function as pre-adaptations which influence the subsequent evolution of a species' group-living patterns.

Communal species in close aggregations utilize vibratory and tactile cues to facilitate aggregation. Communal living serves multiple functions of protection from predators, increased mate availability, communal help in predation (especially on large prey), and freedom from the energy expenditures inherent in defending space. Maternal-social spiders also utilize vibratory and tactile cues in their groups. Offspring of maternal-social species receive increased nourishment and are protected from predators by their web or by concealment on the mother, as well as by the predator-swamping effects of grouping. Juvenile aggregations are influenced both by climatic factors and developmental changes; individuals may be subject to less predation while clumped on the juvenile web.

A variety of techniques have been employed to measure spacing. Block-size analysis, frequency distributions, and variance/mean measures are most useful for field data sampled in quadrat blocks. Interindividual distance measures are useful in quantifying the behavioral spacing responses within groups. Ideally, mean distances to many successive neighbors should be measured in order to evaluate the heterogeneity of the group spatial pattern. Statistical tests may be used in every case to determine the non-randomness of clumped or regular spacing.

Three spatial questions are addressed by the use of new spacing data. It was found that the web-complex builder *Metepeira spinipes* does not have an inflexible spatial pattern, but exhibits different interindividual distances in different microhabitats and can respond to changes in prey availability by varying its spatial arrangement. In another comparison, spacing of the "paradoxically territorial" spider *Oecobius civitas* was found to be significantly more variable than three other species with different social strategies that include defending space. Finally, spacing of a communal species, *Mallos gregalis*, and a solitary species, *Agelenopsis utahena*, were compared under equivalent, controlled, open-field conditions to models of random and mechanical spacing. The communal species exhibited clumping toward close neighbors whereas the solitary species maintained regular spacing; however, at farther neighbor distances, both species exhibited regular spacing.

It is clear that spiders, like other animals, have evolved various ways to share their living space. As we study spiders, we will no doubt learn more about the general nature of group structure and the evolution of all social strategies. The subtle interplay of communication, other behaviors, and the environment can thus provide us with a wealth of examples of the diversity of animal life-styles.

Acknowledgments

Funds for these studies were provided by an NSF grant to Peter N. Witt, who also contributed laboratory space, stimulating discussions, and help with tedious spatial analyses. Ms. Mabel Scarboro helped with computer analysis and Ms. Cher Carrie provided statistical advice. Travel to Mexico and Puerto Rico was made possible by funds from the N. C. Foundation for Mental Health Research, Inc. and *Scientific American* to J. W. Burgess. The National Geographic Society, the American Philosophical Society, and the University of Cincinnati Research Council provided support for George W. Uetz.

Chapter 10 SPIDER FORAGING: BEHAVIORAL RESPONSES TO PREY

Susan E. Riechert
Department of Zoology
University of Tennessee
Knoxville, Tennessee 37916

Jadwiga Łuczak
Institute of Ecology
Dziekanow Lesny
05150 Komianki, Poland

Editors' Note: We decided to include this chapter in the book because it deals with an important part of the spider's relationship with the environment and because it developed from an original contribution to the symposium. Although the contents of the chapter do not fit a generally accepted definition of "communication," they include valuable information on spider behavior and ecology that applies some of the material discussed earlier on sensory physiology and modes of signaling. While the material on predator-prey interactions only touches rather than overlaps with the other chapters, such a summary cannot be found anywhere else.

Introduction

"Communication in the fullest sense implies evolutionary specialization of a mutualistic, cooperative nature . . ." (Marler, 1977). And yet animals are constantly gathering and exchanging information from and with their competitors, predators, and prey. In doing so, they utilize the same vibratory, visual, and chemical cues that are used in social interactions. This chapter deals in part with the degree to which spiders exploit visual, vibratory, and chemical stimuli produced by potential prey either as intended or illegitimate receivers. The information utilized by spiders at different levels of the predator-prey interaction will also be discussed with emphasis placed on the ecological and biological rules that influence the nature of these interactions.

Probability of Encounter

Selection of Foraging Sites

A number of morphological and physiological studies completed in the last decade have presented evidence for the evolution of spider populations under food limitation. Anderson (1970) and Peakall and Witt (1976), for instance, have demonstrated that spiders have very low maintenance energy requirements and Anderson (1974) and others (Ito, 1964; Nakamura, 1972) have presented evidence for a decrease in metabolic rate associated with periods of starvation. When food is available, the highly distensible abdomens of spiders (Anderson, 1970) and their ability to store large amounts of fat (Collatz and Mommsen, 1975) permit them to feed maximally. These findings led Miyashita (1968a) and Anderson (1974) to conclude that spider populations frequently exist under conditions of food deprivation.

Selection of suitable microhabitats which allow maximum foraging time, provide an abundance of prey, and permit efficient handling of this prey appears also to represent an adaptation to a temporally and spatially heterogeneous food supply. The first observation indicating that spiders might actively select microhabitats on the basis of foraging considerations was made by Turnbull (1964). In a simple experiment, he released a number of individuals of the common house spider *Achaearanea tepidariorum* into an empty room. Although *Achaearanea* initially built webs at random locations within the room, these webs were soon abandoned and the majority of the spiders settled in the vicinity of a window where flies were gathered. In a later study (1966), Turnbull presented evidence for the seasonal migration of grassland spiders from an area of decreasing prey density to one offering higher densities. Similarly, Kronk and Riechert (1979) observed seasonal movements of the wolf spider, *Lycosa santrita* from areas of lower prey availability to higher in a desert riparian habitat, while in Poland, Dabrowska-Prot et al. (1973) found sheetline weaving spiders of the species *Linyphia triangularis* and *Enoplognatha lineata* to establish dense populations in an ecotone between an alder wood and mid-forest meadow. The ecotone habitat was less favorable thermally and structurally to these spiders, but provided an abundance of prey. These observations are indicative of the operation of some form of the aggregational response (Readshaw, 1973), in which the behavior of predators leads to their concentration in localized patches of abundant prey.

Other factors associated with prey-capture success have recently been shown to be important determinants of microhabitat association. For instance, a desert spider, *Agelenopsis aperta* associates with microhabitats affording both high prey numbers and long periods of feeding activity (a result of ameliorated thermal conditions) (Riechert and Tracy, 1975; Riechert, 1976). The specific microhabitats utilized in lava bed and grassland areas differed, but the same results were achieved by the associations. In addition, substrate choice has been shown to be related to prey-capture efficiency in two genera of wolf spiders (Lycosidae: Greenquist and Rovner, 1976; Kronk and Riechert, 1979).

The extent to which spiders actively select locations affording greater prey-capture success should vary with the foraging strategies exhibited by the various taxa as well as with the kinds of habitats they typically occupy. The selection mechanism will likewise reflect compromises among the many needs and selection pressures impinging on species populations in different environments. However, some basic components of the selection process can be outlined. For instance, there must be an initial incentive to move from a given location. In the case of spiderlings, motivation for dispersal from the vicinity of the egg sac has been shown to be associated with the onset of agonistic behavior (Riechert, 1978a), with age (Valerio, 1975), and with the presence of specific temperature and wind conditions (Tolbert, 1977). A hunger threshold has been suggested by Turnbull (1964) as providing the stimulus for the migration of older *Achaearanea tepidariorum*, though he felt decreasing humidity caused the migration of both insects and spiders from a pasture in Canada (1966). Hunger probably also resulted in the relocation of the ant specialist *Steatoda fulva*, which is permitted only a single feeding bout at the entrance of an ant mound before the ants change the entrance location (Hölldobler, 1979). An experiment completed by Lindley (personal communication) has, perhaps, provided the best evidence for a causal relationship between hunger and spider migration to date. In the experiment, Lindley divided members of two species of araneids occupying a meadow habitat in England into two groups: one group of individuals received daily food supplementation in the form of *Drosophila* released on their webs, while the other group received no food supplementation. (All individuals were marked as to their group affiliation at the beginning of the experiment.) Lindley found the supplemented group to exhibit a significantly lower emigration rate than the non-supplemented group.

Other factors that have also been shown to stimulate migration in spiders, include: changes in the local thermal environment; web destruction by wind, rain, and marauding animals (Eberhard, 1971; Enders, 1975a, 1976; LeSar and Unzicker, 1978; Marson, 1974); a change in stadium and associated changes in energetic needs (Enders, 1975a; Kronk and Riechert, 1979); production of an egg sac; and loss of a site through agonistic interactions with conspecifics or potential predators (Riechert, 1978b). Several of these factors probably are responsible for the moves made by individuals of a given population, as indicated by the data presented in Table 10.1 for a desert grassland population of the funnel-web spider, *Agelenopsis aperta.*

TABLE 10.1. Relative frequencies of factors responsible for migration in a desert grassland population of the spider *Agelenopsis aperta*. *N* = 518 (initial dispersal of spiderlings not included).

Loss of territorial dispute	19.04%
Following molt	5.26%
Web destruction (wind, rain, animals)	28.07%
Unknown (insufficient prey, unfavorable thermal conditions, etc.)	37.63%

Once migration has been initiated, most workers would agree that random searching occurs, ending perhaps when cues signifying a favorable location are encountered, such as contact with the odor of a preferred prey as in the case of *Steatoda fulva* (Hölldobler, 1979). The specific cues utilized by spiders in locating suitable habitats undoubtedly vary, though the method delineated for the funnel-web spider, *A. aperta* provides some indication of how the more polyphagous (generalist feeders) spiders might accomplish this end. On numerous occasions Riechert has observed *A. aperta* to run out on their sheet-webs when flies were buzzing overhead with no apparent contact made with the sheet or support threads. The hypothesis was made that *Agelenopsis* is responding to the airborne vibrations of this prey, and tapes of buzzing flies were played to spiders both sitting in their funnels and off their webs. Though the data are incomplete, orientation and approach in the direction of the speaker were observed in some runs. These findings, along with those presented in the chapter by Barth, suggest that spiders in search of new locations can orient toward airborne vibrations corresponding to the frequencies of favored prey taxa and may follow a gradient of such stimuli to settle in a patch affording a threshold level of prey.

Once an animal has located such a patch, depending on the environmental pressures impinging on it, it may seek (1) those structural features necessary for construction of a specific type of web (Colebourne, 1974; Enders, 1975a), (2) favorable physical environments (Eberhard, 1971; Riechert and Tracy, 1975), or (3) features providing ease of capture (Greenquist and Rovner, 1976; Kronk and Riechert, 1979). In both experimental and field contexts, *A. aperta* chooses locations that permit it to achieve a preferred body temperature of approximately 31°C. (Though individuals may stop searching anywhere within a favorable range (21-35°C), they frequently will resume the search within a short period, seeking the preferred temperature.) In addition, once a specific site is selected, animals at the more exposed sites will orient their funnels such that the sun's rays extend down the tube for the shortest period of time possible during mid-summer (Riechert and Tracy, 1975).

Obviously this species, as a desert inhabitant, is strongly influenced by its thermal environment and its need to escape heat stress. Where spiders are released from environmental pressures, one often observes the selection of a substrate that will increase the detection of prey (Greenquist and Rovner, 1976) or capture efficiency (Kronk and Riechert, 1979). Both of the studies cited in this case were completed on wolf spiders (Lycosidae), though selection of features requisite to construction of a web probably functions in increasing the efficiency of prey capture as well.

Seasonal and Diel Activity Patterns

Encounter between a spider and a given prey further depends on the temporal activity patterns of each. Seasonal activity delineates the period of overlap between predator and prey occurrences in a community. It can also be used as a predictor of when maximum predatory activity might occur, since spiders are especially active just after the last molt (Haynes and Sisojevic, 1966; Miyashita, 1968a,b; Witt et al., 1972) and before cocoon production (Kajak, 1965; Łuczak, 1970; Vlijm and Richter, 1966).

The life cycles of the spiders that concentrate on specific prey types will probably have undergone evolutionary adjustments corresponding to the activity periods of their prey. Łuczak (1970), for instance, feels that the seasonal and diel activity patterns of *Tetragnatha montana* have developed in response to the activity of its main prey, culicid mosquitoes. For most spiders, however, seasonal activity patterns are likely to be adjusted only to the general temporal variations in prey abundance characteristic of specific habitats. Since this kind of activity pattern does not lend itself to modification by proximate cues, we will forego further discussion of it here.

Diel activity does lend itself to influence by external cues (Edgar, 1970; Riechert and Tracy, 1975; Vlijm and Kessler-Geschiere, 1967); consequently, spiders may modify their behavior according to the activity of associated prey. This influence has been nicely demonstrated for the long-jawed orb weaver *Tetragnatha montana* which occupies the herb stratum of alder forests in Poland. In two studies, the addition of large numbers of mosquitoes to enclosures containing individual *Tetragnatha* stimulated web construction and subsequent feeding activity during the normal midday resting period of this spider (Dabrowska-Prot et al., 1968a,b; Łuczak, 1970).

Optimal Foraging

Numerous contributions have been made to our understanding of the feeding habits of specific spider species, and yet there has been little effort to search for patterns in foraging behavior among spiders or to look for the ultimate factors responsible for the patterns exhibited. Work by Enders (1975b, 1976) and the late Robert Givens (1978) are notable exceptions. We will try to follow their example here by looking for central tendencies in spider foraging behavior and attempting to consider these in terms of optimization.

The Sit and Wait–Active Searcher Continuum

Schoener (1971), in a mathematical treatise on predator specialization toward prey, uses an *Anolis* lizard as a close approximation to a "Sit-and-Wait" predator since "a pure example of this kind of predator is never found in nature." We would like to offer the web-building spider as exemplary of the Sit-and-Wait predator which "expends no time or energy in food search that is not simultaneously used in other activities." Once a suitable location is found and a web constructed, these spiders spend most of their time stationary—waiting for food items to enter within a potential capture range while simultaneously monitoring the vicinity for potential predators, mates, and competitors. Within the web-building group, Enders (1976) feels there exists a continuum from the more sedentary ambushers to the more agile searchers for prey. (According to

Enders, this same continuum should also exist within the short-sighted and long-sighted hunting groups.) Among the web builders he would predict the agelenids (funnel-web spiders), which lack sticky silk and are highly mobile, to engage in more searching activity on and in the vicinity of the web than the often globular-shaped araneids (orb weavers), which rely on the sticky spiral to detain potential prey. Riechert has observed individuals of a desert grassland population of the spider *Agelenopsis aperta* to search the web-sheet between long bouts of sitting at the funnel entrance. This activity results in the collection of leafhoppers and other small prey which would otherwise escape detection. Similar behavior, however, has not been observed in another population of the same species inhabiting a more favorable environment where food is in ready supply. Sit-and-Wait appears to be the dominant strategy even in this agile species, with perhaps a low threshold to web vibrations when hunger levels are high (resulting in apparent searching behavior).

What about the hunting spiders (taxa that do not utilize the web trap in the capture of prey)? Is their categorization as active foragers (Bilsing, 1920; Turnbull, 1973) more myth than fact? As studies are completed with what we thought were the most active searchers, we find that most of their time is spent in the Sit and Wait strategy (Lycosidae: Cragg, 1961; Edgar, 1969, 1970; Ford, 1978; Oxyopidae: Weems and Whitcomb, 1977). There are really only three groups for which there are both behavioral data and dietary evidence documenting an active searching mode: the jumping spiders (Salticidae) (Edwards, 1975; Fitch, 1963; Gardner, 1965; Lincoln et al., 1967; Whitcomb and Tardic, 1963; Whitcomb et al., 1963), philodromid spiders (Turnbull, 1973), and the Loxoscelidae (Bushman et al., 1976). The view of the salticid, for instance, is of an animal that actively patrols a substrate in search of prey (Edwards, 1975); various other spiders—Anyphaenidae, Clubionidae, Gnaphosidae, Loxoscelidae—are considered their nocturnal counterparts. Certainly the fact that members of these groups feed on caterpillars and even on insect eggs (Bushman et al., 1976; Richman and Hemenway, in press) is indicative of active foraging. However, even jumping spiders spend much of their time stationary on plants, orienting toward prey only when they come within the spider's field of vision (Salticidae: Givens, 1978; Hill, 1979b). Though little is known of the predatory behavior of the sac spiders, it is probable that these taxa too utilize the Sit-and-Wait strategy to a large extent.

Why are spiders basically ambush predators? The point Miyashita (1968a) and Anderson (1974) made concerning food limitation suggests an answer. If spiders frequently find themselves under food stress, natural selection should favor those individuals who minimize energy expenditure. This can be accomplished in two ways which are somewhat interdependent: by lowering metabolic rates and by decreasing activity. Spiders have notoriously low metabolic rates (Anderson, 1970; Carrel and Heathcote, 1976; Peakall and Witt, 1976). A Sit-and-Wait foraging strategy is appropriate to this kind of physiology, and coupled with it, permits animals to survive long periods of food deprivation. The foraging cost per food item can be considered to be roughly equivalent to the energy expended in metabolism during the period spent waiting for the prey.

Spiders that utilize web traps have an additional cost—the energy spent in construction of the web. This cost should be divided among the prey items caught during the life of the web, and added to the foraging cost. Ford (1977) estimated the energy expenditure in web construction to be approximately 8 times that required for the basal metabolism of a sheetline-weaving spider over the course of a day. In order for web building to have evolved, then, we would have to assume that the increased perception radius provided by the first webs must have resulted in a greater net benefit than the flexibility to move from patches of low prey density to those affording more prey. In what kinds of situations should we expect to find our web builders as opposed to our hunting spiders? We would expect web builders to be favored in spatially homogeneous habitats where movement from patch to patch is not required. In times of food limitation in heterogeneous habitats, the better utilization of a patch provided by the web trap is probably also favored over movement in search of localized prey. Observations of the desert wolf spider *Lycosa santrita* support this hypothesis. *Lycosa santrita* normally waits on grass clumps and attacks prey as they pass within the vicinity. Under food stress, however, it has been observed both to utilize the web of a funnel spider occupying the same habitat and to build its own webs (Kronk and Riechert, 1979). Carico (1978) discusses this phenomenon in reverse—the factors leading to web reduction in spiders. He feels that selection of microhabitat patches where prey are superabundant has led to the loss of the web trap of the theridiid *Euryopis funebris* (Hentz).

Just as location in the habitat and timing of activity influence predator-prey interactions, it is important to point out that the type of foraging strategy exhibited by spiders determines to a large extent the kinds of prey this predator will encounter. Prominent among the prey of a Sit-and-Wait predator are the more motile forms—active fliers, jumpers, and runners—which constitute 98.8%

of all prey hitting the webs of a sheetline weaver (Turnbull, 1960; Nakamura, 1977). The more stationary prey (larvae, eggs, aphids, mites, etc.) are represented primarily in the diets of spiders that actively search for their food (Bushman et al., 1976; Jackson, 1977e; Richman and Hemenway, in press).

Prey Selection

Specialist–Generalist Feeders

Savory in his 1928 book on the biology of spiders states that they will eat just about anything, showing "no trace of discrimination." While most observers might tend to agree with Savory's view of spiders as generalist feeders, Bristowe (1941) emphasized the large number of invertebrate taxa that are rejected by spiders in his treatise on the subject. Perhaps consideration of spider predatory behavior as a problem in optimization might explain these conflicting views.

Basic to current models of animal foraging behavior is the assumption that individuals should attempt to obtain the greatest possible energy reward while feeding, something which can be achieved through discrimination of the relative profitability of various prey types. (By prey types here we are referring to both size classes and taxonomic affiliations.) Profitability is usually expressed in terms of costs and benefits: the greater the benefit relative to the cost, the more profitable the prey (Krebs, 1978). Within the range of available prey, then, there exists an optimal set of prey types for which it is economically feasible to attempt capture.

Prey falling outside this optimal set generally have a higher cost than benefit, making it uneconomical to attempt their capture. Such forbidden prey for spiders might include animals that are beyond the size range of the trophic apparatus (the front grasping legs or chelicerae), distasteful prey, and those predatory insects which have the capacity to reverse the direction of the predator-prey interaction. Thus, many coleopterans with their impenetrable elytra (hardened front wings) and heavy bodies are outside the optimal diet of the majority of spider species. Energy expenditure would be great in attempted capture of these beetles, while eventual escape is probable. Likewise, most spiders do not attempt capture of distasteful taxa (chemically noxious) including lysid, chrysomellid, and coccinellid beetles; milkweed and stink bugs; and some lepidopteran larvae and adults. If the individual spider does find these taxa to be noxious or distasteful, capture of them is wasted energy expenditure. Predatory wasps and ants are also rejected as prey by many spiders, probably because of the potential for injury to the predator; and very large invertebrates, regardless of taxonomic affiliation, are routinely rejected as prey. This is despite the fact that even hunting spiders readily attack prey that are larger than themselves (Morse, 1979; Robinson and Valerio, 1972; Rovner, 1980a).

Nevertheless, the optimal set of prey is quite large for most spiders relative to that exhibited by other animal groups. This fact is demonstrated by the data presented in Table 10.2. Although there is some dietary evidence to suggest that spiders actively maintain a mixed diet for nutritional reasons (Greenstone, 1979), the great diversity of prey types taken by this predatory group primarily reflects an adaptation to temporally heterogeneous food supplies and adoption of the Sit-and-Wait foraging strategy. By virtue of its stationary hunting method, the Sit-and-Wait predator will encounter fewer prey than the mobile predators which can actively seek out specific prey types; and if its prey are in limited supply, the spider will need to act as an opportunist—attack any prey type that falls within the optimal set.

The degree to which an individual animal will specialize on the more profitable prey within the optimal set, then, must depend on local prey abundance and the relative availability of specific prey types to it as well as on its internal state (e.g., pre- or post-molt, gravid, degree of satiation) at the time a potential prey is encountered. If we assume that animals tend to harvest their food efficiently as a result of selection pressures on them (MacArthur and Pianka, 1966), then, given saturation levels of prey, spiders should choose among the prey and try only for those taxa and size classes that are most profitable (Bristowe's 1941 observation). On the other hand, under less favorable conditions, individual spiders should be opportunistic—try for all prey that are economically feasible to capture (Savory's 1928 observation). Exemplifying this decision rule are the acceptance rates of different prey taxa observed for two populations of *Agelenopsis aperta* afforded different levels of prey (Figure 10.1). Members of a population occupying a stringent grassland environment with low numbers of potential prey attempted capture of 99% of the prey taxa hitting their webs, including aposematic taxa which were subsequently consumed (Riechert and Tracy, 1975). In a comparative study recently completed by Riechert (unpublished data) of another population of *A. aperta* occupying a riparian habitat in southeastern Arizona, spiders were found to attempt capture of only 59% of the prey taxa encountered, and

TABLE 10.2. Relative taxonomic composition of the diets of various spiders (mean and standard deviations).

Diet Items	*Lycosidae (Pardosa)*	*Salticidae (Phidippus)*	*Hypochilidae (Hypochilus)*	*Dictynidae (Dictyna, Mallos)*	*Theridiidae (Achaearanea)*	*Agelenidae (Agelenopsis, Coelotes)*	*Tetragnathidae (Tetragnatha)*	*Araneidae (Argiope, Araneus, Nephila)*
Misc. Arachnida			8.1		1.3	1.6 ± 2.9		2.4 ± 2.1
Misc. Insecta			5.4	.7 ± .8	.9	2.5 ± 4.7	8.4 ± 11.4	2.5 ± 3.0
Araneae	31.1 ± 6.3	21	9.9	.3 ± .5	14.8	3.5 ± 3.2	1.0 ± 1.4	
Coleoptera		2	8.1	3.1 ± 2.3	14.4	2.1 ± 0.5	2.9 ± 1.6	7.7 ± 14.7
Collembola	4.3 ± 2.8							
Dermaptera		2						
Diptera	29.7 ± 11.5	40	18.0	89.5 ± .2	31.7	47.3 ± 27.4	38.9 ± 24.2	48.8 ± 39.0
Hemiptera	12.0 ± 14.2	14	1.0			0.9 ± 0.8	8.9 ± 12.5	3.1 ± 2.6
Homoptera	2.0 ± 2.4	8	9.9	1.5 ± .8	5.6	17.6 ± 18.9	12.5 ± 1.6	5.7 ± 5.5
Hymenoptera	3.8 ± 4.8	6	8.1	1.7 ± 1.1	10.2	16.8 ± 5.0	1.9 ± 2.7	14.2 ± 27.5
Isoptera						0.2 ± 0.3		
Lepidoptera Adults	2.3 ± 2.7	4	3.6	2.6 ± 3.5	3.3	3.1 ± 3.9		8.4 ± 13.4
Lepidoptera Larvae	0.4 ± .8	2			2.0	0.7 ± 4.2		
Odonata					4.2	2.2 ± 3.8		0.2 ± 0.4
Opiliones	.4 ± .8	2	4.5		6.5	1.4 ± 2.5		0.3 ± 0.6
Orthoptera			22.5	.3 ± .5	3.0	10.2 ± 9.5		2.7 ± 4.7
Plecoptera	.2 ± 0.4				1.3			

Lycosidae: Edgar (1969, 1970), Hallander (1970); Salticidae: Jackson (1977c); Hypochilidae: Riechert; Dictynidae: Jackson (1977e); Theridiidae: Riechert; Agelenidae: Riechert and Tracy (1975), Riechert (1978a,b, 1979); Tetragnathidae: Dabrowska-Prot and Luczak (1968), LeSar and Unzicker (1978); Araneidae: Kajak (1965), Robinson and Robinson (1970, 1973a), Riechert .

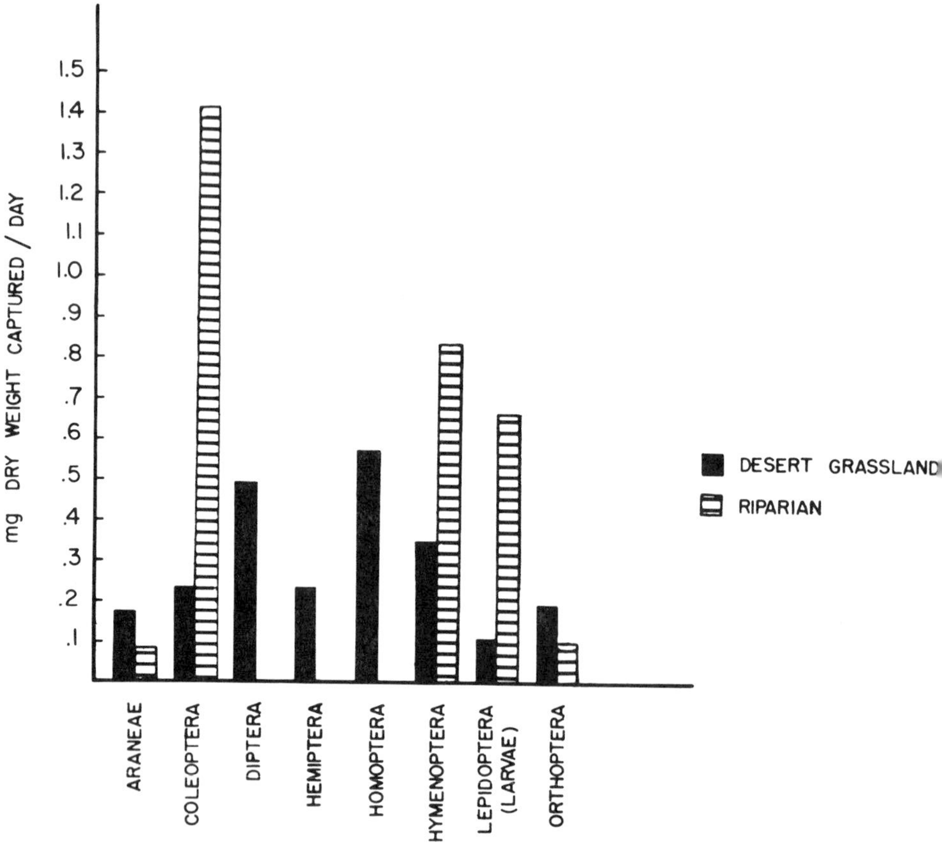

FIGURE 10.1. Between-population comparison of capture rates of *Agelenopsis aperta* with different prey types. Riparian population from southeastern Arizona, desert grassland population from south-central New Mexico.

aposematic taxa were usually rejected. Members of this latter population receive almost threefold the prey numbers encountered by the desert grassland individuals, weigh significantly more than their cousins, and in general live under optimal conditions. That degree of specificity can be adjusted to local conditions is further indicated in the results of an experiment completed on members of the desert grassland population. Following four weeks of food supplementation in the field, the average attack rate exhibited by individuals used in the experiment decreased from 59.7% of all prey encountered while active to 36.4%. (The riparian population exhibits an average attack rate of 25.5%.)

For situations where some choice can be exhibited toward specific prey, how is this prey selection achieved? For the web builder, prey selection may occur at two levels. Initially prey may be filtered via the web trap according to size and taxon. Following entanglement a given prey item may then either be ignored, approached and attacked, or approached and rejected. Further, rejection can take two forms in that a prey item may be left on the web or physically removed. Riechert, for instance, has observed *Agelenopsis aperta* to cut large rhinoceros beetles, *Dynastes* sp., from their sheet-webs and then repair the rifts made during the process. This is the active phase of prey choice where the individual spider might utilize visual, vibratory, and/or chemical cues to discern the size and type of intruder present. Its choice between acceptance and rejection is then based on this information and on either prior experience with or some predisposed "view" of organisms exhibiting similar characteristics to the intruder just encountered.

Passive Selection

Spiders use silk for a variety of purposes including the construction of egg sacs, retreats, safety or drag lines, and travel lines for aerial dispersal (Levi, 1978). Its best-known use, however, is in the construction of the web trap.

Bristowe (1941) stated that spiders exhibit different catching traps in avoidance of competition for food. Spider numbers and species diversity are higher as a result of this form of niche partitioning. Associated with web structure diversification is the evolutionary development of the web as an effective trap, culminating with the orb-web, a model of effectiveness in economy of construction (Kullmann, 1972b; Levi, 1978; Robinson, 1977b; Witt, 1965).

The major web types (e.g., scattered, sheet, and orbs) are believed by some workers to be specifically adapted to particular habitat structures and to the capture of specific prey, with the orb-web capable of capturing a wider range of taxa than other web types (Bilsing, 1920; Turnbull, 1973). Data are available on the passive selection of specific web structures for a number of species populations (Dabrowska-Prot and Łuczak, 1968; Kajak, 1965; Kiritani et al., 1972; Nyffeler and Benz, 1978, 1979; Riechert and Tracy, 1975; Robinson and Robinson, 1970, 1971; Turnbull, 1960). Unfortunately, none of these studies includes species exhibiting different web structures while occupying the same habitat; and a between habitat comparison of these data cannot be made, since the range of prey available frequently varies with habitat, season, etc. (Table

10.2), and since different methodologies have been used in the various studies. Riechert and Cady, however, have recently completed a comparison of the fundamental niches of spiders occupying sandstone rock faces in the Cumberland Plateau, east Tennessee (1978-79 field studies). Prominent in this community are: a scattered line weaver, *Achaearanea tepidariorum*; a sheetline weaver and funnel-web builder, *Coelotes montanus*; and an orb-weaver, *Araneus cavaticus*. Figure 10.2 shows a comparison of the catches of these webs (encounter and attack frequencies with various taxonomic groups). Important to our discussion here is the degree to which the web types vary in their catches. A chi-square test performed on the data shows marked differences between the prey compositions associated with these webs ($\chi^2 = 116$, d.f. $= 14$, $P < 0.001$). However, if one looks at the prey orders responsible for this variation, we see that location on the cliff face rather than web type itself is responsible for the filtering. The major differences in prey association, for instance, are between the orb-web which hangs off ledges and is often a meter or more away from the cliff face, and the other two web types which have much of the catching area in contact with the cliff face itself. Thus, crawling prey such as other spiders, harvestmen, beetles, and crickets more frequently encounter the *Coelotes* and *Achaearanea* webs, while the Lepidoptera, which flutter in the vicinity of the cliff face but do not huddle under its ledges, contact the *Araneus* webs. These data indicate that structure plays little or no role in the passive selection of prey taxa, exceptions occurring only in a few special cases (e.g., ladder webs: Eberhard, 1975). What usually does occur is that certain web types require specific habitat features for their construction which place them in different microhabitats from other web types. It is primarily the microhabitat association that results in the partitioning of prey taxa; and microhabitat and web size influence the numbers of prey contacting webs.

Any filtering that might occur is more likely to involve prey size differences within a given class of webs than prey type. Mesh size differences in orb-weavers, for instance, has often been considered a method by which different species and maybe even different instars of the same species partition food (Risch, 1977). A good example of this phenomenon is given by Uetz et al. (1978) for the North American araneids, *Argiope aurantia* and *A. trifasciata*, which frequently share the same habitat. Differences in prey size are statistically significant for the two species, with *A. aurantia*, as the larger spider, exhibiting a larger mesh size and feeding on larger prey.

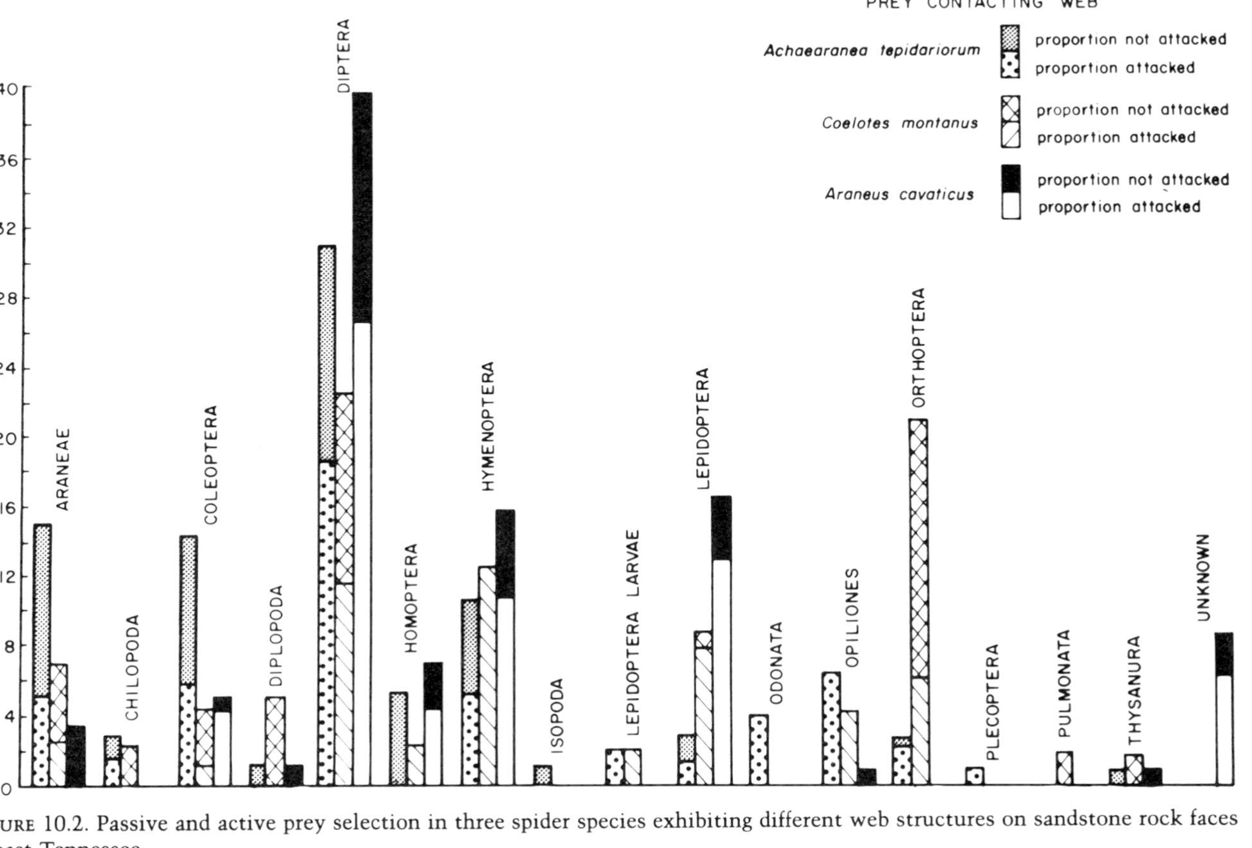

FIGURE 10.2. Passive and active prey selection in three spider species exhibiting different web structures on sandstone rock faces in east Tennessee.

Active Selection

We know that the species composition of prey eaten in a web frequently does not reflect the full range available to the spiders (Kajak, 1965; Riechert and Tracy, 1975), despite the fact that some workers have largely viewed spiders as passive filterers, consuming specific prey taxa in proportion to their numbers hitting the webs (Pointing, 1966). When prey are sufficiently numerous, spiders appear actively to discriminate among the taxa and size classes available to them. What are the criteria that are used in the selection process? Keeping profitability in mind, the spider should consider its own level of hunger and the size and type of potential prey encountered. All three parameters are important if the individual is to decide whether it should wait for a more profitable (preferred) prey or consume the one at hand.

PREY SIZE Both Nielsen (1932) and Bristowe (1941) considered size to be an important determinant of the acceptability of a given prey item. According to Bristowe, an acceptable prey must be smaller than its predator, but not smaller than one-sixth of its size. In economic terms, the lower size limit reflects the point at which greater energy expenditure is spent retrieving and handling the prey than will be gained from its consumption. At the higher end of the range, the probability of escape is such that large energy expenditure may be met with no gain whatsoever.

PREY TYPE In addition to the ease of handling specific prey types, there is also evidence to suggest that the relative familiarity of the prey type is important to the profitability decision. Spiders have been shown experimentally to be capable of both habituation (Szlep, 1964) and learning (LeGuelte, 1969). Neither of these studies involved prey capture specifically, but Turnbull (1960), in his field study of the predator-prey relations of the sheetline weaver *Linyphia triangularis* found that the probability of attack increases with familiarity with a specific prey type. Riechert (unpublished data) has seen similar behavior in the predator-prey interactions of *Agelenopsis aperta*. An example of this phenomenon was the change in response of members of a riparian population to honeybees, associated with variation in bee abundance. *Agelenopsis* rejected all bees offered experimentally to them early in the field season and prior to bee abundance in the habitat. Following the bloom of sweet clover in the study area, these pollinators frequently encountered the webs of this spider (relative frequency of contact = 27.2%) and the spider's capture success with them increased from 0 to 71%.

PHYSIOLOGICAL STATE Turnbull (1960) alludes to the extreme vagary of the selection process as evidenced by the results of a study of the prey relations of *Linyphia triangularis.* Of 153 prey types contacting the webs of this spider, 33 of the species were accepted sometimes and rejected at other times. Contributing to this variance was the experiential component just mentioned; but Turnbull also noted that the more frequent the contact of a given prey type with the webs, the greater was the variability in spider behavior toward it. Although this observation appears to contradict the idea that probability of attack increases with familiarity, it actually reflects an unrelated factor—variation in the physiological state of the predator: the greater the frequency of encounter the greater the probability that contact with the webs will be made during periods when individual spiders are very receptive, less receptive, and totally unreceptive to food as a result of such factors as stage in the molt cycle, body temperature, time since last feeding, etc.

PREY AVAILABILITY Finally, the relative availability of different prey types is likely to affect the probability of attack. Dabrowska-Prot and Łuczak (1968), for instance, found that *Tetragnatha montana* would not prey upon the dipteran *Tricholauxania* sp. in the presence of culicid mosquitoes. Only the culicids were eaten. However, after the spiders had spent a few days in the presence only of *Tricholauxania*, Tarwid (personal communication) found that *Tetragnatha montana* would accept this dipteran as food. Permitted to choose among two abundant prey types, the spider exhibited a strong preference, though it accepted as food the less preferred prey when only this was available.

The Predatory Sequence

Let us now look at the predatory sequence itself and try to distinguish among the cues that spiders might utilize in making their profitability decisions and in handling specific prey in the most efficient manner possible. Figure 10.3 shows a schematic representation of the basic stages in the predator-prey interaction for both web-building and hunting spiders. Once a suitable hunting location is found, the majority of a spider's foraging time is spent in a Sit-and-Wait position. Following contact with or receipt of some stimulus from a potential prey, the predatory sequence is initiated with the spider's orientation toward the prey, through which its movements can be monitored or its location determined. The sequence

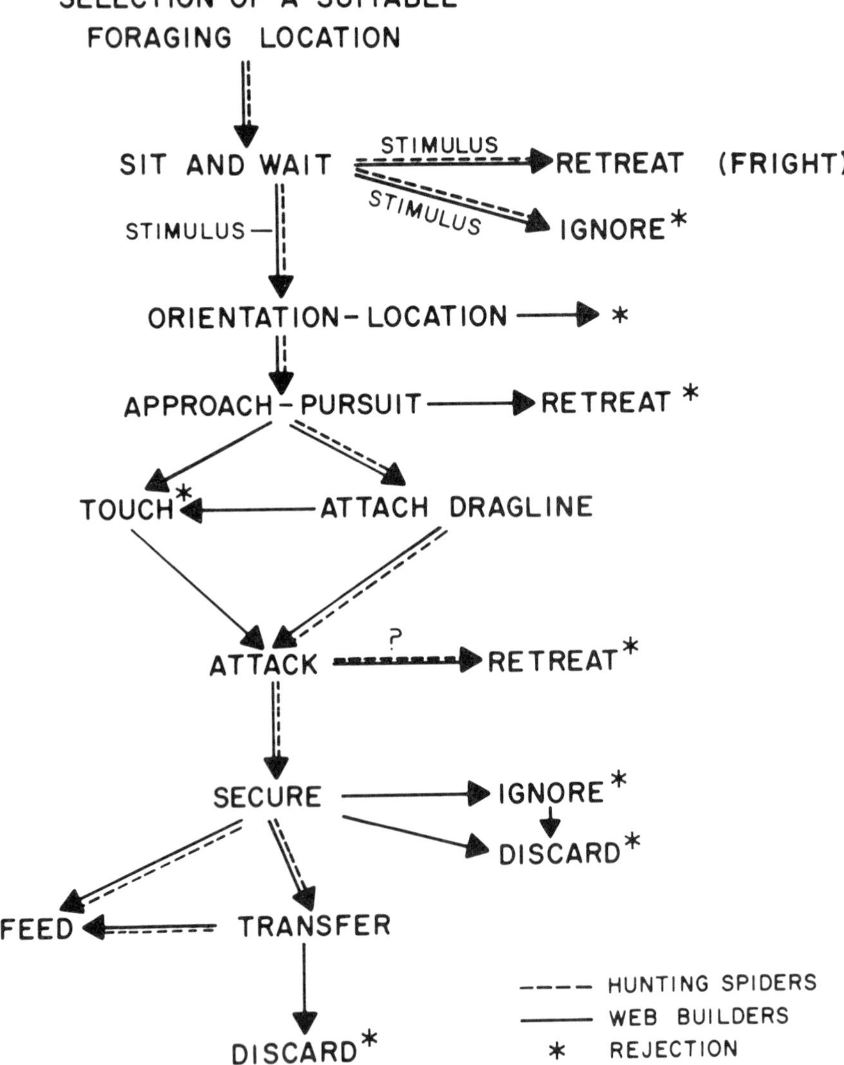

FIGURE 10.3. Generalized spider predatory sequence, showing major transitions and stages at which selection has been known to occur.

continues with an approach or pursuit followed by attack and the capture or securing of the prey item. The prey may subsequently be eaten at the location of its capture or transferred to a retreat or web hub.

Note that at every stage in the sequence, the behavior of the prey can elicit appropriate responses in the predator, and that although the previous behavior exhibited by the predator may trigger subsequent behavior to some extent, alternative behavior patterns and variability in their duration, frequency, and amplitude are determined largely by stimuli provided by the prey (Harwood, 1974; Peters, 1931, 1933a,b; Robinson, 1969; Robinson and Olazarri, 1971).

Selection Process

Figure 10.3 also shows the points at which rejection of a potential prey item is known to occur in spiders (i.e., just about any time, including following capture). The principles of economics dictate that the rejection process should be completed as early as possible in the predatory sequence to avoid wasteful time and energy expenditures. An efficient predator is one that perhaps can avoid even the orientation or approach phases when unprofitable prey are contacted. Though *Agelenopsis aperta* has been observed to discard or ignore secured prey, most prey rejection takes place early in the encounter (Table 10.3).

Table 10.3. Probability of prey rejection taking place at various stages of the prey capture sequence of *Agelenopsis aperta* (desert grassland habitat, south-central New Mexico). $N = 534$ natural capture sequences.

Following	*Total Rejection*	*Ignore*	*Frighten*
Sit-and-Wait	12.03%	9.60%	2.43%
Orientation	1.71%	1.71%	
Approach	0.57%		
Touch	1.52%		
Bite	3.42%		
Secure-Transfer	1.33%		
Total frequency of prey rejected	20.58%		

Among the hunting spiders, prey selection has only been cited for the Sit-and-Wait and Orientation-Location stages of the capture sequence, though we expect some rejection to occur following the initial attack (Figure 10.3). Most of our understanding of the predatory behavior of this group comes from work with its highly visual member, the jumping spiders (Salticidae). Although tactile and vibratory information might be important to other hunters, salticids primarily use visual recognition of a set of specified characteristics in the selection of prey of appropriate size, color, velocity, and distance (Carthy, 1958; Crane, 1949a; Dill, 1975; Drees, 1952; Land, 1969). The set of visual cues necessary to elicit a capture response in the salticids varies with the physiological state of the spider (Drees, 1952). In general, however, prey must be moving to elicit the capture sequence, an exception being very small objects in close proximity to the predator (Drees, 1952).

Despite the fact that the jumping spider relies strongly on visual perception in its prey selection process, it may monitor other sources of information as well. For instance, Hollis and Branson (1964) successfully habituated *Phidippus audax* (Hentz) to auditory and vibratory stimuli. Spider legs in general are equipped with extremely sensitive vibratory receptors which permit some spiders to monitor airborne vibrations (see chapter by Barth) and all web builders to discriminate among prey vibrating at different frequencies and amplitudes on their webs (Frings and Frings, 1966; Parry, 1965; Walcott, 1969). As Witt (1975) states, the web "constitutes an essential route of access to the spider's perceptual systems."

The information the web conveys permits the individual spider to exhibit appropriate responses at a safe distance: attack to a preferred prey, alarm and retreat to a potential predator, and indifference to a less preferred prey (Barrows, 1915; Bays, 1962; Boys, 1880; Cloudsley-Thompson, 1960; Emerit, 1967; Kaston, 1937; Suter, 1978; Szlep, 1964). It is to an individual's benefit, for example, to discriminate between acceptable and unacceptable prey through web vibrations, since this prevents the spider from wasting energy in unprofitable pursuits (Suter, 1978). Recognition of the nature of a prey item that has contacted the web before beginning the approach has another advantage to the spider in that this can signal the kind of capture sequence dictated by the prey type. Thus the orb-web spider *Argiope argentata* has processed size information concerning the entangled prey before it leaves its resting position at the orb hub, or at least during its initial approach, and its approach is modified according to this information—slower for larger prey, often with the front legs raised (Robinson and Olazarri, 1971).

Accurate processing of vibratory information can also mean the difference between *A. aperta*'s success with or loss of a prey item (Riechert, unpublished observations). It may even be important in avoiding possible injury. Flies and moths must be approached and pounced on quickly before they recover flight, while large grasshoppers and ants need to be approached more cautiously. Their escape is less imminent, though damage from a kick or bite are finite possibilities. Both of these prey are approached from the front with *A. aperta* in a reared-up position, while flies and moths are approached from the thorax side on the run.

Toward the specialist end of the spider feeding continuum are the dictynids, which primarily feed on dipterans (Table 10.2; Jackson, 1977e). Burgess (1975) has shown that these spiders respond to a fairly narrow range of vibration frequencies corresponding to the wing-beat frequency of flies. The web design also transmits this frequency better than others, so it is difficult to distinguish which factor is responsible for the response. At the other end of the continuum are the spiders that show less discrimination of vibratory cues. The lampshade spider, *Hypochilus gertschi* (Hypochilidae), responds to a potential prey item only if it hits the shade portion of the web and struggles continuously (Shear, 1969), while many spiders discriminate only between potential prey (smaller items) and potentially damaging items (large) on the basis of weight information transmitted through the web. This is true of the scytodid *Drymusa dinora* which flattens on its sheet in the presence of very large prey until they leave the web (Valerio, 1974a). An orb weaver of the family Tetragnathidae, *Tetragnatha laboriosa*, also shows little discrimination, responding to most intruders as prey with the exception of large beetles (LeSar and Unzicker, 1978).

Following location of the prey object via orientation and approach, many spiders will touch or palpate a prey item, especially if it is unfamiliar to them (Robinson, 1969; Robinson and Olazarri, 1971; Shear, 1969). Some hunting spiders, in fact, actually require tactile stimulation under most circumstances to elicit attack (e.g., *Philodromus rufus* (Thomisidae): Haynes and Sisojevic, 1966; and *Euryopis funebris* (Theridiidae): Carico, 1978).

The bite, however, is the ultimate test of a prey's palatability (Peters, 1933a; Robinson, 1969; Robinson and Robinson, 1973a). *Nephila maculata*, for example, pulls and discards distasteful insects from its orb-web following biting (Robinson and Robinson, 1973a); and *Agelenopsis aperta* retreats and ignores unsuitable prey following biting (Table 10.3). If unsuitable prey are among the aposematic taxa, rubbing of the chelicerae against plant material may even follow the short bite (Riechert, unpublished observations). It is less clear as to whether the visually hunting spiders may reject prey following contact. Reference is made only to rejection or attack following orientation and retinal scanning of the object to determine its size relative to that of the spider and its characterization as potential prey, a conspecific, predator, etc. (Dill, 1975; Edwards et al., 1974).

Finally, Riechert (unpublished data) has observed *Agelenopsis* to ignore subdued prey and eventually to remove these from the web. This has only occurred in the presence of abundant prey, and in 100% of the cases, a prey captured shortly after the securing of the eventually discarded item is given priority in consumption. Haynes and Sisojevic (1966) have noted similar behavior in the crab spider *Philodromus rufus*.

Prey Capture

VARIATION RELATED TO PREY CHARACTERISTICS Turnbull in his 1973 review of spider ecology expressed the view that prey rarely escape a determined attack. This phenomenon is largely related to two facts: (1) following the rules of profitability, spiders tend to reject prey that are difficult to handle, and (2) they have developed methods of attack suitable to particular prey (Bilsing, 1920; Eberhard, 1967; Harwood, 1974; Levi, 1967; Robinson, 1969; Robinson and Olazarri, 1971; Robinson et al., 1969). The first of these features has already been discussed in the previous section. The variability in the handling of different prey taxa by *Agelenopsis aperta* is exemplary of the second point.

Figure 10.4 shows the placement of 534 predatory sequences of *A. aperta* from a desert grassland habitat relative to one another along two axes of variability. Apparent from the map, based on multiple discriminant analysis results, is the fact that various prey taxa are handled differently. The method of capture utilized in subduing coleopterans and hymenopterans differs maximally from that utilized in subduing dipterans and lepidopterans. These two major groupings of sequences are located at opposite ends of the second axis, reflecting differences in the method of capture (Figure 10.4). The potentially dangerous beetles, ants, and wasps are captured via numerous short bites and the binding of biting or stinging body parts to the web with silk. On the other hand, dipterans and lepidopterans are pounced upon and a sustained bite to the thorax or thorax side delivered, often while the spider receives a free ride around the web-sheet. Successful capture of these latter prey re-

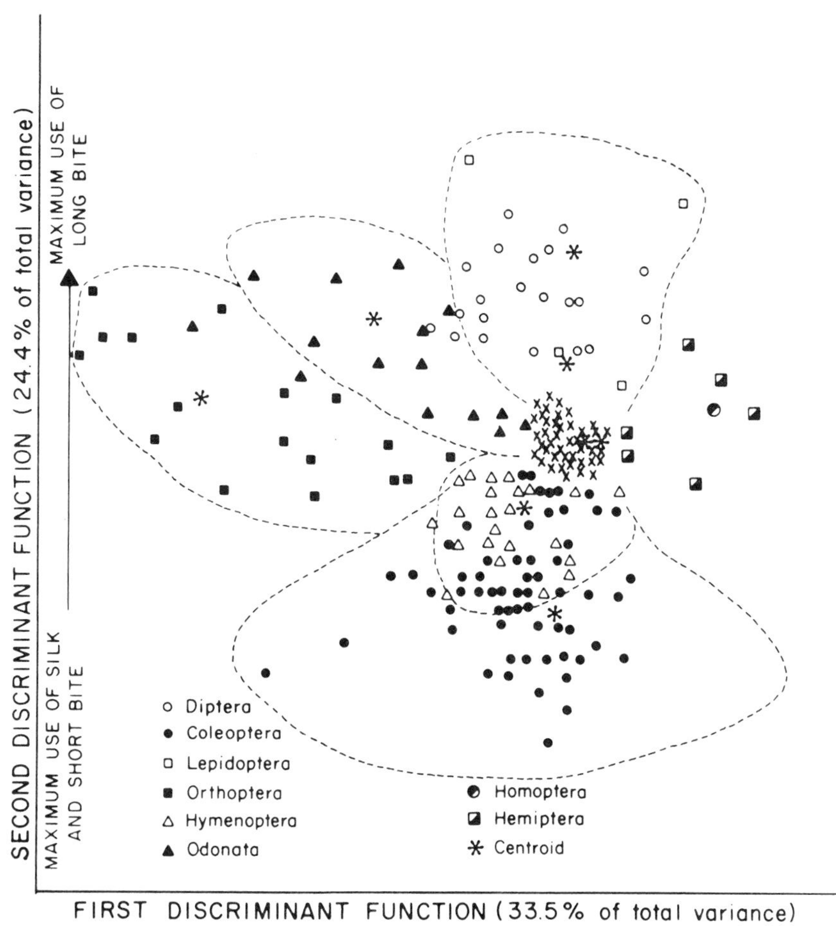

FIGURE 10.4. Results of discriminant analyses completed on predatory behavior exhibited by *Agelenopsis aperta*, desert grassland populations, in 534 encounters with diverse prey. (Sample size: minimum of 20 sequences for each prey type.)

quires immediate action, while close-quarter contact is not dangerous to the spider. The capture technique used on grasshoppers and crickets is actually most similar to that applied to beetles. The placement of the orthopteran encounters on the graph closer to the sustained-bite group reflects the fact that *A. aperta* frequently dismembers one of the large hind legs of these prey during their capture. When this happens, the spider ignores the remainder of the prey while it feeds on the limb. Because this feeding occurs on the web-sheet rather than in the funnel (the case for prey in general), the behavior is recorded as a sustained bite.

The first axis, accounting for the greatest amount of the among-sequence variability, reflects differences in the handling time and hence energy expenditure involved with different prey types. Large grasshoppers, crickets, roaches, and dragonflies require considerably more time to subdue than the smaller prey represented at the right side of the graph by the leafhoppers and mirid bugs. The cluster of sequences near the centroids for Hemiptera and Homoptera, in fact, represents captures involving small prey of various taxonomic affiliations. Small prey items, in general, are searched for, seized, and carried to the funnel, where they are consumed while the spider waits for additional prey.

We would expect *Agelenopsis* as a sheetline weaver to have a fairly complex predatory repertoire. Its web provides little retention of prey hitting it, serving primarily for prey detection and localization. Yet its large area and continuous nature frequently place the spider in contact with a diversity of prey sizes and taxa. Efficient handling by the spider is required to make maximum use of the available nutriment. If Witt's (1965) idea of "a steady evolution toward efficiency of the web as a trap with decreasing participation of the trapper" is correct, we would expect spiders utilizing sticky traps which actually restrain movement to exhibit less variability in their capture techniques.

Though not as pronounced as in the agelenids, the araneids do vary their capture techniques with different prey (Peters, 1933a; Harwood, 1974; Robinson and Olazarri, 1971; Robinson and Robinson, 1969; Robinson et al., 1969). The model of predatory behavior developed for *Argiope argentata* by Robinson and Olazarri (1971) is exemplary of the capture techniques used by the araneids, and as such is shown in Figure 10.5. Small prey have limited surface contact with the sticky spiral of the orb-web. Their escape is prevented by a fast approach and immediate seizure in the chelicerae. Most larger prey are first secured by wrapping or enswathing in silk and then bitten. This capture strategy has been suggested to be an energy, time, and injury conservation measure in that it frees the spider to engage in other captures while it also permits it to subdue the prey at a safe distance (Eberhard, 1967; Robinson, 1969; Robinson et al., 1969). There seems to be something wrong with this argument, however, since the short bite which is used as the test of palatability comes after much energy is expended in securing the prey (Figure 10.5). If the prey is subsequently found to be

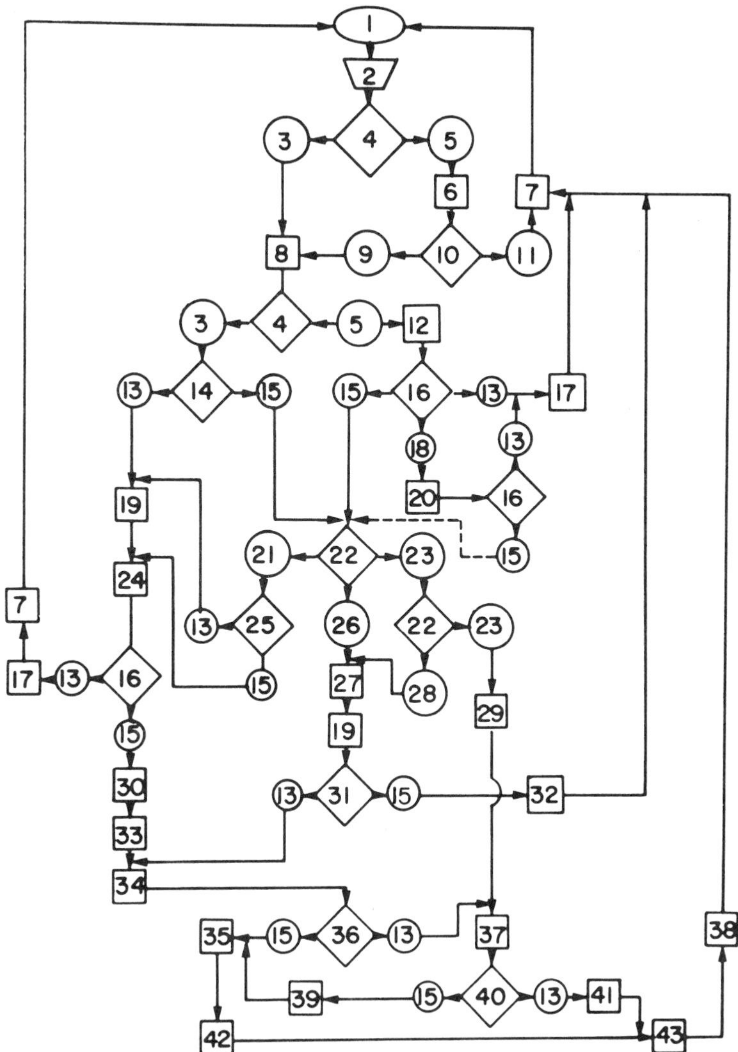

FIGURE 10.5. Model of the predatory behavior of *Argiope argentata*, modified from Robinson and Olazarri (1971). Numbers represent the following: 1.Spider at hub waiting. 2.Alerted by impact. 3.Sustained. 4.Type of vibration? 5.None or erratic. 6.Pluck. 7.Return to hub. 8.Run to prey. 9.Prey in hub. 10.Response indicates? 11.No prey. 12.Touch and palpate. 13.No. 14.Further checks. 15.Yes. 16.Is prey edible? 17.Reject. 18.Unsure. 19.Wrap. 20.Brief bites. 21.Not very light or vibrating fly or lepidopteran. 22.Prey is? 23.Very light or a weakly vibrating fly. 24.Short bite. 25.Silk on prey? 26.A lepidopteran. 27.Long bite. 28.Fly strongly vibrating? 29.Seize in jaw. 30.Rest at hub. 31.Is there prey at hub? 32.Fix to web. 33.Return to prey. 34.Cut out. 35.Carry on silk. 36.Prey is heavy. 37.Carry in jaws. 38.Feed. 39.Wrap in transport. 40.Prey sticks during transport? 41.Wrap at hub. 42.Hang at hub. 43.Manipulate.

distasteful or otherwise unsatisfactory, this energy expenditure has been wasted. To have evolved, then, rejection of a wrapped item must be an infrequent occurrence relative to the number of prey that would be lost as a result of having failed initially to secure the prey. Unfortunately, a resolution of this question is not available in the literature at this time.

At any rate, the wrap–short bite order of events is reversed among the araneids in the capture of lepidopterans as an adaptation for their quick capture. This is a necessary counter-adaptation to the scales of moths and butterflies, which have developed to permit rapid escape from adhesive surfaces (i.e., spider webs; Eisner et al., 1964). Prey belonging to this category are recognized as such through touching, and then biting precedes wrapping (Figure 10.5).

Prey characteristics also determine whether araneids thoroughly wrap prey on location or carry them to the hub for post-immobilization wrapping (Robinson et al., 1969). Prey small enough to be carried in the chelicerae are not wrapped until reaching the hub, whereas larger prey are wrapped in situ and then tethered and dragged to the hub.

Prey capture in the hackled-band weavers (cribellate spiders) which construct a sticky sheet is far more stereotyped than in the orb weavers. Dictynids, for instance, wait until a prey item ceases its struggling before they even make their approach. A sustained bite to a leg or antenna is then made (Bristowe, 1941; Jackson, 1977e; Witt et al., 1978). It is interesting that while most spiders attack the anterior half of the prey—presumably in a location where the venom will have a rapid effect—the dictynids, which have a highly efficient web trap, bite an extremity where there is minimal chance of injury to the predator. *Hypochilus gertschi*, another hackled-band weaver, also exhibits a single method of attack to all prey. Following orientation, the spider approaches the prey, touches it, and then repeatedly bites it as it pulls in the side of the web-shade on which the prey is caught (Shear, 1969).

The hunting spiders also exhibit a fairly short and invariable predatory sequence as a consequence of their greater selectivity toward prey within a limited size range. These capture sequences involve orientation, approach, crouch, and pounce with a sustained bite (Cragg, 1961; Drees, 1952; Edgar, 1969; Gardner, 1965; Precht and Freytag, 1958). Using highspeed cinematography, Melchers (1967) and Rovner (1980a) have provided further insight into the hunter's capture sequence, showing that preceding the bite, the prey is grasped in the spider's front legs. It is then brought by the legs into contact with the cheliceral fangs. Important to capture success, then, is the ability to maintain a hold on the prey prior to cheliceral contact. While lacking webs, the hunting spiders nevertheless possess a tool which increases their grasping ability—tarsal scopulae or claw tufts made up of adhesive hairs (Rovner, 1978).

There is some evidence for variation in the attack behavior of hunters, corresponding to different prey. *Phidippus* (Gardner, 1965) and *Salticus* (Dill, 1975) will circle larger prey and attack from behind instead of exhibiting the normal frontal attack characteristic of salticids; and when attacking ants, salticids tend to hold them "at arms length" in avoidance of a bite or sting (Edwards et al., 1974).

RESPONSES TO PREY ACTIONS In addition to the responses elicited largely by the behavioral or structural characteristics of different prey, within a chain of events a succession of stimuli from the prey determine whether alternative behavior patterns will be used and when and to what extent.

The relationship between prey behavior and its cuing of successive stages of the predatory sequence of *Hypochilus gertschi* is shown in Figure 10.6 (after Shear, 1969). Especially significant in this fairly simple capture routine is the required constant movement of the prey for continuation of the capture sequence. A much more complex system is shown in Figure 10.5 for the orb weaver *Argiope argentata*. At the initiation of an interaction, no or little movement by the prey elicits web plucking and subsequent approach. If the prey is struggling vigorously, however, plucking is bypassed and there is an immediate approach. At various stages along the predatory sequence of this spider, movement or lack of movement by the prey results in the assumption of different predatory pathways.

In *A. aperta*, too, specific predatory behaviors are triggered by prey actions. For instance, there is a significant probability for the transition "prey moves–spider approaches" to occur as well as the transition "prey motionless–spider searches" at the initiation of an encounter (Riechert, unpublished data). Later in the interaction, struggling by the prey elicits short bites by the spider, while attempted kicking and biting lead to encirclement of the offending body part with silk. In addition, prey that are slowly escaping off the sheet are often pulled back toward the funnel, and silk used to hinder further escape. The degree to which *A. aperta* behavior is modified by that of its prey is best shown by an example. Riechert once observed a carabid beetle to catch the front tarsus of the spider with its large mandibles during an attack sequence. It took several

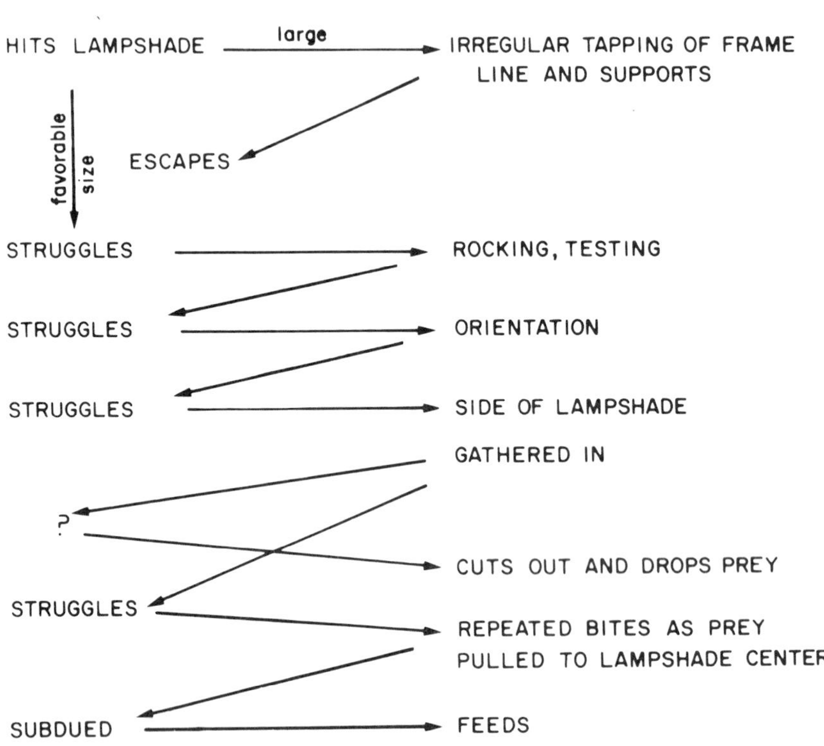

FIGURE 10.6. Response of *Hypochilus gertschi* to potential prey. Modified from Shear (1969).

minutes for the spider to free its leg from this bite and when it did, it picked the prey up by the abdomen and forced its mandibles into the web-sheet. While holding the prey thus, the spider laid silk over the mandibles, gluing them to the web. Following preening of the injured leg, the spider resumed the capture sequence, though bites were administered to the abdomen of the prey as opposed to the thorax where bites are usually directed.

The actions just discussed represent behavior patterns triggered by stimuli from the prey in specific contexts. It is hard to tell from the literature on the subject whether the raising of the front legs by salticids, thomisids, and segestriids in response to bees and ants (Bristowe, 1958; Edwards et al., 1974) is elicited by snapping mandibles and raised stingers or is an innate response triggered by recognition of the prey type. The velocity at which a prey is moving, however, does determine the speed of a salticid's approach to the prey (Gardner, 1964). This certainly is an example of a vagrant spider's response to the behavior of its prey.

Evolutionary Trends

Robinson (1977b) makes a nice analogy between the selection pressures impinging on spiders and their prey when he says, "Clearly there has been a long evolutionary arms race between spiders and their prey, and we are just beginning to understand some of the weapons systems that have emerged." The evolution of spider predatory behavior is a complex problem, requiring much additional research. We can only attempt here to outline some tendencies which have surfaced.

From Generalization toward Specialization

It appears that capture efficiency tends to lead away from filter feeding toward specialization on particular prey. An initial step in this direction is represented by the development of methods which increase the spider's capture success with a specific prey (e.g., araneids and moths). This may be followed with selection of foraging locations that provide for maximum capture of a particular prey type and restriction of foraging bouts to periods coinciding with the activity of the prey, as noted for *Tetragnatha montana* and its major prey, mosquitoes, in Poland (Łuczak, 1970). Finally there may be further changes in web construction, the character of tarsal hairs, or the method of attack, which permit maximum efficiency in the handling of a particular prey to the detriment of capture success with other prey types.

Specialization provides for greater capture success with specific prey but decreased success with the majority of prey types. For specialization to have evolved, a given prey type must be sufficiently abundant to support the predator specializing on it. With the exception of ant predators, most of our spider specialists are restricted to tropical areas where prey are present in greater numbers and more constant supply than in temperate areas. In addition, directional selection favoring dietary specialization is most likely to have occurred in the context of competition for food. Through specialization, various spider species are avoiding competition with the more successful feeding generalists.

The following two cases exemplify the phenomenon of specialized feeding in spiders. Both have evolved in the tropics where prey numbers and spider diversity are high (prime templates for the operation of directional selection). The ladder web has evolved independently in at least two tropical spider species (Robinson, 1977b). It is specialized for catching moths, which, hitting the top of the vertically expanded and laterally compressed orb-web, slide toward the bottom, losing scales along the way. At some point along the descent, the epidermis of the moth is exposed enough to adhere to the sticky spiral and the spider merely has to climb up or down to the prey and secure it (Eberhard, 1975; Robinson, 1977b). The nocturnal hunting activity of the ladder-web spiders coincides with moth activity, while its elongate ladder structure requires greater energy expenditure in retrieval of prey, making it inefficient for the capture of other prey types. Hence, species using this structure fit criteria for feeding specialization.

The ogre-faced spider *Dinopis longipes* (Dinopidae) is an ant specialist (Robinson and Robinson, 1971). It settles in a location opposite (above or across from) an ant trail, builds a rectangular web which it holds between its front legs, and upon visual stimulation reaches down with its web trap and glues the top of an ant to it. The ant is then lifted above the ground with its legs dangling, where it is wrapped and subdued at arm's length. This method of attack is important for a couple of reasons: (1) it removes the ant from the vicinity of the trail, hence avoiding retaliation by other members of the ant colony, and (2) by taking the ant off its feet and holding it at a distance, a potentially dangerous bite or sting is avoided (Robinson and Robinson, 1971).

Avoidance of Predation

To a large extent the behavioral characteristics of prey determine their probability of encounter with spiders (Turnbull, 1960). Strong fliers, including the wasps and flies, for example, often fly up to and hover in the vicinity of spider webs, but they usually change their course, avoiding the webs altogether (Rypstra, 1979; Turnbull, 1960), or carefully fly through them avoiding contact with their silk lines (Turnbull, 1960). Usually, it is only their tremendous abundance that makes them frequent prey of spiders (Table 10.2); however, some spiders have evolved adaptations to attract at least the flies. Jackson (1977e) points to the extension lines of dictynid webs as a specialization to attract dipterans in search of perching sites and to Ackerman's (1932) study of another cribellate spider whose web is reduced to only one sticky line on which it captures primarily dipterans. *Agelenopsis aperta* also catches a large number of dipterans (Riechert and Tracy, 1975). It builds its webs in close proximity to fecal material and flowering plants—places where flies and wasps tend to assemble (Riechert, 1976). In addition to changes in web structure and hunting locations, spiders have also been observed to change the timing of their activity to counter avoidance adaptations of their prey. For instance, Rypstra (1979) feels that the nocturnal foraging activity of some colonial spiders is a response to the active avoidance of their conspicuous webs by diurnal flies, wasps, and lepidopterans.

By the nature of their curved flight paths, beetles and bugs are more likely to contact spider webs than are flies and wasps (Turnbull, 1960). The jumpers (grasshoppers) should be included in this susceptible group as well. Perhaps the thick integuments of the coleopterans and the aposematic tendencies of many hemipterans, homopterans, orthopterans and coleopterans lacking hard elytra are a response to their high probability of encounter with spider webs. The kicking behavior characteristic of grasshoppers caught in spider webs must also represent a substantial deterrent to their capture, as does the defensive spray emitted by some beetles (Eisner and Dean, 1976). Many of the stronger flying insects even exhibit defensive adaptations which are employed following capture. Once captured, the sweat bee *Halictus*, for instance, behaves as if it were a piece of debris in the web and deliberately cleanses its wings of silk with its legs before attempting escape (Suter, 1978). Other flies and wasps struggle irregularly, an action that tends also to facilitate their escape (Eisner et al., 1964; Suter, 1978).

We have pretty much limited this discussion to antipredator ploys exhibited by potential prey. Numerous examples of spider counterploys exist, many of which have already been mentioned in other sections of this chapter. Only one example will be presented here. Dill (1975) discusses an antipredatory behavior used against the hunting spiders which appears to have led to the development of a counterploy. Prey have been shown to initiate escape behavior when a predator exceeds a threshold rate of change in the angle that the predator subtends at the eyes of the prey (Dill, 1973). Both Gardner (1964) and Dill (1975) note that salticids tend to reduce the velocity of their approach to prey as they near the object, an action that inhibits the exhibition of the escape response of the prey. Though this behavior of the approaching spider may actually represent conflicting drives between retreat and attack, its exhibition has a beneficial effect in this context.

Do Spiders Exhibit a Functional Response to Their Prey?

Spiders have received much attention for their potential in the control of insect numbers. It is perhaps fitting to conclude this chapter with a discussion of the degree to which they exhibit a functional response (capture more prey as the prey numbers increase), since such a response is important to a significant limiting effect.

Holling (1959) has demonstrated a density-dependent functional response (sigmoid) in small mammals, and a similar response has been observed in birds (Tinbergen, 1960). Based on his work with the wolf spider *Pardosa ramulosa* inhabiting saltwater ponds in California, Greenstone (1978) states that spiders do not display a functional response. There appears to be conflicting evidence, however; Dabrowska-Prot and Łuczak (1968) in field experiments with *Tetragnatha montana* have observed a sigmoid functional response up to a threshold level, above which consumption did not increase. Nakamura (1977) and Haynes and Sisojevic (1966) have obtained similar results in laboratory experiments with lycosids and philodromids respectively. The latter workers (1966) say of their findings of a sigmoid response, "Predators which capture prey by ambush seem to have a functional response that is acutely sensitive to changes in prey."

The increases in capture activity of the spiders in these two studies were mediated by increases in the activity of their prey associated with higher prey numbers. The experiential differences in capture success exhibited by *A. aperta* may also result in the exhibition of a sigmoid functional response, only mediated by increasing familiarity with specific prey types. Consequently, as a prey becomes more numerous and spider encounter with the type increases, spider effectiveness in handling this prey should also increase with an associated increase in their rate of capture.

Kajak (1978) and Kiritani and Kakiya (1975) report cases in which unlimited functional responses have been exhibited by spiders in which their capture of prey has continued far in excess of their needs. This is the overkill phenomenon which has frequently been observed in spiders (Edwards et al., 1974; Givens, 1978; Riechert, 1974b). Miyashita (1968a) considers overkill to represent an adaptation to take a great amount of food over short periods of time when prey are in abundant supply. Although actual consumption of this captured prey continues to a point (50 times base level in the orb weaver *Araneus quadratus*; Kajak, 1978), many of the prey captured are eventually abandoned (Haynes and Sisojevic, 1966; Riechert, herein) and less nutriment is taken from each individual food item that is consumed (Givens, 1978; Haynes and Sisojevic, 1966).

The evidence presented in this chapter overwhelmingly supports the conclusion that spiders are extremely efficient prey-capture machines, largely resulting from their evolution under conditions of limited prey availability. Many behavioral features and ecological dependencies of their predator-prey interaction are known, though conflicting data involving important elements of the system have been presented. It is through thorough investigation of these exceptions to central tendencies that we can hope to gain enough understanding of spider predatory behavior to permit manipulation of spiders in a biological control effort.

Acknowledgments

The authors gratefully acknowledge support from the Polish Academy of Sciences (Institute of Ecology) and the National Geographic Society. Some of the data reported were collected under grants from the National Science Foundation, DEB 80-02882 and DEB 74-23817.

EPILOGUE

Jerome S. Rovner
Department of Zoology
Ohio University
Athens, Ohio 45701

In the spider, evolution has produced a predator no less interesting than any of the large, carnivorous mammals which attract much attention from laymen and biologists alike. But due to problems of scale, we do not as easily explore the spider's world nor, due to differences in sensory systems, do we readily enter its Umwelt. However, as the various authors in this book show us, it is well worth the extra effort needed to make the transition in perspective that permits analyses of the systems associated with behavior in the spider, particularly those involved in communication.

We find that various receptors which evolved in spiders for prey capture have also taken on a role in receiving signals from conspecifics. Unlike most other predators, spiders tend to rely on solid-borne, mechanical cues for prey capture and communication. Krafft repeatedly brings this idea home to us in his chapter, placing special emphasis on the essential role of silk in the most communicative of all the arachnids, the social spiders. Barth, on the other hand, reminds us that silk is not necessary for such purposes in the wandering spiders; the slit sense organs and trichobothria hairs provide for detection of prey crawling on the same surface, on nearby surfaces, or flying close by. It is likely that the presence of the slit sense organs in the ancestral wandering spiders provided pre-adaptations for the first web-weavers; the early web-builders then used these mechanoreceptors for detecting prey-caused vibrations of their silk threads and subsequently for receiving signals from conspecifics.

The slit sense organs also enable spiders to receive acoustic signals transmitted through the air or through solid media. As discussed by Uetz and Stratton, a surprising number of spiders use stridulation, percussion, or appendage vibration to produce sounds that, like those well known in crickets and other insects, probably serve to achieve reproductive isolation, and play an important role in spacing mechanisms and in defense against predators.

The highly developed visual system that underlies the salticids' principal, but for spiders, untypical communicatory channel is a major subject of Lyn Forster's review. Readers familiar with insect vision probably are intrigued by the very different path taken by evolution to produce the visual system of the jumping spiders. In another vein, those interested in measurements of information exchange during interactions between animals may find Forster's transition analyses of salticid display a useful addition to the literature on communication. Approaching the jumping spiders from another perspective, Jackson reminds us that even members of this group of spiders use more than one channel of communication, although their elaborate badges and displays tend to draw the attention of human observers and thereby cause us to emphasize visual aspects in our studies.

The least understood communication channel in spiders, chemical signaling (reviewed by Tietjen and Rovner), is essentially closed to our own direct perception; furthermore, the full measure of available methods of investigation has yet to be brought to bear. Unlike our rapidly growing knowledge of chemical signals in the insects, we do not yet know the molecular identity or the sites of production and release of any of the pheromones in spiders.

Approaching spiders as a behavioral ecologist, population biologist, and sociobiologist, Riechert offers us a review of her findings which illustrate how desert funnel-web spiders use communication for coercion and thus for selfish gain in competition for limited resources. Her data are of interest relative to the theoretical work of Maynard Smith, Dawkins, and others, involving the application of game theory to the prediction of Evolutionarily Stable Strategies. Likewise of interest to a broad audience is the review by Burgess and Uetz on one of the major effects of the latter competitive strategies, i.e., the patterns of social spacing in spiders.

Choosing to expand the definition of communication to its most general sense, Riechert and Łuczak seek to elucidate behaviors that have evolved in spiders to optimize their food-getting efforts. The primary strategy turns out to be that of using sit-and-wait behavior, which apparently arose as a means of limiting energy expenditure under a frequent condition of limited prey availability.

In looking over the chapters of this book, one becomes aware of the value of experimental approaches, as a result of which cause-effect relationships can be placed on firm ground. Most early work on spider communication was entirely descriptive, important exceptions being studies like those of Peters (1933a), Kaston (1936), and Crane (1949b). Obviously, there will always be a need for descriptive work, such as papers detailing courtship displays, since the latter information, for example, may be important for comparative studies having the purpose of elucidating phylogenetic relationships. However, a further increase in experimentation by those studying spider communication will be a welcome development. Without such an approach by past and present workers, there would have been far fewer conclusions reached by the contributors to this volume.

In our survey of various facets of spider communication, major gaps in our knowledge come to mind. For example, only little has been said about the ontogeny of signaling behavior; not surprisingly, adult behaviors have attracted the interest of most observers. However, the presumably simpler signaling systems of juvenile spiders merit our examination as well. Inhibition of cannibalism, territorial defense, and other interactions are no less important among the juveniles of many spiders than in the adults.

While the important vibratory channel is relatively well understood at the sensory level, largely through the research conducted by Barth and by Görner, certain components of this system, such as the effect of the transmission medium on the signal, are just beginning to receive attention. One expects to see increasing emphasis on such factors as the tuning of web structures and the influence of various substrates on vibratory signaling by wandering spiders.

Another aspect of the mechanical channel that should be studied but which presents even greater difficulties for the researcher is that of tactile communication. Even though the spider's contact chemosensitive hairs may transduce both chemical and mechanical stimuli, separating the two modalities is not a great problem for the sensory physiologist measuring action potentials in spiders mounted beneath the microscope and presented with controlled stimuli. However, how does one determine which modality is involved when freely interacting spiders touch one another? One cannot always assume contact chemoreception, although it is much involved in spider communication. For example, the attachment response of lycosid spiderlings to their mother requires a mechanical stimulus—the presence of specialized hairs on the mother's abdomen. However, the spiderling's behavior may turn out not to be exclusively dependent on a mechanoreceptive cue; a chemical signal may augment the response. In another vein, when a male spider contacts and mounts a female, is the information used by both partners for species and sex recognition transmitted entirely by a chemical signal, or are identifying mechanical cues involved as well? The design of experimental approaches to such questions presents a challenge to future investigators.

An area that is now under increasing study by spider behaviorists is that of interspecific communication, as described in the sections on parasitism and commensalism in Krafft's chapter and in Barth's chapter. However, except for the study by Tretzel (1959), the nature of non-symbiotic interspecific communication occurring when spiders sharing the same habitat encounter one another has received too little attention. Lyn Forster's recent work on interspecific communication in salticid spiders, summarized in her chapter in this volume, is a step in the right direction. While meetings between conspecifics are more likely to occur (due to microhabitat selection and the use of pheromones), the diversity of species present and their population densities in any patch of forest or field, as revealed by the work of spider taxonomists and ecologists, suggest that interspecific meetings must be frequent, thus leading to the evolution of generalized signals that serve to inhibit predation by other spiders. Nonetheless, as with behavioral biologists examining other taxa of animals, those studying spiders have dealt largely with courtship, agonistic, or parent-young signaling, i.e., conspecific communication.

The most challenging void in our knowledge of spider communication is the role of the central nervous system. While we are beginning to know much about the receptors that spiders use to detect signals in the environment, a complete understanding of the spider's Umwelt must await the onset of efforts by investigators with the appropriate skills to tackle this most difficult problem of research on the central integration of sensory information, i.e., the nature of the circuitry that has been organized according to genetically transmitted information and possibly modified through individual experience. Likewise, the components of the central nervous system regulating the output of signals, as well as the influence of endocrine tissues on the timing of the signaling behavior, are essentially unknown in spiders.

As to what we do know, the major topics treated in this volume have been presented by authors whose knowledge was gained through their own research in their respective areas and their consequent immersion in the literature of those areas. Thus, the work of each investigator in the laboratory or field is placed within the broader context of a review of the topic.

When one considers the information only recently available to fill the various chapters of this book, it seems that the study of communication in spiders has at last begun to blossom. Using techniques unavailable to earlier workers and developing hypotheses within conceptual frameworks that have only recently been put forth, a number of investigators with diverse backgrounds and interests converge on this complex subject to offer new perspectives, new insights, and several surprises. As a consequence, for those of us already devoted to this group of animals, our enthusiasm for studying the biology of spiders has been heightened. As to those for whom the reading of this book represents a first detailed encounter with behavior in spiders, you may now better understand why increasing numbers of biologists find themselves "hooked" on these fascinating animals. And for those who read these pages seeking information that may be useful to them in their search for principles of animal communication, perhaps some of you will be encouraged by what is attempted here.

LITERATURE CITED

Ackerman, C. On the spider *Miagrammopes* sp. which constructs a single line snare. Ann. Natal. Mus., *5*: 411-422, 1932.

Alexander, C. P. Insects of Micronesia, Diptera: Tipulidae. B. P. Bishop Museum Press, Hawaii, *12*: 8, 1972.

Alexander, R. D. Sound production and associated behavior in insects. Ohio J. Sci., *57*: 101-113, 1957.

Alexander, R. D. Acoustical communication in arthropods. Ann. Rev. Entomol., *12*: 495-526, 1967.

Allard, H. A. The drumming spider (*Lycosa gulosa* Walckenaer). Proc. Biol. Soc. Wash., *49*: 67-68, 1936.

Anderson, J. F. Metabolic rates of spiders. Comp. Biochem. Physiol., *33*: 51-72, 1970.

Anderson, J. F. Responses to starvation in the spiders *Lycosa lenta* (Hentz) and *Filistata hibernalis* (Hentz). Ecology, *55*: 576-585, 1974.

Andrew, R. J. The information potentially available in mammal display. In *Non-Verbal Communication*, R. A. Hinde (ed.), Cambridge University Press, pp. 179-206, 1972.

Aspey, W. P. Agonistic behavior and dominance-subordinance relationships in the wolf spider *Schizocosa crassipes*. Proc. 6th Int. Arachn. Congr., 102-106, 1974.

Aspey, W. P. Ontogeny of display in immature *Schizocosa crassipes* (Araneae: Lycosidae). Psyche, *82*: 174-180, 1975.

Aspey, W. P. Response strategies of adult male *Schizocosa crassipes* (Araneae: Lycosidae). Psyche, *83*: 94-105, 1976.

Aspey, W. P. Wolf spider sociobiology: I. Agonistic display and dominance-subordinance relationships in adult male *Schizocosa crassipes*. Behaviour, *62*: 103-141, 1977a.

Aspey, W. P. Wolf spider sociobiology: II. Density parameters influencing agonistic behaviors in *Schizocosa crassipes*. Behaviour, *62*: 142-163, 1977b.

Baker, R. R. Territorial behavior of the nymphalid butterflies *Aglais urticae* (L.) and *Inachis io* (L.). J. Anim. Ecol., *41*: 453-469, 1972.

Bali, Geetha and F. G. Barth. Tuning curves of individual slits in a spider vibration receptor (in prep.).

Barash, D. P. *Sociobiology and Behavior*. Elsevier, 1977.

Barlow, G. W. Modal action patterns. In *How Animals Communicate*, T. A. Sebeok (ed.), Indiana University Press, pp. 98-134, 1977.

Barlow, G. W. and R. F. Green. The problems of appeasement and of sexual roles in the courtship behavior of the blackchin mouthbreeder, *Tilapia melanotheron* (Pisces: Cichlidae). Behaviour, *36*: 84-115, 1970.

Barrows, W. M. The reactions of an orb-weaving spider, *Epeira sclopetaria* (Cl.) to rhythmic vibrations of its web. Biol. Bull., *29*: 316-332, 1915.

Barth, F. G. Ein einzelnes Spaltsinnesorgan auf dem Spinnentarsus: seine

Erregung in Abhängigkeit von den Parametern des Luftschallreizes. Z. vergl. Physiol., *55*: 407-449, 1967.

Barth, F. G. Der sensorische Apparat der Spaltsinnesorgane (*Cupiennius salei* Keys., Araneae). Z. Zellforsch., *112*: 212-246, 1971.

Barth, F. G. Die Physiologie der Spaltsinnesorgane. I. Modellversuche zur Rolle des cuticularen Spaltes beim Reiztransport. J. comp. Physiol., *78*: 315-336, 1972a.

Barth, F. G. Die Physiologie der Spaltsinnesorgane. II. Funktionelle Morphologie eines Mechanoreceptors. J. comp. Physiol., *81*: 159-186, 1972b.

Barth, F. G. Sensory information from strains in the exoskeleton. In *The Insect Integument*, H. R. Hepburn (ed.). Elsevier, pp. 445-473, 1976.

Barth, F. G. Slit sense organs: "strain gauges" in the arachnid exoskeleton. Symp. zool. Soc. London, *42*: 439-448, 1978.

Barth, F. G. and J. Bohnenberger. Lyriform slit sense organ: threshold and stimulus amplitude ranges in a multi-unit mechanoreceptor. J. comp. Physiol., *125*: 37-43, 1978.

Barth, F. G. and W. Libera. Ein Atlas der Spaltsinnesorgane von *Cupiennius salei* Keys. Chelicerata (Araneae). Z. Morph. Tiere, *68*: 343-369, 1970.

Barth, F. G. and N. Rehner. Zur Orientierung einer Jagdspinne beim Beutefang. Staatsexamensarbeit Universität Frankfurt, unpublished, 1978.

Barth, F. G. and E.-A. Seyfarth. *Cupiennius salei* Keys. (Araneae) in the highlands of central Guatemala. J. Arachnol., *7*: 255-263, 1979.

Barth, F. G. and J. Stagl. The slit sense organs of arachnids. A comparative study of their topography on the walking legs. Zoomorph., *86*: 1-23, 1976.

Barth, F. A. and M. Wadepuhl. Slit sense organs on the scorpion leg (*Androctonus australis*, L. Buthidae). J. Morph., *145*: 209-227, 1975.

Bastock, M. *Courtship: An Ethological Study.* Aldine, 1967.

Bateman, A. J. Intra-sexual selection in *Drosophila*. Heredity, *2*: 349-368, 1948.

Baylis, J. R. A quantitative study of long-term courtship: II. A comparative study of the dynamics of courtship in two New World cichlid fishes. Behaviour, *59*: 117-161, 1976.

Bays, S. M. A study of the training possibilities of *Araneus diadematus* Cl. Experientia, *18*: 423, 1962.

Beer, C. G. What is a display? Am. Zool., *17*: 155-165, 1977.

Bennet, T. J. and R. D. Lewis. Visual orientation in the Salticidae. N. Z. Entomol., *7*: 58-63, 1979.

Bentzien, M. M. Biology of the spider *Diguetia imperiosa* (Araneida, Diguetidae). Pan-Pacific Entomol., *49*: 110-123, 1973.

Beranek, L. *Acoustics.* McGraw-Hill, 1954.

Berestynska-Wilczek, M. Investigations of the sensitivity of the spiders *Pirata piraticus* (Clerck) to vibrations of the water surface. Acta Biologica Cracoviensia Zoologia, *5*: 263-277, 1962.

Berland, L. Observations sur l'accouplement des araignées. Arch. Zool. Exp. et Gen., Notes et Revues, *5*: 47-53, 1912.

Berland, L. Utilisation pour la capture des mouches des nids de l'Araignée mexicaine *Coenothele gregalis*. Bull. Mus., 432-433, 1913.

Berland, L. Nouvelles observations d'accouplements d'araignées. Arch. Zool. Exp. et Gen., Notes et Revues, *5*: 109-119, 1914.

Berland, L. Contributions à l'étude de la biologie des arachnides (Premier Mémoire). Ann. Soc. Entomol. France, *91*: 193-208, 1923.

Berland, L. Contributions à l'étude de la biologie des arachnides (Deuxième Mémoire). Arch. Zool. Exp. et Gen., Notes et Revues, *66, 2*: 7-29, 1927.

Berland, L. Les Arachnides. *Encyclopédie entomologique*. Lechevalier, p. 485, 1932.

Beroza, M. *Chemicals Controlling Insect Behavior*. Academic Press, 1979.

Bhattacharya, G. C. Observations on some peculiar habits of the spider (*Marpissa melanognathus*). J. Bombay Nat. Hist. Soc., *39*: 142-144, 1936.

Bick, G. H. and J. C. Bick. Demography and behavior of the damselfly *Argia apicalis* (Say) (Odonata: Coenagriidae). Ecology, *46*: 461-472, 1965.

Billaudelle, H. Zur Biologie der Mauerspinne *Dictyna civica* (Łuczak). Z. Angew. Ent., *41*: 475-512, 1957.

Bilsing, S. W. Quantitative studies in the food of spiders. Ohio J. Sci., *20*: 215-260, 1920.

Bishop, D. T. and C. Cannings. A generalized war of attrition. J. Theor. Biol., *70*: 85-124, 1978.

Bishop, S. C. Singing spiders. N.Y. St. Mus. Bull., *260*: 65-69, 1925.

Blanke, R. Untersuchungen zur Ökophysiologie und Ökethologie von *Cyrtophora citricola* Forskal (Araneae, Araneidae) in Andalusien. Forma et Functio, *5*: 125-206, 1972.

Blanke, R. Nachweis von Pheromonen bei Netzspinnen. Naturwissenschaften, *60*: 481, 1973.

Blanke, R. Das Sexualverhalten der Gattung *Cyrtophora* als Hilfsmittel für Phylogenetische Aussagen. Proc. 6th Int. Arachn. Congr., 116-119, 1975a.

Blanke, R. Untersuchungen zum Sexualverhalten von *Cyrtophora cicatrosa* (Stoliczka) (Araneae, Araneidae). Z. Tierpsychol., *37*: 62-74, 1975b.

Blest, A. D. The concept of ritualization. In *Current Problems in Animal Behavior*, W. H. Thorpe and O. L. Zangwill (eds.), Cambridge University Press, pp. 102-124, 1961.

Blest, A. D. The rapid synthesis and destruction of photoreceptor membrane by a dinopid spider. Proc. R. Soc. Lond. B., *196*: 198-222, 1978.

Blest, A. D. and M. F. Land. The physiological optics of *Dinopis subrufus* L. Koch: a fish-lens in a spider. Proc. R. Soc. Lond. B., *196*: 197-222, 1977.

Blest, A. D., D. S. Williams, and L. Kao. The posterior median eyes of the dinopid spider *Menneus*. Cell Tissue Res., *211*: 391-403, 1980.

Blum, M. S. The source and specificity of trail pheromones in *Termitopone, Monomorium* and *Huberia*, and their relation to those of some other ants. Proc. R. Ent. Soc. Lond. (A), *41*: 155-160, 1966.

Blum, M. S. Pheromonal bases of social manifestations in insects. In *Pheromones*, M. C. Birch (ed.), *Frontiers of Biology*, 32, North Holland, pp. 190-199, 1974.

Blum, M. S. and N. A. Blum (eds.). *Sexual Selection and Reproductive Competition in Insects.* Academic Press, 1979.

Blum, M. S. and J. M. Brand. Social insect pheromones: their chemistry and function. Am. Zool., *12*: 553-576, 1972.

Blumenthal, H. Untersuchungen über das "Tarsalorgan" der Spinnen. Z. Morph. Ökol. Tiere, *29*: 667-719, 1935.

Bohnenberger, J. On the transfer characteristics of a lyriform slit sense organ. Symp. zool. Soc. London, *42*: 449-455, 1978.

Bonner, J. T. How slime molds communicate. Sci. Am., *209*: 84-93, 1963.

Bonnet, P. L'instinct maternel des araignées à l'épreuve de l'expérimentation. Bull. Soc. Hist. Nat. Toulouse, I, *81*: 185-250, 1940.

Bonnet, P. *Bibliographia Araneorum*, Vol. 1 (publ. by author), 1945.

Bossert, W. J. and E. O. Wilson. The analysis of olfactory communication among animals. J. Theor. Biol., *5*: 443-469, 1963.

Bouissou, M. F. and J. P. Signoret. La hiérarchie sociale chez les mammifères. Rev. Comp. Anim., *4*: 43-61, 1970.

Boys, C. V. The influence of a tuning fork on the garden spider. Nature, *23*: 149, 1880.

Brach, V. The biology of the social spider *Anelosimus eximius* (Araneae: Theridiidae). Bull. So. Calif. Acad. Sci., *74*: 37-41, 1975.

Brach, V. Subsocial behavior in the funnel-web wolf spider *Sosippus floridanus* (Araneae: Lycosidae). Florida Entomol., *59*: 225-229, 1976.

Brach, V. *Anelosimus studiosus* (Araneae: Theridiidae) and the evolution of quasisociality in theridiid spiders. Evolution, *31*: 154-161, 1977.

Bradoo, B. L. Some observations on the ecology of social spider *Stegodyphus sarasinorum* Karsch (Araneae: Eresidae) from India. Oriental Insects, *6*: 193-203, 1972.

Braun, R. Zur Biologie von *Teutana triangulosa.* Z. wiss. Zool., *159*: 255-318, 1956.

Brignoli, P. M. Ragni d'Italia. XX Note Sugli Hahniidae. Fragmentata Entomol., *8*: 265-274, 1973.

Bristowe, W. S. The mating habits of spiders, with special reference to the problems surrounding sex dimorphism. Proc. zool. Soc. London, *1929*: 309-358, 1929.

Bristowe, W. S. *The Comity of Spiders*, Vol. 2. London, Ray Society, 1941.

Bristowe, W. S. *The World of Spiders*. Collins, 1958.

Bristowe, W. S. and G. H. Locket. The courtship of British lycosid spiders and its probable significance. Proc. zool. Soc. London, *1926*: 317-347, 1926.

Brock, V. E. and R. H. Riffenburgh. Fish schooling: a possible factor in reducing predation. J. Conseil. Perm. Internat. Explor. Mer., *25*: 307-317, 1960.

Brockmann, H. J., A. Grafen, and R. Dawkins. Evolutionarily Stable Strategy in a digger wasp. J. Theor. Biol., *77*: 472-496, 1979.

Brown, J. L. *The Evolution of Behavior.* Norton, 1975.

Brownell, P. H. Compressional and surface waves in sand: used by desert scorpions to locate prey. Science, *197*: 4302-4304, 1977.

Brownell, P. H. and R. D. Farley. Detection of vibrations in sand by tarsal sense organs of the nocturnal scorpion, *Paruroctonus mesaensis.* J. comp. Physiol., *131*: 23-30, 1979a.

Brownell, P. H. and R. D. Farley. Prey-localizing behaviour of the nocturnal desert scorpion, *Paruroctonus mesaensis*: orientation to substrate vibrations. Anim. Behav., *27*: 185-193, 1979b.

Bryant, E. B. Some new and little known species of New Zealand spiders. Records Canterbury Museum, *IV*: 71-94, 1935.

Buchli, H.H.R. Hunting behavior in the Ctenizidae. Am. Zool., *9*: 175-193, 1969.

Buckle, D. J. Social behavior in spiders with special reference to maternal behavior in *Clubiona kulezynskii* (Lessert). Proc. 19th Ann. Meeting Entomol. Soc. Saskatchewan, 1971.

Buckle, D. J. Sound production in the courtship of 2 lycosid spiders: *Schizocosa avida* (Walck.) and *Tarentula aculeata* (Cl). Blue Jay, *30*: 110-113, 1972.

Burch, T. L. The importance of communal experience to survival for spiderlings of *Araneus diadematus.* J. Arachnol., *7*: 1-18, 1979.

Burgess, J. W. The sheet web as a transducer, modifying vibration signals in social spider colonies of *Mallos gregalis.* Neurosci. Abstr., 557, 1975.

Burgess, J. W. Social spiders. Sci. Am., *234*: 100-106, 1976.

Burgess, J. W. Social behavior in group-living spider species. Symp. zool. Soc. London, *42*: 69-78, 1978.

Burgess, J. W. Measurement of spatial behavior: methodology applied to rhesus monkeys, neon tetras, and social and communal spiders, cockroaches and gnats in open fields. Behav. Neur. Biol., *26*: 132-160, 1979a.

Burgess, J. W. Web-signal processing for tolerance and group predation in the social spider *Mallos gregalis* Simon. Anim. Behav., *27*: 157-164, 1979b.

Burgess, J. W. The spider species *Mallos gregalis* Simon (Araneae: Dictynidae): Ph.D. Thesis: Department of Zoology, N. C. State University, 1979c.

Burgess, J. W. and P. N. Witt. Spider webs: design and engineering. Interdisc. Sci. Rev., *1*: 322-335, 1976.

Burghardt, G. M. Defining "Communication." In *Communication by Chemical Signals,* J. W. Johnston, Jr., D. G. Moulton, and A. Turk (eds.), Appleton-Century-Crofts, pp. 241-308, 1970.

Burghardt, G. M. Ontogeny of communication. In *How Animals Communicate,* T. A. Sebeok (ed.), Indiana University Press, 1977.

Burroughs, J. *Pepacton.* Houghton Mifflin, 1881.

Bushman, L. L., W. H. Whitcomb, R. C. Hemenway, D. L. Mays, Ru Nguyen, N. C. Leppla, and B. J. Smittle. Predators of the velvetbean caterpillar eggs in Florida soybeans. Environ. Entomol., *6*: 403-407, 1976.

Buskirk, R. E. Coloniality, activity patterns and feeding in a tropical orb-weaving spider. Ecology, *56*: 1314-1328, 1975a.

Buskirk, R. E. Aggressive display and orb defense in the colonial spider *Metabus gravidus*. Anim. Behav., *23*: 560-567, 1975b.

Busnel, R. G. (ed.). *Acoustic Behaviour of Animals*. Elsevier, 1963.

Butler, C. G., D.J.C. Fletcher, and D. Walter. Nest-entrance marking with pheromones by the honeybee, *Apis mellifera* L. and by a wasp *Vespula vulgaris* L. Anim. Behav., *17*: 142-147, 1969.

Cade, W. The evolution of alternative male reproductive strategies in field crickets. In *Sexual Selection and Reproductive Competition in Insects*, M. S. Blum and N. A. Blum (eds.), Academic Press, pp. 343-379, 1979.

Cady, A. B. Microhabitat selection and locomotor activity of *Schizocosa ocreata* (Araneae: Lycosidae) as viewed at a woodland site. M.S. Thesis, Ohio University, 1978.

Callahan, P. S. *Tuning in to Nature: Solar Energy, Infrared Radiation and the Insect Communication System*. Deven-Adair, 1975.

Cameron, E. A., C. P. Schwalbe, M. Beroza, and E. F. Knipling. Disruption of gypsy moth mating with microencapsulated disparlur. Science, *183*: 972-973, 1974.

Campanella, P. J. and L. L. Wolf. Temporal leks as a mating system in a temperate zone dragonfly (Odonata, Anisoptera). I. *Plathemis lydia* (Drury). Behaviour, *51*: 41-87, 1974.

Campbell, B. (ed.). *Sexual Selection and the Descent of Man*. Aldine, 1972.

Campbell, F. M. On supposed stridulating organs of *Steatoda guttata* Wider and *Liniphia tenebricola* Wider. J. Linn. Soc. London, *15*: 152-155, 1881.

Carico, J. E. The Nearctic species of the genus *Dolomedes* (Araneae: Pisauridae). Bull. Mus. Comp. Zool., *144*: 435-488, 1973.

Carico, J. E. Predatory behavior in *Euryopis funebris* (Hentz) (Araneae: Theridiidae) and the evolutionary significance of web reduction. Symp. zool. Soc. London, *42*: 51-58, 1978.

Carmichael, J. H. Jumping spiders. Nat. Hist., New York, *78*: 28-35, 1969.

Carpenter, F. L. and R. E. MacMillen. Threshold model of feeding territoriality and test with a Hawaiian honeycreeper. Science, *194*: 639-642, 1976.

Carpenter, G. H. The smallest of stridulating spiders. Nat. Sci., *12*: 319-322, 1898.

Carrel, J. E. and R. D. Heathcote. Heart rate in spiders: influence of body size and foraging energetics. Science, *193*: 148-150, 1976.

Carthy, J. D. *An Introduction to the Behavior of Invertebrates*. George Allen and Unwin, 1958.

Caryl, P. G. Communication by agonistic displays: what can games theory contribute to ethology? Behaviour, *68*: 136-152, 1979.

Chamberlin, R. V. The spider fauna of the shores and islands of the gulf of California. Proc. Calif. Acad. Sci., *12*: 561-694, 1924.

Chase, R. The mentalist hypothesis and invertebrate neurobiology. Perspect. Biol. Med., *23*: 103-117, 1979.

Chauvin, R. Le comportement social chez les animaux, P.U.F., 1961.

Chauvin, R. Les lois de l'ergonomie chez les fourmis au cours du transport d'objets. C. R. Acad. Sc. Paris, *273*: 1862-1865, 1971.

Chopard, L. Sur les bruits par certaines araignées. Bull. Soc. Zool. France, *59*: 132-134, 1934.

Christenson, T. E. and K. C. Goist. Costs and benefits of male-male competition in the orb weaving spider, *Nephila clavipes*. Behav. Ecol. Sociobiol., *5*: 87-92, 1979.

Christian, U. Zur Feinstruktur der Trichobothrien der Winkelspinne *Tegenaria derhami* (Scopoli) (Agelenidae, Araneae). Cytobiol., *4*: 172-185, 1971.

Christian, U. Trichobothrien, ein Mechanoreceptor bei Spinnen. Elektronenmikroskopische Befunde bei der Winkelspinne *Tegenaria derhami* (Scopoli) (Agelenidae, Araneae). Verh. dtsch. Zool. Ges., *66*: 31-36, 1972.

Chrysanthus, Fr. Hearing and stridulation in spiders. Tijdschr. voor Entomol., *96*: 57-83, 1953.

Clark, P. J. and F. J. Evans. Distance to nearest neighbor as a measure of spatial relationships in populations. Ecology, *35*: 445-452, 1954.

Cloudsley-Thompson, J. L. The life-histories of British cribellate spiders of the genus *Ciniflo* Bl. (Dictynidae). Ann. Mag. Nat. Hist., *12*: 787-794, 1955.

Cloudsley-Thompson, J. L. Notes on Arachnida. 34. The sense of hearing in *Dolomedes fimbriatus* (Clerck). Entomol. Mon. Mag., *95*: 216, 1960.

Cole, L. C. A study of the cryptozoa of an Illinois woodland. Ecol. Monogr., *16*: 49-86, 1946.

Colebourn, P. H. The influence of habitat structure on the distribution of *Araneus diadematus* Clerck, J. Anim. Ecol., *43*: 401-409, 1974.

Collatz, K. G. and T. Mommsen. Physiological conditions and variations of body constituents during the moulting cycle of the spider *Tegenaria atrica* (Agelenidae). Comp. Biochem. Physiol. (A), *52*: 465-476, 1975.

Comfort, A. Likelihood of human pheromones. Nature, *230*: 432-434, 1971.

Cooke, J.A.L. A contribution to the biology of the British spiders belonging to the genus *Dysdera*. Oikos, *16*: 20-25, 1965.

Coulson, J. C. and E. White. A study of the colonies of the kittiwake *Rissa tridactyla* (L.), Ibis, *98*: 63-79, 1956.

Cragg, J. B. Some aspects of the ecology of moorland animals. J. Anim. Ecol., *30*: 205-234, 1961.

Crane, J. Comparative biology of salticid spiders at Rancho Grande, Venezuela, Part I. Systematics and life histories in *Corythalia*. Zoologica, *33*: 1-38, 1948.

Crane, J. Comparative biology of salticid spiders at Rancho Grande, Venezuela, Part III. Systematics and behavior in representative species. Zoologica, *34*: 31-52, 1949a.

Crane, J. Comparative biology of salticid spiders at Rancho Grande, Venezuela, Part IV. An analysis of display. Zoologica, *34*: 159-215, 1949b.

Crane, J. Basic patterns of display in fiddler crabs (Ocypodidae, Genus *Uca*). Zoologica, *42*: 69-82, 1957.

Crosby, C. R. Two new species of Theridiidae. Canad. Entomol., *38*: 308-310, 1906.

Cullen, J. M. Some principles of animal communication. In *Non-Verbal Communication*, R. A. Hinde (ed.), Cambridge University Press, pp. 101-125, 1972.

Curtis, J. T. *The Vegetation of Wisconsin*, University of Wisconsin Press, 1959.

Cushing, D. H. and F. R. Hardin-Jones. Why do fish school? Nature, *218*: 918-920, 1968.

Dabrowska-Prot, E. and J. Łuczak. Studies on the incidence of mosquitoes in the food of *Tetragnatha montana* Simon and its food activity in the natural habitat. Ekol. Pol. A., *16*: 843-853, 1968.

Dabrowska-Prot, E., J. Łuczak, and K. Tarwid. Prey and predator density and their reactions in the process of mosquito reduction by spiders in field experiments. Ekol. Pol. A., *16*: 773-819, 1968a.

Dabrowska-Prot, E., J. Łuczak, and K. Tarwid. The predation of spiders on forest mosquitoes in field experiments. J. Med. Entomol., *5*: 252-256, 1968b.

Dabrowska-Prot, E., J. Łuczak, and Z. Wójcik. Ecological analysis of two invertebrate groups in the wet alder wood and meadow ecotone. Ekol. Pol. A., *21*: 753-812, 1973.

Dahl, F. Über die Hörhaare bei den Arachnoiden. Zool. Anz., *6*: 267-270, 1883.

Dahl, F. Die Sinneshaare der Spinnentiere. Zool. Anz., *51*: 215-219, 1911.

Darchen, R. Ethologie de quelques araignées sociales. L'interattraction, la construction et la chasse. C. R. du Vème Congrès de l'U.I.E.I.S., Toulouse, 333-345, 1965a.

Darchen, R. Ethologie d'une Araignée sociale, *Agelena consociata* Denis. Biologia Gabonica, *1*: 117-146, 1965b.

Darchen, R. Une nouvelle Araignée sociale du Gabon *Agelena republicana* Darchen. Biologia Gabonica, *3*: 31-42, 1967.

Darchen, R. Ethologie d'*Achaearanea disparata* Denis, Aranea, Therididae, Araignée sociale du Gabon. Biologia Gabonica, *3*: 5-25, 1968.

Darchen, R. La fondation de nouvelle colonies d'*Agelena consociata* et d'*Agelena republicana*. C. R. Cong. Arachnol., France, *1976*: 20-39, 1976.

Darchen, R. and J. C. Ledoux. *Achaearanea disparata*, Araignée sociale du Gabon, synonyme ou espèce jumelle d'*A. tessellata*, solitaire. Rev. Arachnol., *1*: 121-132, 1978.

Darling, F. F. *Bird Flocks and the Breeding Cycle: A Contribution to the Study of Avian Sociality.* Cambridge University Press, 1938.

Darwin, C. *The Descent of Man, and Selection in Relation to Sex.* John Murray, 1871.

David, F. N. and P. G. Moore. Notes on contagious distributions in plant populations. Ann. Bot. London, N.S., *18*: 47-53, 1954.

Davis, W. T. Spider calls. Psyche, *2*: 120, 1904.

Dawkins, R. *The Selfish Gene,* Oxford University Press, 1976.

Dawkins, R. and J. R. Krebs. Animal signals: information or manipulation? In *Behavioral Ecology,* J. R. Krebs and N. B. Davies (eds.), Sinauer, 1978.

Denny, M. The physical properties of spider's silk and their role in the design of orb webs. J. Exp. Biol., *65*: 483-506, 1976.

Den Otter, C. J. Setiform sensilla and prey detection in the bird-spider *Sericopelma rubronitens* Ausserer (Araneae, Theraphosidae). Neth. J. Zool., *24*: 219-235, 1974.

Dethier, V. G. *Nebraska Symposium of Motivation.* University of Nebraska Press, 1966.

DeVoe, R. D. Ultraviolet and green receptors in principal eyes of jumping spiders. J. gen. Physiol., *66*: 193-207, 1975.

Dewsbury, D. A. *Comparative Animal Behavior.* McGraw-Hill, 1978.

Dietlein, W. Ein Beitrag zur Sexualbiologie und Brutfürsorge bei Lycosiden mit besonderer Berücksichtigung von *Pirata piraticus* (Clerck). Thesis, Nuremberg, 1967.

Diguet, L. Sur l'araignée mosquéro. C. R. Acad. Sc. Paris, *148*: 735-736, 1909a.

Diguet, L. Le mosquéro. Bull. Soc. Natur. Acclim. France, *56*: 368-375, 1909b.

Dijkgraaf, S. Untersuchungen über die Funktion der Seitenorgane von Fischen. Z. vergl. Physiol., *20*: 162-214, 1934.

Dijkstra, H. Comparative research of the courtship behaviour in the genus *Pardosa.* III. Agonistic behaviour in *Pardosa amentata.* Bull. Mus. Natl. Hist. Nat., 2 ème Serie, *41*: Suppl. 1., 91-97, 1969.

Dijkstra, H. Searching behaviour and tactochemical orientation in males of the wolf spider *Pardosa amentata* (Cl.) (Araneae, Lycosidae). Proc. kon. ned. Akad. Wet., Ser. C, *79*: 235-244, 1976.

Dijkstra, H. Dynamics of dominance in the wolf spider *Pardosa amentata* (Cl.) (Lycosidae, Araneae). Symp. zool. Soc. London, *42*: 483-484, 1978.

Dill, L. M. An avoidance learning submodel for a general predation model. Oecologia (Berl.), *13*: 291-312, 1973.

Dill, L. M. Predatory behavior of the zebra spider, *Salticus scenicus* (Araneae; Salticidae). Can. J. Zool., *53*: 1284-1289, 1975.

Dingle, H. and R. L. Caldwell. The aggressive and territorial behaviour of the mantis shrimp *Gonodactylus bredini* Manning (Crustacea: Stomatopoda). Behaviour, *33*: 115-136, 1969.

Dondale, C. D. Sexual behavior and its application to a species problem in the spider genus *Philodromus* (Araneae, Thomisidae). Can. J. Zool., *42*: 817-827, 1964.

Dondale, C. D. Sexual behavior and the classification of the *Philodromus rufus* complex in North America (Araneae, Thomisidae). Can. J. Zool., *45*: 453-459, 1967.

Dondale, C. D. and B. M. Hegdekar. The contact sex pheromone of *Pardosa lapidicina* Emerton. Can. J. Zool., *51*: 400-401, 1973.

Dondale, C. D. and J. H. Redner. Revision of the Nearctic wolf spider genus *Schizocosa* (Araneida: Lycosidae). Canad. Entomol., *110*: 143-181, 1978.

Drees, O. Untersuchungen über die angeborenen Verhaltensweisen bei Springspinnen (Salticidae). Z. Tierpsychol., *9*: 169-207, 1952.

Drews, C. D. and R. A. Bernard. Electrophysiological responses of chemosensitive sensilla in the wolf spider. J. exp. Zool., *198*: 423-428, 1976.

Duelli, P. Movement detection in the posterolateral eyes of jumping spiders (*Evarcha arcuata*, Salticidae). J. comp. Physiol., *124*: 15-26, 1978.

Dumais, J., J. M. Perron, and C. D. Dondale. Eléments du comportement sexuel chez *Pardosa xerampelina* (Keyserling) (Araneida: Lycosidae). Can. J. Zool., *51*: 265-271, 1973.

Dumortier, B. Morphology of sound emission apparatus in Arthropoda. In R. G. Busnel (ed.), *Acoustic Behavior of Animals*. Elsevier, 1963.

Dumpert, K. Spider odor receptor: electrophysiological proof. Experientia, *34*: 754-755, 1978.

Dumpert, K. and W. Gnatzy. Cricket combined mechanoreceptors and kicking response. J. comp. Physiol., *122*: 9-25, 1977.

Dunn, R. A. The peacock spider. Walkabout (April), 38-39, 1957.

Dzimirski, I. Untersuchungen über Bewegungssehen und Optomotorik bei Springspinnen (Salticidae). Z. Tierpsychol., *16*: 385-402, 1959.

Eakin, R. M. and J. L. Brandenburger. Fine structure of the eyes of jumping spiders. J. Ultrastruct. Res., *37*: 618-663, 1971.

Eberhard, W. G. Attack behavior of diguetid spiders and the origin of prey wrapping in spiders. Psyche, *74*: 173-181, 1967.

Eberhard, W. G. The ecology of the web of *Uloborus diversus* (Araneae: Uloboridae). Oecologia (Berl.), *6*: 328-342, 1971.

Eberhard, W. G. The "inverted ladder" orb web of *Scoloderus* sp. and the intermediate orb of *Eustala*(?) sp. Araneae: Araneidae. J. Nat. Hist., *9*: 93-106, 1975.

Eberhard, W. G. Aggressive chemical mimicry by a bolas spider. Science, *198*: 1173-1175, 1977.

Edgar, W. D. Prey and predators of the wolf spider *Lycosa lugubris*. J. Zool. Lond., *159*: 405-411, 1969.

Edgar, W. D. Prey and feeding behavior of adult females of the wolf spider *Pardosa amentata* (Clerck). Neth. J. Zool., *20*: 487-491, 1970.

Edmunds, M. On the association between *Myrmarachne* sp. (Salticidae) and ants. Bull. Brit. Arachn. Soc., *4*: 149-160, 1978.

Edwards, G. B. Biological studies on the jumping spider *Phidippus regius* C. L. Koch. M.Sc. Thesis, University of Florida, Gainesville, 1975.

Edwards, G. B. Sound production by courting males of *Phidippus mystaceus* (Araneae: Salticidae) Psyche (in press).

Edwards, G. B., J. F. Carroll, and W. H. Whitcomb. *Stoidis aurata* (Araneae: Salticidae) a spider predator of ants. Florida Entomol., *57*: 337-345, 1974.

Eibl-Eibesfeldt, I. *Ethology: The Biology of Behavior*, Holt, Rinehart and Winston, 1970.

Eisner, T., R. Alsop, and G. Ettershank. Adhesiveness of spider silk. Science, *146*: 1058-1061, 1964.

Eisner, T. and J. Dean. Ploy and counterploy in predator-prey interactions: orb-weaving spiders versus bombardier beetles. Proc. Nat. Acad. Sci., *73*: 1365-1367, 1976.

Emerit, M. Innervation trichobothricale et axiale de la patte de l'aranéida, *Gasteracantha versicolor* (Walck.) (Argiopidae). C. R. Acad. Sc. Paris, Ser. D, *265*: 1134-1137, 1967.

Enders, F. Effects of prey capture, web destruction and habitat physiognomy on web-site tenacity of *Argiope* spiders (Araneidae). J. Arachnol., *3*: 75-82, 1975a.

Enders, F. The influence of hunting manner on prey size particularly in spiders with long attack distances. Am. Nat., *109*: 737-763, 1975b.

Enders, F. Airborne pheromone probable in orb web spider *Argiope aurantia* (Araneidae). British Arachnol. Soc. News, *13*: 5-6, 1975c.

Enders, F. Clutch size related to hunting manner of spider species. Ann. Ent. Soc. Amer., *69*: 991-998, 1976.

Engelhardt, W. Die mitteleuropäischen Arten der Gattung *Trochosa* C. L. Koch, 1848 (Araneae, Lycosidae). Z. Morph. Ökol. Tiere, *54*: 219-392, 1964.

Esch, H. and D. Wilson. The sounds produced by flies and bees. Z. vergl. Physiol. *54*: 256-267, 1967.

Estes, R. D. Behaviour and life history of the wildebeest (*Connochaetes taurinus* Burchell). Nature, *212*: 999-1000, 1966.

Exline, H. and H. Levi. American spiders of the genus *Argyrodes* (Theridiidae). Bull. Mus. Comp. Zool., *127*: 75-204, 1962.

Fabre, J. H. *The Life of the Spider*. Dodd Mead, 1913.

Falconer, W. Notes on *Eboria caliginosa* Falc. Naturalist, pp. 253-254, 1910.

Farkas, S. R. and H. H. Shorey. Mechanisms of orientation to a distant pheromone source. In *Pheromones*, M. C. Birch (ed.), *Frontiers of Biology*, 32, North Holland, 1974.

Farley, C. and W. A. Shear. Observations on the courtship behaviour of *Lycosa carolinensis*. Bull. Brit. Arachn. Soc., *2*: 153-158, 1973.

Farr, J. A. Social behavior of the golden silk spider, *Nephila clavipes*. J. Arachnol., *4*: 137-144, 1977.

Finck, A. Vibration sensitivity in an orb-weaver. Am. Zool., *12*: 539-543, 1972.

Fine, M. L., H. W. Winn, and B. L. Olla. Communication in fishes. In *How*

Animals Communicate, T. A. Sebeok (ed.), Indiana University Press, pp. 472-518, 1977.

Fishelson, L., D. Popper, and N. Gunderman. Diurnal cyclic behaviour of *Pempheris aualensis* Cuv, & Vol. (Pempheridae: Teleostei). J. Nat. Hist., *5*: 503-506, 1971.

Fisher, J. Evolution and bird sociality. In *Evolution as a Process,* J. Huxley, A. C. Hardy, and E. B. Ford (eds.), George Allen and Unwin, pp. 71-83, 1954.

Fitch, H. A. Spiders of the University of Kansas Natural History Reservation and Rockefeller Experimental Tract. Misc. Pub. Univ. Kansas Mus. Natur. Hist., *33*: 1-202, 1963.

Fitzgerald, T. D. Trial marking by larvae of the eastern tent caterpillar. Science, *194*: 961-963, 1976.

Foelix, R. F. Chemosensitive hairs in spiders. J. Morph., *132*: 313-334, 1970.

Foelix, R. F. and A. Choms. Fine structure of a spider joint receptor and associated synapses. Europ. J. Cell. Biol., *19*: 149-159, 1979.

Foelix, R. F. and I.-W. Chu-Wang. The morphology of spider sensilla. I. Mechanoreceptors. Tissue and Cell, *5*: 451-460, 1973a.

Foelix, R. F. and I.-W. Chu-Wang. Morphology of spider sensilla. II. Chemoreceptors. Tissue and Cell, *5*: 461-478, 1973b.

Ford, M. J. Energy costs of the predation strategy of the web-spinning spider *Lepthyphantes zimmermanni* Bertkau (Linyphiidae). Oecologia (Berl.), *28*: 341-349, 1977.

Ford, M. J. Locomotory activity and the predation strategy of the wolf spider *Pardosa amentata* (Cl.) (Lycosidae). Anim. Behav., *26*: 31-35, 1978.

Forster, L. M. A qualitative analysis of hunting behaviour in jumping spiders (Araneae: Salticidae). N. Z. J. Zool., *4*: 51-62, 1977a.

Forster. L. M. Some factors affecting feeding behaviour in young *Trite auricoma* spiderlings (Araneae: Salticidae). N. Z. J. Zool., *4*: 435-443, 1977b.

Forster, L. M. Mating behaviour in *Trite auricoma,* a New Zealand jumping spider. Peckhamia, *1*: 35-36, 1977c.

Forster, L. M. Visual mechanisms of hunting behaviour in *Trite planiceps,* a jumping spider. N. Z. J. Zool., *6*: 79-93, 1979a.

Forster, L. M. Comparative aspects of the behavioural biology of New Zealand jumping spiders (Salticidae: Araneae). Ph.D. dissertation, University of Otago, New Zealand, 1979b.

Forster, R. R. The spiders of New Zealand, Part 1. Otago Museum Bull. No. 1, Dunedin, 1967.

Forster, R. R. and L. M. Forster. *Small Land Animals of New Zealand,* McIndoes, Dunedin, 1970.

Forster, R. R. and L. M. Forster. *New Zealand Spiders.* Collins, 1973.

Forster, R. R. and L. M. Forster. Jumping Spiders. In *New Zealand's Heritage,* R. Knox (ed.), *6*: 2114-2117, 1976.

Fowler, H. G. and J. Diehl. Biology of a Paraguayan colonial orb-weaver *Eriophora bistriata* (Rengger) (Araneae, Araneidae). Bull. Brit. Arachn. Soc., *4*: 241-250, 1978.

Frings, H. and M. Frings. Reactions of orb weaving spiders (Argiopidae) to airborne sounds. Ecology, *47*: 578-588, 1966.

Frings, H. and M. Frings. Other invertebrates. In *Animal Communication*, T. A. Sebeok (ed.), Indiana University Press, 1968.

Gadgil, M. Male dimorphism as a consequence of sexual selection. Am. Nat., *106*: 574-580, 1972.

Galton, F. Gregariousness in cattle and men. Macmillan's Magazine, London, *23*: 353, 1871.

Gardner, B. T. Hunger and sequential responses in the hunting behavior of salticid spiders. J. comp. Physiol. Psych., *58*: 167-173, 1964.

Gardner, B. T. Observations on three species of *Phidippus* jumping spiders (Araneae: Salticidae). Psyche, *72*: 133-147, 1965.

Gardner, B. T. Hunger and characteristics of the prey in the hunting behavior of salticid spiders. J. comp. Physiol. Psych., *62*: 475-478, 1966.

Gaston, L. K., R. S. Kaae, H. H. Shorey, and D. Sellers. Controlling the pink bollworm by disrupting the sex pheromone communication between adult males. Science, *196*: 904-905, 1977.

Gatlin, L. L. *Information Theory and the Living System.* Columbia University Press, 1972.

Gerhardt, U. Weitere Untersuchungen zur Biologie der Spinnen. Z. Morph. Ökol. Tiere, *6*: 1-77, 1926.

Gerhardt, U. and A. Kaestner. Araneae. In *Handbuch der Zoologie*, Vol. 3, W. G. Kükenthal (ed.). Berlin: De Gruyter, pp. 394-656, 1937.

Gertsch, W. J. *American Spiders* (2nd ed.). Van Nostrand Reinhold, 1979.

Gervet, J. L'interaction entre individus dans un groupement animal. Essai de classification. Année Psychologique, *2*: 397-410, 1965.

Gervet, J. Interactions entre individus et phénomène social. Neth. J. Zool., *18*: 205-252, 1968.

Gettmann, W. W. Beutefang bei Wolfspinnen der Gattung *Pirata* (Arachnida; Araneae: Lycosidae). Ent. Germ., *3*: 93-99, 1976.

Ghiselin, M. T. *The Economy of Nature and the Evolution of Sex.* University of California Press, 1974.

Gilbert, L. W. Postmating female odor in *Heliconius* butterflies: a male-contributed antiaphrodisiac? Science, *196*: 419-420, 1976.

Giulio, L. Optomotor responses of the jumping spider *Heliophanus muscorum* Walck. (Araneae: Salticidae) elicited by turning spiral. Monitore zool. ital. (N.S.), *13*: 143-157, 1979.

Givens, R. P. Dimorphic foraging strategies of a salticid spider (*Phidippus audax*). Ecology, *59*: 309-321, 1978.

Glatz, L. Zur Biologie und Morphologie von *Oecobius annulipes* Lucas (Araneae, Oecobiidae). Z. Morph. Tiere, *61*: 185-214, 1967.

Gnatzy, W. Die Feinstruktur der Fadenhaare auf den Cerci von *Periplaneta americana* L. Verh. dtsch. Zool. Ges., *66*: 37-42, 1972.

Gnatzy, W. The ultrastructure of the threadhairs on the cerci of the cockroach *Periplaneta americana* L.: The intermoult phase. J. Ultrastr. Res., *54*: 124-134, 1976.

Gnatzy, W. and K. Schmidt. Die Feinstruktur der Sinneshaare auf den Cerci von *Gryllus bimaculatus* Deg. (Saltatoria, Gryllidae). Z. Zellforsch., *122*: 190-209, 1971.

Goodman, L. A. The analysis of cross-classified data: independence, quasi-independence and interactions in contingency tables with or without missing entries. J. Amer. Stat. Assoc., *63*: 1091-1131, 1968.

Görner, P. Die Orientierung der Trichterspinne nach polarisiertem Licht. Z. vergl. Physiol., *45*: 307-314, 1962.

Görner, P. A proposed transducing mechanism for a multiple-innervated mechanoreceptor (trichobothrium) in spiders. Cold Spr. Harb. Symp. Quant. Biol., *30*: 69-73, 1965.

Görner, P. and P. Andrews. Trichbothrien, ein Ferntastsinnesorgan bei Webspinnen (Araneen). Z. vergl. Physiol., *64*: 301-331, 1969.

Graeser, K. and H. Markl. Vibrations in a spider's web (in prep.).

Grassé, P. P. Sociétés animales et effet de groupe. Experientia, *2*: 77-82, 1946.

Grassé, P. P. Le fait social, ses critères biologiques, ses limites. Coll. intern. C.N.R.S., Paris, Structure et Physiologie des Sociétiés animales, 7-17, 1952.

Grassé, P. P. and R. Chauvin. L'effet de groupe et la survie des neutres dans les sociétés d'insectes. Rev. Sc., *82*: 461-464, 1944.

Grasshoff, M. Die Kreuzspinne *Araneus pallidus*—ihr Netzbau und ihre Paarungsbiologie. Natur und Museum, *94*: 305-314, 1964.

Greenquist, E. A. and J. S. Rovner. Lycosid spiders on artificial foliage: stratum choice, orientation preferences and prey wrapping. Psyche, *83*: 196-209, 1976.

Greenstone, M. H. The numerical response to prey availability of *Pardosa ramulosa* (McCook) (Araneae: Lycosidae) and its relationship to the role of spiders in the balance of nature. Symp. zool. Soc. London, *42*: 183-193, 1978.

Greenstone, M. H. Spider feeding behavior optimises dietary essential amino acid composition. Nature, *282*: 501-503, 1979.

Greenwalt, C. H. Dimensional relationships for flying animals. Smithson. Misc. Collect., *144*: 1-46, 1962.

Greig-Smith, P. The use of random and contagious quadrats in the study of the structure of plant communities. Ann. Bot. London, N.S. *16*: 293-216, 1952.

Greig-Smith, P. The use of pattern in ecological investigations. Recent Advances Bot., *2*: 1354-1358, 1961.

Griffin, D. R. The importance of atmospheric attenuation for the echolocation of bats (Chiroptera). Anim. Behav., *19*: 55-61, 1971.

Griffin, D. R. Expanding horizons in animal communication behavior. In *How Animals Communicate*, T. A. Sebeok (ed.), Indiana University Press, 1977.

Griffin, D. R. Prospects for a cognitive ethology. Behav. Brain Sciences, *4*: 527-538, 1978.

Griswold, C. E. Biosystematics of *Habronattus* in California. M.Sc. Thesis, University of California, Berkeley, 1977.

Grünbaum, A. A. Über das Verhalten der Spinne (*Epeira diademata*) besonders gegenüber vibratorischen Reizen. Psychol. Forsch. *9*: 275-299, 1927.

Gueldner, R. L. and G. L. Wiygul. Rhythms in pheromone production of the male boll weevil. Science, *199*: 984-986, 1978.

Guibe, J. P. Présence d'un appareil stridulatoire chez le mâle de deux espèces d'Araignées *Theridium ovatum* Cl. et *Gongylidiellum vivum* O. P. Cambr. Bull. Soc. Zool. France, *63*: 65-67, 1943.

Gwinner-Hanke, H. Zum Verhalten zweier stridulierender Spinnen, *Steatoda bipunctata* Linné und *Teutana grossa* Koch (Theridiidae, Araneae), unter besonderer Berücksichtigung des Fortpflangzungsverhaltens. Z. Tierpsychol., *27*: 649-678, 1970.

Hailman, J. P. Uses of the comparative study of behavior. In *Evolution, Brain, and Behavior: Persistent Problems*, R. B. Masterton, W. Hodos, and H. Jerison (eds.), Erlbaum, 181-189, 1976.

Hailman, J. P. Communication by reflected light. In *How Animals Communicate*, T. A. Sebeok (ed.), Indiana University Press, pp. 184-210, 1977a.

Hailman, J. P. *Optical Signals: Animal Communication and Light*. Indiana University Press, 1977b.

Haldane, J.B.S. Animal communication and the origin of human language. Sci. Prog., *43*: 385-401, 1955.

Hall, J. R. Synchrony and social stimulation in colonies of the black-headed weaver *Ploceus cucullatus* and Viellots black weaver *Melanopteryx nigerrimus*. Ibis, *112*: 93-104, 1970.

Hallander, H. Courtship display and habitat selection in the wolf spider *Pardosa chelata* (O. F. Müller). Oikos, *18*: 145-150, 1967.

Hallander, H. Prey, cannibalism, and microhabitat selection in the wolf spider *Pardosa chelata* O. F. Muller and *P. pullata* Clerck. Oikos, *21*: 337-340, 1970.

Hamilton, W. D. Geometry for the selfish herd. J. Theor. Biol., *31*: 295-311, 1971.

Hamilton, W. D. Innate social aptitudes in man, an approach from evolutionary genetics. In *Biosocial Anthropology*, R. Fox (ed.), Wiley, pp. 133-157, 1975.

Hammerstein, P. The role of asymmetries in animal contests. Anim. Behav., *29*: 193-205, 1981.

Hardie, R. C. and P. Duelli. Properties of single cells in posterior lateral eyes of jumping spiders. Z. Naturforsch., *33*: 156-158, 1978.

Harris, G. G. and W. A. van Bergeijk. Lateral-line organ response to near-field displacements of sound sources in water. J. Acoust. Soc. Amer., *34*: 1831-1841, 1962.

Harris, D. J. and P. J. Mill. Observations on the leg receptors of *Ciniflo* (Araneida: Dictynidae). I. External mechanoreceptors. J. comp. Physiol., *119*: 37-54, 1977a.

Harris, D. J. and P. J. Mill. Observations on the leg receptors of *Ciniflo*

(Araneida: Dictynidae) II. Chemoreceptors. J. comp. Physiol., *119*: 55-62, 1977b.

Harrison, J. B. Acoustic behavior of a wolf spider, *Lycosa gulosa*. Anim. Behav., *17*: 14-16, 1969.

Hartshorne, C. *Born to Sing: An Interpretation and World Survey of Bird Song*. Indiana University Press, 1973.

Harwood, R. H. The predatory behavior of *Argiope aurantia* (Lucas). Am. Midl. Nat., *91*: 130-139, 1974.

Hasselt, A.W.M. van. Stridulatie-organen bij Spinnen. Tidjschr. Ent. *19*: 101, 1876.

Haupt, J. Beitrag zur Kenntnis der Sinnesorgane von Symphylen: I. Elektronenmikroskopische Untersuchung des Trichobothriums von *Scutigerella immaculata* Newport. Z. Zellforsch., *110*: 588-599, 1970.

Haupt, J. Ultrastruktur der Trichobothrien von *Allopauropus (Decapauropus)* (Pauropoda). Abh. Verh. Naturwiss. Ver. Hamburg, NF *21/22*: 271-277, 1978.

Haupt, J. and Y. Coineau. Trichobothrien und Tastborsten der Milbe *Microcaeculus* (Acari, Prostigmata, Caeculidae). Z. Morph. Tiere, *81*: 305-322, 1975.

Haynes, D. L. and P. Sisojevic. Predatory behavior of *Philodromus rufus* Wackenaer (Araneae: Thomisidae). Can. Ent., *98*: 113-133, 1966.

Hazlett, B. A. Factors affecting the aggressive behavior of the hermit crab *Calcinus tibicen*. Z. Tierpsychol., *23*: 655-671, 1966.

Hazlett, B. A. Size relationship and aggressive behavior in the hermit crab *Clibanarius vittatus*. Z. Tierpsychol., *25*: 608-614, 1968.

Hazlett, B. A. Responses to agonistic postures by the spider crab *Microphrys bicornutus*. Mar. Behav. Physiol., *1*: 85-92, 1972.

Hazlett, B. A. Ritualization in marine crustacea. In *Behavior of Marine Animals*, Vol. 1, H. E. Winn and B. L. Olla (eds.), Plenum, pp. 97-125, 1972.

Hazlett, B. A. (ed.). *Quantitative Methods in the Study of Animal Behaviour*, Academic Press, 1977.

Hazlett, B. A. and W. H. Bossert. A statistical analysis of the aggressive communication systems of some hermit crabs. Anim. Behav., *13*: 367-373, 1965.

Hazlett, B. A. and G. F. Estabrook. Examination of agonistic behavior by character analysis. II. Hermit crabs. Behaviour, *49*: 88-110, 1974.

Hegdekar, B. M. and C. D. Dondale. A contact sex pheromone and some response parameters in lycosid spiders. Can. J. Zool., *47*: 1-4, 1969.

Heil, K. H. Beiträge zur Physiologie und Psychologie der Springspinnen. Z. vergl. Physiol., *23*: 1-25, 1936.

Helsdingen, P. J. van. A stridulatory organ in *Antistea elegans*. (Araneida, Hahniidae). Entomol. Berichten, *23*: 143-145, 1963.

Hendry, L. B. Insect pheromones: diet related? Science, *192*: 143-145, 1976.

Higashi, G. A. and J. S. Rovner. Post-emergent behaviour of juvenile lycosid spiders. Bull. Brit. Arachn. Soc., *3*: 113-119, 1975.

Hill, D. E. The mating of *Phidippus princeps*. Peckhamia, *1*: 5-7, 1977a.

Hill, D. E. The pretarsus of salticid spiders. Zool. J. Linn. Soc., *60*: 319-338, 1977b.

Hill, D. E. Modified setae of the salticid palp. Peckhamia, *1*: 7-9, 1977c.

Hill, D. E. The behavior of *Eris marginata* (Araneae: Salticidae). Peckhamia, *1*: 63-70, 1978.

Hill, D. E. The scales of salticid spiders. Zool. J. Linn. Soc., *65*: 193-218, 1979a.

Hill, D. E. Orientation by jumping spiders of the genus *Phidippus* (Araneae: Salticidae) during the pursuit of prey. Behav. Ecol. Sociobiol., *5*: 301-322, 1979b.

Hinde, R. A. Interaction of internal and external factors in integration of canary reproduction. In *Sex and Behavior*, F. A. Beach (ed.), Wiley, pp. 381-415, 1965.

Hinde, R. A. *Animal Behaviour: A Synthesis of Ethology and Comparative Psychology.* McGraw-Hill, 1970.

Hinde, R. A. *Non-Verbal Communication.* Cambridge University Press, 1972.

Hinde, R. A. *Biological Bases of Human Social Behaviour.* McGraw-Hill, 1974.

Hinde, R. A. The concept of function. In *Function and Evolution in Behaviour*, G. Baerends, C. Beer, and A. Manning (eds.), Clarendon Press, pp. 3-15, 1975.

Hines, W.G.S. Competition with an Evolutionarily Stable Strategy. J. Theor. Biol., *67*: 141-153, 1977.

Hinton, H. E. and R. S. Wilson. Stridulatory organs in spiny orb-weaver spiders. J. Zool. Lond. *162*: 482-484, 1970.

Hirai, K., H. H. Shorey, and L. K. Gaston. Competition among courting male moths: male-to-male inhibitory pheromone. Science, *202*: 644-645, 1978.

Hirschberg, D. Beiträge zur Biologie, insbesondere zur Brutpflege einiger Theridiiden. Zeit. für Wissenschaftliche Zoologie, *179*: 189-252, 1969.

Hoffmann, C. Bau und Funktion der Trichobothrien von *Euscorpius carpathicus.* Z. vergl. Physiol., *54*: 290-352, 1967.

Holden, W. Behavioral evidence of chemoreception on the legs of the spider *Araneus diadematus* Cl. J. Arachnol., *3*: 207-210, 1977.

Hollander, J. Species barriers in the *Pardosa pullata* group (Araneae, Lycosidae). Proc. 5th Int. Arachn. Congr., Brno, 129-142, 1971.

Hollander, J. and H. Lof. Differential use of the habitat of *Pardosa pullata* (Clerck) and *Pardosa prativaga* (L. Koch) in a mixed population (Araneae, Lycosidae). Tijdschrift voor Entomologie, *115*: 205-215, 1972.

Hollander, J., H. Dijkstra, H. Alleman, and L. Vlijm. Courtship behaviour as species barrier in the *Pardosa pullata* group (Araneae, Lycosidae). Tijdschrift voor Entomologie, *116*: 1-22, 1973.

Hollander, J. and H. Dijkstra. *Pardosa vlijmi* sp. nov. a new ethospecies sibling *Pardosa proxima* (C. L. Koch, 1848), from France, with description of courtship display (Araneae, Lycosidae). Beaufortia, *289*: 57-65, 1974.

Hölldobler, B. *Steatoda fulva* (Theridiidae). A spider that feeds on harvester ants. Psyche, *77*: 202-208, 1979.

Holling, C. S. The components of predation as revealed by a study of small mammal predation on the European pine sawfly. Can. Ent., *91*: 293-320, 1959.

Hollis, J. H. and B. A. Branson. Laboratory observations on the behavior of the salticid spider *Phidippus audax* (Hentz), Trans. Kansas Acad. Sci., *67*: 131-148, 1964.

Holzapfel, M. Die Bedeutung der Netzstarrheit für die Orientierung der Trichterspinne *Agelena labyrinthica.* Rev. Suisse Zool., *40*: 247-250, 1933a.

Holzapfel, M. Die nichtoptische Orientierung der Trichterspinne *Agelena labyrinthica* (Cl.) Z. vergl. Physiol., *20*: 55-116, 1933b.

Homann, H. Beiträge zur Physiologie der Spinnenaugen. I. Untersuchungsmethoden, II. Das Sehvermögen der Salticiden. Z. vergl. Physiol., *7*: 201-268, 1928.

Homann, H. Beiträge zur Physiologie der Spinnenaugen. III. Das Sehvermögen der Lycosiden. Z. vergl. Physiol., *14*: 40-67, 1931.

Homann, H. Die Augen der *Araneae.* Anatomie, Ontogenie und Bedeutung für die Systematik. Z. Morph. Tiere, *69*: 201-272, 1971.

Honjo, S. Social behavior of *Dictyna follicola* Bos et Str. (Araneae: Dictynidae). Acta Arachnol., *27*: 213-219, 1977.

Horch, K. An organ for hearing and vibration sense in the ghost crab *Ocypode.* Z. vergl. Physiol., *73*: 1-21, 1971.

Horel, A., C. Roland, and R. Leborgne. Mise en évidence d'une tendance au groupement chez les jeunes de l'araignée solitaire *Coelotes terrestris.* Rev. Arachnol., *2*: 157-164, 1979.

Horton, C. C. Apparent attraction of moths by the webs of araneid spiders. J. Arachnol., *7*: 88, 1979.

Howe, N. R. and Y. M. Sheich. Anthopleurine: a sea anemone alarm pheromone. Science, *189*: 386-388, 1975.

Huber, F. The insect nervous system and insect behaviour. Anim. Behav., *26*: 969-981, 1978.

Hukusima, S. and M. Miyafugi. Life histories and habits of *Misumenops tricuspidatus* Fabricius (Araneae: Thomisidae). Ann. Rep. Soc. Plant Prot. N. Jap., *21*: 5-12, 1970.

Humphries, D. A. and P. M. Driver. Erratic display as a device against predators. Science, *156*: 1767-1768, 1967.

Humphries, D. A. and P. M. Driver. Protean defence by prey animals. Oecologia (Berl.), *5*: 285-302, 1970.

Huntingford, F. A. The relationship between inter- and intraspecific aggression. Anim. Behav., *24*: 485-497, 1976.

Hutchinson, H. F. The hunting spider. Nature, *20*: 581, 1879.

Huxley, J. Darwin's theory of sexual selection and the data subsumed by it, in the light of recent research. Am. Nat., *72*: 416-433, 1938.

Ito, Y. Preliminary studies on the respiratory energy loss of a spider, *Lycosa pseudoannulata.* Res. Pop. Ecol., *4*: 13-21, 1964.

Jackson, R. R. The evolution of courtship and mating tactics in the jumping spider, *Phidippus johnsoni*. Ph.D. dissertation. University of California, Berkeley, 1976a.

Jackson, R. R. Predation as a selection factor in the mating strategy of the jumping spider *Phidippus johnsoni* (Salticidae, Araneae), Psyche, *83*: 243-255, 1976b.

Jackson, R. R. An analysis of alternative mating tactics of the jumping spider *Phidippus johnsoni* (Araneae: Salticidae). J. Arachnol., *5*: 185-230, 1977a.

Jackson, R. R. Courtship versatility in the jumping spider, *Phidippus johnsoni* (Araneae: Salticidae). Anim. Behav., *25*: 953-957, 1977b.

Jackson, R. R. Prey of the jumping spider *Phidippus johnsoni* (Araneae: Salticidae). J. Arachnol., *5*: 145-149, 1977c.

Jackson, R. R. Web sharing males and females of dictynid spiders. Bull. Brit. Arachn. Soc., *4*: 109-112, 1977d.

Jackson, R. R. Comparative studies of *Dictyna* and *Mallos* (Araneae, Dictynidae). III. Prey and predatory behavior. Psyche, *83*: 267-280, 1977e.

Jackson, R. R. Male mating strategies of dictynid spiders with differing types of social organization. Symp. zool. Soc. London, *42*: 79-88, 1978a.

Jackson, R. R. The mating strategy of *Phidippus johnsoni* (Araneae: Salticidae): I. Pursuit time and persistence. Behav. Ecol. Sociobiol., *4*: 123-132, 1978b.

Jackson, R. R. Life history of *Phidippus johnsoni* (Araneae: Salticidae). J. Arachnol., *6*: 1-29, 1978c.

Jackson, R. R. Comparative studies of *Dictyna* and *Mallos* (Araneae, Dictynidae). I. Social organization and web characteristics. Revue Arachnologique, *1*: 133-164, 1978d.

Jackson, R. R. Nests of *Phidippus johnsoni* (Araneae, Salticidae): characteristics, pattern of occupation, and functions. J. Arachnol., *7*: 47-58, 1979a.

Jackson, R. R. Predatory behavior of the social spider *Mallos gregalis*: is it cooperative? Insectes Sociaux, *26*: 300-312, 1979b.

Jackson, R. R. Comparative studies of *Dictyna* and *Mallos* (Araneae, Dictynidae). II. The relationships between courtship, mating, aggression and cannibalism in species with differing types of social organization. Rev. Arachnol., *2*: 103-132, 1979c.

Jackson, R. R. The mating strategy of *Phidippus johnsoni* (Araneae, Salticidae): II. Sperm competition and the function of copulation. J. Arachnol., *8*: 217-240, 1980a.

Jackson, R. R. The mating strategy of *Phidippus johnsoni* (Araneae, Salticidae): III. A cost-benefit analysis. J. Arachnol., *8*: 241-249, 1980b.

Jackson, R. R. The mating strategy of *Phidippus johnsoni* (Araneae, Salticidae): IV. Interpopulational variation in courtship persistence. Behav. Ecol. Sociobiol., *6*: 257-263, 1980c.

Jackson, R. R. Cannibalism as a factor in the mating strategy of the jumping spider *Phidippus johnsoni* (Araneae: Salticidae). Bull. Brit. Arachn. Soc., *5*: 129-133, 1980d.

Jackson, R. R. Nest-mediated sexual discrimination by a jumping spider, *Phidippus johnsoni* (Araneae, Salticidae). J. Arachnol., *9*: 87-92, 1981a.

Jackson, R. R. The courtship behavior of *Phidippus femoratus* (Araneae, Salticidae). Southwest Nat., *104*: 284, 1981b.

Jackson, R. R. The relationship between reproductive security and intersexual selection in the jumping spider *Phidippus johnsoni*. Evolution, *35*: 601-604, 1981c.

Jackson, R. R. and S. E. Smith. Aggregations of *Mallos* and *Dictyna* (Araneae, Dictynidae): population characteristics. Psyche, *85*: 65-79, 1978.

Jacobson, M. Chemical insect attractants and repellents. Ann. Rev. Entomol., *11*: 403-422, 1966.

Jacobson, M., M. Beroza, and R. T. Yamamoto. Isolation and identification of the sex attractant of the American cockroach. Science, *139*: 48-49, 1963.

Jacson, C. C. and K. J. Joseph. Life history, bionomics and behaviour of the social spider *Stegodyphus sarasinorum* Karsch. Insectes Sociaux, *20*: 189-204, 1973.

Jennings, D. T. Plant association of *Misumenops coloradensis* Gertsch (Araneae: Thomisidae) in central New Mexico. Southwest Nat., *16*: 201-207, 1971.

Jennings, D. T. An overwintering aggregation of spiders (Araneae) on cottonwood in New Mexico. Ent. News, *83*: 61-67, 1972.

Job, W. Beiträge zur Biologie der fangnetzbauenden Wolfsspinne *Aulonia albimana* (Walck, 1805). Zool. Jb. Syst., *101*: 560-608, 1974.

Johnson, C. The evolution of territoriality in the Odonata. Evolution, *18*: 89-92, 1964.

Kaestner, A. Reaktionen der Hüpfspinnen (Salticidae) auf unbewegte farblose und farbige Gesichtsreize. Zool. Beitr., *1*: 13-50, 1950.

Kaestner, A. *Invertebrate Zoology*. Wiley, 1969.

Kaissling, K. E. Insect olfaction. In *Handbook of Sensory Physiology*, Vol. IV. Chemical Senses, Part I. Olfaction. Springer-Verlag, 1971.

Kajak, A. An analysis of food relations between spiders *Araneus cornutus* Clerck and *Araneus quadratus* Clerck and their prey in a meadow. Ekol. Polska A., *13*: 717-764, 1965.

Kajak, A. Analysis of consumption by spiders under laboratory and field conditions. Ekol. Pol., *26*: 409-427, 1978.

Kajak, A. and J. Łuczak. Clumping tendencies in some species of meadow spiders. Bull. Acad. Polonaise Sc., *11*: 471-476, 1961.

Karlson, P. and A. Butenandt. Pheromones (ectohormones) in insects. Ann. Rev. Entomol., *4*: 49-58, 1959.

Kaston, B. J. The senses involved in the courtship of some vagabond spiders. Ent. Amer., *16*: 97-167, 1936.

Kaston, B. J. The slit sense organs of spiders. J. Morph., *58*: 189-209, 1937.

Kaston, B. J. *Spiders of Connecticut*. Conn. St. Geol. and Nat. Hist. Survey, *70*: 1-874, 1948.

Kaston, B. J. The evolution of spider webs. Am. Zool., *4*: 191-207, 1964.

Kaston, B. J. Some little known aspects of spider behavior. Am. Midl. Nat., *73*: 336-356, 1965.

Katzenellenbogen, J. S. Insect pheromone synthesis: new methodology. Science, *194*: 139-148, 1976.

King, J. A. The ecology of aggressive behavior. Ann. Rev. Ecol. System., *4*: 117-138, 1973.

Kinsler, L. E. and A. R. Frey. *Fundamentals of Acoustics*. Wiley, 1962.

Kiritani, J. and N. Kakiya. An analysis of the predator-prey system in the paddy field. Research Popul. Ecol., Kyoto Univ., *17*: 29-38, 1975.

Kiritani, K., S. Kawahare, T. Sasaba, and F. Nakasuji. Quantitative evaluation of predation of spiders on the green rice leaf hopper, *Nephotettix cincticeps* Uhler, by a sight-count method. Research Popul. Ecol., Kyoto Univ., *13*: 187-200, 1972.

Kirschfeld, K. The resolution of lens and compound eyes. In *Neural Principles in Vision*, F. Zettler and R. Weiler (eds.), Springer-Verlag, pp. 354-372, 1976.

Kittredge, J. S. and F. T. Takahashi. The evolution of sex pheromone communication in the Arthropoda. J. Theor. Biol., *35*: 467-471, 1972.

Klopfer, P. H. and J. J. Hatch. Experimental considerations. In *Animal Communication*, T. A. Sebeok (ed.), Indiana University Press, pp. 31-43, 1968.

Kolosvary, G. Biologische Studien über einige Spinnenarten von Szeged. 10 Congr. Intern. Zoologie, Budapest, 1929.

Kolosvary, G. Neue Daten zur Lebensweise der *Trochosa (Hogna) singoriensis* (Laxm.). Zool. Anz., *98*: 307-311, 1932.

Koomans, M. J., S.W.F. van der Ploeg, and H. Dijkstra. Leg wave behavior of wolf spiders of the genus *Pardosa* (Lycosidae, Araneae). Bull. Brit. Arachn. Soc., *3*: 53-61, 1974.

Krafft, B. Sur une possibilité d'échanges de substance entre les individus chez l'araignée sociale *Agelena consociata* Denis. C. R. Acad. Sc. Paris, *260*: 5376-5378, 1965.

Krafft, B. Various aspects of the biology of *Agelena consociata* Denis when bred in the laboratory. Am. Zool., *9*: 201-210, 1969.

Krafft, B. Contribution à la biologie et à l'éthologie d'*Agelena consociata* Denis (araignée sociale du Gabon.) I. Biologia Gabonica, *3*: 197-301, 1970a.

Krafft, B. Contribution à la biologie et à l'éthologie d'*Agelena consociata* Denis (araignée sociale du Gabon). II. Etude expérimentale de certains phénomènes sociaux. Biologia Gabonica, *4*: 307-369, 1970b.

Krafft, B. Les rhythmes d'activité d'*Agelena consociata* Denis. Activité de tissage et activité locomotrice. Biologia Gabonica, *4*: 99-130, 1970c.

Krafft, B. Contribution à la biologie et à l'éthologie d'*Agelena consociata* Denis (araignée sociale du Gabon). III. Etude expérimentale de certains phénomènes sociaux (suite). Biologia Gabonica, *7*: 3-56, 1971a.

Krafft, B. Les interactions entre les individus chez *Agelena consociata*,

Araignée sociale du Gabon. Proc. 5th Int. Arachn. Congr., Brno, 159-164, 1971b.

Krafft, B. Les interactions limitant le cannibalisme chez les araignées solitaires et sociales. Bull. Soc. Zool. France, *100*: 203-221, 1975a.

Krafft, B. La tolérance réciproque chez l'Araignée sociale *Agelena consociata* Denis. Proc. 6th Int. Arachn. Congr., 107-112, 1975b.

Krafft, B. The recording of vibratory signals performed by spiders during courtship. Symp. zool. Soc. London, *42*: 59-67, 1978.

Krafft, B. Organisation et évolution des sociétés d'araignées. Journal de Psychologie, *1*: 23-51, 1979.

Krafft, B., C. Roland, D. Mielle, and R. Leborgne. Contribution à l'étude des signaux intervenant lors du comportement sexuel d'araignées du genre *Amaurobius, Coelotes* et *Tegenaria.* 103ème Congrès National des Sociétés Savantes, Nancy, Sciences, *3*: 19-29, 1978.

Krafft, B. and A. Horel. Le comportement maternel et les relations mères-jeunes chez les araignées. Reprod. Nutr. Dévelop., *20*: 747-758, 1979.

Krafft, B. and R. Leborgne. Perception sensorielle et importance des phénomènes vibratoires chez les araignées. J. Psychologie, *3*: 299-334, 1979.

Krafft, B. and C. Roland. Un labyrinthe appliqué à l'étude des attractions sociaux et sexuelles et de leur specificité chez les araignées. Rev. Arachnol., *2*: 165-171, 1979.

Krebs, J. R. Optimal foraging. Decision rules for predators. In J. R. Krebs and N. B. Davies (eds.), *Behavioral Ecology*, Sinauer, 1978.

Kronestedt, T. Study of a stridulatory apparatus in *Pardosa fulvipes* (Collett) (Araneae, Lycosidae) by scanning electron microscopy. Zool. Scripta., *2*: 43-47, 1973.

Kronestedt, T. Study on chemosensitive hairs in the wolf spiders (Araneae, Lycosidae) by scanning electron microscopy. Zool. Scripta., *8*: 279-285, 1979.

Kronk, A. W. and S. E. Riechert. Parameters affecting the habitat choice of *Lycosa santrita* Chamberlin and Ivie. J. Arachnol., *7*: 155-166, 1979.

Kuenzler, E. J. Niche relations of three species of lycosid spiders. Ecology, *39*: 494-500, 1958.

Kulczynski, W. Fragmentata arachnologica III. De organo stridenti nonullorum Theridiidarum. Bull. Acad. Cracovie, pp. 564-568, 1905.

Kullmann, E. Beobachtungen des Netzbaues und Beiträge zur Biologie von *Cyrtophora citricola* Forskal. Zool. Jb. Syst., *86*: 181-216, 1958.

Kullmann, E. Beobachtungen und Betrachtungen zum Verhalten der Theridiide *Conopistha argyrodes* Walckenaer (Araneae). Mitt. Zool. Mus. Berlin, *35*: 276-296, 1959.

Kullmann, E. Beobachtungen an *Theridium tepidariorum* C. L. Koch als Mitbewohner von *Cyrtophora*-Netzen. Deutsche Entomol. Z., *7*: 146-163, 1960.

Kullmann, E. Neue Ergebnisse über Netzbau und das Sexualverhalten einiger Spinnenarten. Zeitsch. Zool. Syst. Evolutionsforschung, *2*: 41-122, 1964.

Kullmann, E. Soziale Phänomene bei Spinnen. Insectes Sociaux, *15*: 289-298, 1968.

Kullmann, E. Beobachtungen zum Sozialverhalten von *Stegodyphus sarasinorum* Karsch. Bull. Mus. Hist. Nat., *2*: Suppl. 1, 76-81, 1969.

Kullmann, E. Evolution of social behavior in spiders (Araneae, Eresidae and Theridiidae). Am. Zool., *12*: 419-426, 1972a.

Kullmann, E. The convergent development of orb-webs in cribellate and ecribellate spiders. Am. Zool., *12*: 395-405, 1972b.

Kullmann, E., S. Nawabi, and W. Zimmermann. Neue Ergebnisse zur Brutbiologie cribellater Spinnen aus Afghanistan und der Serengeti. Z. Kölner Zoo, *3*: 87-108, 1972.

Kullmann, E., F. Otto, T. Braun, and R. Raccanello. Grundlagen und Ordnung—Übersicht der Netzkonstruktionen der Spinnen. Mitt. Inst. f. Leichte Flächentragwerke, Stuttgart, 1975.

Kullmann, E. and W. Zimmermann. Versuche zur Toleranz bei der permanent sozialen Spinnenart *Stegodyphus sarasinorum* Karsch (Eresidae). Proc. 5th Int. Arachn. Congr., Brno, 175-182, 1971.

Kullmann, E. and W. Zimmermann. Regurgitationsfütterungen als Bestandteil der Brutfürsorge bei Haubennetz und Röhrenspinnen (Araneae, Theridiidae und Eresidae). Proc. 6th Int. Arachn. Congr., 120-124, 1974.

Lahee, F. H. The calls of spiders. Psyche, *2*: 74, 1904.

Land, M. F. Structure of the principal eyes of jumping spiders (Salticidae: Dendryphantinae) in relation to visual optics. J. exp. Biol., *51*: 443-470, 1969a.

Land, M. F. Movements of the retinae of jumping spiders (Salticidae: Dendryphantinae) in response to visual stimuli. J. exp. Biol., *51*: 471-493, 1969b.

Land, M. F. Orientation by jumping spiders in the absence of visual feedback. J. exp. Biol., *54*: 119-139, 1971.

Land, M. F. Stepping movements made by jumping spiders during turns mediated by the lateral eyes. J. exp. Biol., *57*: 133-151, 1972a.

Land, M. F. Mechanisms of orientation and pattern recognition by jumping spiders (Salticidae). In *Information Processing in the Visual Systems of Arthropods*, R. Wehner (ed.), Springer-Verlag, pp. 231-247, 1972b.

Land, M. F. A comparison of the visual behaviour of a predatory arthropod with that of a mammal. In *Invertebrate Neurons and Behaviour*, C.A.G. Wiersma (ed.), MIT Press, pp. 411-418, 1974.

Land, M. F. and T. S. Collet. Chasing behaviour of houseflies (*Fannia canicularis*): a description and analysis. J. comp. Physiol., *89*: 331-357, 1974.

Lang, H. H. "Hörorgane" bei Insekten. Umschau, *79*: 257-258, 1979.

Langer, R. M. Elementary physics and spider webs. Am. Zool., *9*: 81-89, 1969.

Leborgne, R. and B. Krafft. Techniques d'enregistrement et d'analyse des signaux vibratoires intervenant dans les comportements des araignées sédentaires. Rev. Arachnol., *2*: 173-182, 1979.

Leborgne, R., C. Roland, and A. Horel. Quelques aspects de la communication chimique et vibratoire chez certaines Agelenidae. Proc. 8th Int. Arachn. Congr., 215-220, 1980.

Legendre, R. Recherches sur l'olfaction des araignées. Ann. Sci. Nat. Zool., *11*: 141-155, 1958.

Legendre, R. Quelques remarques sur le comportement des *Argyrodes malgaches*. Ann. Sci. Nat. Zool., *12*: 507-512, 1960.

Legendre, R. Etudes sur les Archaea. II. La capture des proies et la prise de nourriture. Bull. Soc. Zool. France, *86*: 316-319, 1961.

Legendre, R. L'audition et l'émission de sons chez les Aranéides. Ann. Biol., *2*: 371-390, 1963.

Legendre, R. Un organe stridulant nouveau chez les Archaeidae (Aranéides). Bull. Soc. Zool. France, *95*: 29-30, 1970.

Legendre, R. and A. Lopez. Variations morphologiques sexuelles des glandes gnathocoxales chez les Araneidae (Araneae), C. R. Acad. Sc. Paris, *279*: 1769-1771, 1974a.

Legendre, R. and A. Lopez. Etude histologique de quelques formations glandulaires chez les araignées du genre *Argyrodes* (Theridiidae) et description d'un nouveau type de glande: la glande clypéale des mâles. Bull. Soc. Zool. France, *99*: 453-460, 1974b.

Legendre, R. and A. Lopez. Ultrastructure de la glande clypéale des mâles d'araignées appartenant au genre *Argyrodes* (Theridiidae). C. R. Acad. Sc. Paris, *281*: 1101-1103, 1975.

LeGuelte, L. Learning in spiders. Am. Zool., *9*: 145-152, 1969.

Lehrman, D. S. The reproductive behavior of ring doves. Sci. Am., November, 1964.

Le Masne, G. Classification et caractéristiques des principaux types de groupements sociaux réalisés chez les Invertébrés. Coll. intern. C.N.R.S., Paris, Structure et Physiologie des Sociétés animales, 19-70, 1952.

LeSar, C. D. and J. D. Unzicker. Life history, habits and prey preferences of *Tetragnatha laboriosa* (Araneae: Tetragnathidae). Environ. Ent., *7*: 879-884, 1978.

Levi. H. W. Predatory and sexual behavior of the spider *Sicarius* (Araneae: Sicariidae). Psyche, *76*: 29-40, 1967.

Levi, H. W. Orb weaving spiders and their webs. Amer. Sci., *66*: 734-742, 1978.

Levi, H. W., L. R. Levi, and H. S. Zim. *Spiders and Their Kin*. Golden Press, 1968.

Liesenfeld, F. J. Untersuchungen am Netz und über den Erschütterungssinn von *Zygiella x-notata* (Cl.) (Araneidae). Z. vergl. Physiol., *38*: 563-592, 1956.

Liesenfeld, F. J. Über Leistung und Sitz des Erschütterungssinnes von Netzspinnen. Biol. Zbl. *80*: 465-475, 1961.

Lin, N. Territorial behavior in cicada killer wasp. *Sphecius speciosus* (Drury) (Hymenoptera: Sphecidae). Behaviour, *20*: 115-113, 1963.

Lincoln, C. J., R. Phillips, W. H. Whitcomb, G. C. Dowell, W. P. Boyer, K. O. Bell, C. L. Dean, E. J. Matthews, J. B. Graves, L. D. Newsom, D. F. Clower, J. R. Bradley, and J. L. Bagnet. The bollworm-tobacco budworm problem in Arkansas and Louisiana. Agric. Exp. Sta. Div. Agr. Univ. Arkansas Bull., *720*: 1-66, 1967.

Lindauer, M. Social behaviour and mutual communication. In *The Physiology of Insects*, M. Rockstein (ed.), Academic Press, pp. 123-186, 1965.

Linsenmair, K. E. Anemomenotaktische Orientierung bei Scorpionen (Chelicerata, Scorpiones). Z. vergl. Physiol., *60*: 445-449, 1968.

Lloyd, J. E. Bioluminescence and communication. In *How Animals Communicate*, T. A. Sebeok (ed.), Indiana University Press, pp. 164-183, 1977.

Lloyd, M. Mean crowding. J. Anim. Ecol., *36*: 1-30, 1967.

Locket, G. H. Observations on the mating habits of some web-spinning spiders. Proc. zool. Soc. London, *1926*: 1125-1146, 1926.

Lopez, A. Morphologie et rapports particuliers des glandes épigastriques dans deux familles d'Aranéides: les Dysderidae et les Clubionidae. Bull. Soc. Zool. France, *97*: 113-119, 1972.

Lopez, A. Les glandes hypodermiques monocellulaires et canaliculées des Aranéides. Forma et Functio, *7*: 317-326, 1974.

Lorenz, K. *Studies in Animal and Human Behaviour*, Vol. 1 (translated by R. Martin). Harvard University Press, 1970.

Lubin, Y. D. Adaptive advantages and the evolution of colony formation in *Cyrtophora* (Araneae: Araneidae). Zool. J. Linn. Soc., *54*: 321-339, 1974.

Lucas, F. Spiders and their silks. Discovery, *25*: 20-26, 1964.

Łuczak, J. Behavior of a spider population in the presence of mosquitoes. Ekol. Pol., *18*: 625-634, 1970.

Łuczak, J. Skupienia srodowiskowe I socjalne pajakow (habitat and social concentrations of spiders). Wiadomosci Ekologiczne *17*: 404-412, 1971.

MacArthur, R. H. and E. R. Pianka. On the optimal use of a patchy environment. Am Nat., *100*: 603-609, 1966.

Machado, A. de B. Araignées nouvelles pour la faune Portugaise (II). Mem. Est. Mus. Zool. Univ. Coimbra. Ser. I, *117*: 1-60, 1941.

Magni, F., F. Papi, H. E. Savely, and P. Tongiorgi. Research on the structure and physiology of the eyes of a lycosid spider. II. The role of different pairs of eyes in astronomical navigation. Archs. ital. Biol., *103*: 146-158, 1964.

Main, B. Y. Adaptive radiation of trapdoor spiders. Aust. Mus. Mag., *12*: 160-163, 1957.

Main, B. Y. *Spiders*. The Australian Naturalist Library, Collins, 1976.

Manly, B.F.J. and L. M. Forster. A stochastic model for the behaviour of naive spiderlings (Araneae: Salticidae). Biometrical Journal, *21*: 115-122, 1979.

Manning, A. *Drosophila* and the evolution of behaviour. Viewpoints in Biology, *4*: 124-169, 1965.

Manning, A. *An Introduction to Animal Behaviour*. 3rd ed., Addison-Wesley, 1979.

Markl, H. Die Verständigung durch Stridulationssignale bei Blattschneiderameisen. II. Erzeugung und Eigenschaften der Signale. Z. vergl. Physiol., *60*: 103-150, 1968.

Markl, H. Leistungen des Vibrationssinnes bei wirbellosen Tieren. Fortsch. Zool., *21*: 100-120, 1973.

Markl, H. and S. Fuchs. Klopfsignale mit Alarmfunktion bei Roßameisen (*Camponotus*, Formicidae, Hymenoptera). Z. vergl. Physiol., *76*: 204-225, 1972.

Markl, H. and B. Hölldobler. Recruitment and food-retrieving behavior in *Novomessor* (Formicidae, Hymenoptera). Behav. Ecol. Sociobiol., *4*: 183-216, 1978.

Markl, H. and J. Tautz. The sensitivity of hair receptors in caterpillars of *Barathra brassicae* L. (Lepidoptera, Noctuidae) to particle movement in a sound field. J. comp. Physiol., *99*: 79-87, 1975.

Markl, H. and K. Wiese. Die Empfindlichkeit des Rückenschwimmers *Notonecta glauca* L. für Oberflächenwellen des Wassers. Z. vergl. Physiol., *62*: 413-420, 1969.

Marler, P. Visual system. In *Animal Communication*, T. A. Sebeok (ed.), Indiana University Press, pp. 103-126, 1968.

Marler, P. The evolution of communication. In *How Animals Communicate*, T. A. Sebeok (ed.), Indiana University Press, 1977.

Marson, J. Some observations on the variation in the camouflage used by *Cyclosa insulana* (Costa), an Asiatic spider in its web. Proc. zool. Soc. London, *11*: 598-605, 1974.

Martin, H. Leistungen des topochemischen Sinnes bei der Honigbiene. Z. vergl. Physiol., *50*: 254-292, 1965.

Marx, J. L. Insect control. I: Use of pheromones. Science, *181*: 736-737, 1973.

Masters, W. M. Insect disturbance stridulation: its defensive role. Behav. Ecol. Sociobiol., *5*: 187-200, 1979.

Matthews, R. W. and J. R. Matthews. *Insect Behavior*. Wiley, 1978.

Maynard Smith, J. Game theory and the evolution of fighting. In *On Evolution*, J. Maynard Smith (ed.), Edinburgh University Press, pp. 3-23, 1972.

Maynard Smith, J. The theory of games and the evolution of animal conflict. J. Theor. Biol., *47*: 209-221, 1974.

Maynard Smith, J. Evolution and the theory of games. Amer. Sci., *46*: 41-45, 1976.

Maynard Smith, J. and G. R. Price. The logic of animal conflict. Nature, *246*: 15-18, 1973.

Maynard Smith, J. and G. A. Parker. The logic of asymmetric contests. Anim. Behav., *24*: 159-175, 1976.

Mayr, E. *Animal Species and Evolution*. Harvard University Press, 1963.

Mayr, E. Teleological and teleonomic, a new analysis. Boston Studies in the Philosophy of Science. *14*: 91-117, 1974a.

Mayr, E. *Populations, Species, and Evolution*. Harvard University Press, 1974b.

McClintock, M. K. Menstrual synchrony and suppression. Nature, *229*: 244-245, 1971.

McCook, H. C. *American Spiders and their Spinningwork*, Vol. 1. Philadelphia, 1889.

McKeown, K. C. *Spider Wonders of Australia*. Angus and Robertson, 1936.

McKeown, K. C. *Australian Spiders*, 2nd ed. Angus and Robertson, 1952.

Meijer, J. A glandular secretion in the ocular area of certain erigonine spiders (Araneae, Linyphiidae). Bull. Brit. Arachn. Soc., *3*: 251-252, 1976.

Meinwald, J., G. Prestwich, K. Nakanishi, and I. Kubo. Chemical ecology: studies from East Africa. Science, *199*: 1167-1173, 1978.

Melchers, M. Zur Biologie und zum Verhalten von *Cupiennius salei* (Keyserling), einer amerikanischen Ctenide. Zool. Jb. Abt. Syst., Ökol. u. Georgr. *91*: 1-90, 1963.

Melchers, M. Der Beutefang von *Cupiennius salei* Keyserling (Ctenidae). Z. Morph. Ökol. Tiere, *58*: 321-346, 1967.

Meyer, E. Neue sinnesbiologische Beobachtungen an Spinnen. Z. Morph. Ökol. Tiere, *12*: 1-69, 1928.

Michelsen, A. Sound reception in different environments. In *Sensory Ecology*, M. A. Ali (ed.), Plenum, 345-373, 1978.

Michelsen, A. and H. Nocke. Biophysical aspects of sound communication in insects. Adv. Insect Physiol., *10*: 247-297, 1974.

Michener, C. D. Comparative social behavior of bees. Ann. Rev. Entomol., *14*: 299-342, 1969.

Middendorf, G. A. Resource partitioning by an iguanid lizard: thermal and density influences. Ph.D. dissertation. University of Tennessee, 1979.

Mielle, D. Contribution à l'étude de comportement prédateur et des mécanisms de tolérance dans le genre *Tegenaria*. Thesis, Nancy, 1978.

Mill, P. J. and D. J. Harris. Observations on the leg receptors of *Ciniflo* (Araneida: Dictynidae). J. comp. Physiol., *119*: 63-72, 1977.

Miller, W. H. and A. W. Snyder. The tiered vertebrate retina. Vis. Res., *17*: 239-255, 1977.

Millot, J. Sens chimique et sens visuel chez les araignées. Ann. Biol. Anim., *22*: 1-21, 1946.

Millot, J. Ordre des Aranéides (Araneae). In *Traité de Zoologie*, P. Grassé (ed.), *6*: 589-743, 1949.

Millot, J. and P. Bourgin. Sur la biologie des *Stegodyphus* solitaires. Bull. Biol. France et Belg., *76*: 298-313, 1942.

Miyashita, K. Quantitative feeding biology of *Lycosa T-insignata* Boes. et Str. (Araneae: Lycosidae). Bull. Nat. Inst. Agri. Sci. (Jap.), Ser. C., *22*: 329-344, 1968a.

Miyashita, K. Changes of the daily food consumption during adult stage of *Lycosa pseudoannulata* Boes. et Str. (Araneae: Lycosidae). Appl. Ent. Zool., *3*: 203-204, 1968b.

Monath, M. Barking spiders: *Solenocosmia crassipes*-Aviculariidae. N. Qd. Nat., *26*: 118, 1957.

Montgomery, T. H. Studies on the habits of spiders, particularly those of the mating period. Proc. Acad. Nat. Sci. Phil., *1*: 59-149, 1903.

Montgomery, T. H. Further studies on the activities of araneids. Am. Nat., *42*: 697-709, 1908.

Montgomery, T. H. Further studies on the activities of Araneids. II. Proc. Acad. Nat. Sci., Phil., *61*: 548-569, 1909.

Montgomery, T. H. Significance of the courtship and secondary sexual characters of araneids. Am. Nat., *44*: 151-177, 1910.

Morse, D. H. Prey capture by the crab spider *Misumena calycina* (Araneae: Thomisidae). Oecologia (Berl.), *39*: 309-319, 1979.

Morse, P. M. *Vibrations and Sound*. McGraw-Hill, 1948.

Moulton, D. G. Olfaction in mammals. Am. Zool., *7*: 421-249, 1967.

Moynihan, M. Control, suppression, decay, disappearance and replacement of displays. J. Theor. Biol., *29*: 85-112, 1970.

Nakamura, K. The ingestion in wolf spiders II. The expression of degree of hunger and amount of ingestion in relation to spider's hunger. Research Popul. Ecol., Kyoto Univ., *14*: 82-96, 1972.

Nakamura, K. A model for the functional response of a predator to varying prey densities, based on the feeding ecology of wolf spiders. Bull. Nat. Inst. Agri. Sci. (Jap.), Ser. C., *31*: 28-89, 1977.

Nappi, A. J. Notes on the courtship and mating habits of the wolf spider *Lycosa helluo* Walckenaer. Am. Midl. Nat., *74*: 368-373, 1965.

Neumann, J. von and O. Morgenstern. *Theory of Games and Economic Behavior*. Princeton University Press, 1944.

Nicklaus, R. Die Erregung einzelner Fadenhaare von *Periplaneta americana* in Abhängigkeit von der Grösse und Richtung der Auslenkung. Z. vergl. Physiol., *50*: 331-362, 1965.

Nicklaus, R. Zur Richtcharakteristik der Fadenhaare von *Periplaneta americana*. Z. vergl. Physiol., *54*: 434-437, 1967.

Nielsen, E. *The Biology of Spiders*, Vol. 1. Copenhagen, 1932.

Nocke, H. Biophysik der Schallerzeugung durch die Vorderflügel der Grillen. Z. vergl. Physiol., *74*: 272-314, 1971.

Nørgaard, E. Environment and behavior of *Theridion saxatile*. Oikos, *7*: 159-192, 1956.

Nyffeler, M. and G. Benz. Die Beutespektren der Netzspinnen *Argiope bruennichi* (Scop.), *Araneus quadratus* Cl. und *Agelena labyrinthica* Cl. in Ödlandwiesen bei Zürich. Rev. Suisse Zool., *85*: 747-757, 1978.

Nyffeler, M. and G. Benz. Zur ökologischen Bedeutung der Spinnen der Vegetationsschicht von Getreide und Rapsfeldern bei Zürich (Schweiz). Zeit. Angew. Entomol., *87*: 348-376, 1979.

Opell, B. D. Revision of the genera and tropical American species of the spider family Uloboridae. Bull. Mus. Comp. Zool., *148*: 443-549, 1979.

Opell, B. D. and J. A. Beatty. The Nearctic Hahniidae (Arachnida: Araneae). Bull. Mus. Comp. Zool., *147*: 393-433, 1976.

Orians, G. H. Social stimulation within blackbird colonies. Condor, *63*: 330-337, 1961.

Painter, T. S. On the dimorphism of the males of *Maevia vittata*. Zool. Jahrb. Abt. Syst., *35*: 625-636, 1913.

Palmgren, P. Über die Brutpflegeinstinkthandlungen der Wolfspinnen (Lycosidae). Soc. Scientiarium Fennica, Commentationes Biologicae, *9*: 1-29, 1944.

Papi, F. Astronomische Orientierung bei der Wolfspinne *Arctosa perita* (Latr.). Z. vergl. Physiol., *37*: 230-233, 1955.

Parker, G. A. Assessment strategy and the evolution of fighting behavior. J. Theor. Biol., *47*: 223-243, 1974.

Parker, G. A. Selfish genes, evolutionary games and the adaptiveness of behaviours. Nature, *274*: 849-865, 1978.

Parker, G. A. and R. A. Stuart. Animal behavior as a strategy optimizer: evolution of resource assessment strategies and optimal emigration threshold. Am. Nat., *110*: 1055-1076, 1976.

Parry, D. A. The signal generated by an insect in a spider's web. J. exp. Biol., *43*: 185-192, 1965.

Payne, T. L. Pheromone perception. In *Pheromones,* M. C. Birch (ed.), *Frontiers of Biology,* 32, North Holland, 1974.

Peakall, D. B. and P. N. Witt. The energy budget of an orb web-building spider. Comp. Biochem. Physiol., *54*: 187-190, 1976.

Peckham, G. W. and E. G. Peckham. Observations on sexual selection in spiders of the family Attidae. Occ. Pap. Wisconsin Nat. Hist. Soc., *1*: 3-60, 1889.

Peckham, G. W. and E. G. Peckham. Additional observations on sexual selection in spiders of the family Attidae. Occ. Pap. Wisconsin Nat. Hist. Soc., *1*: 117-151, 1890.

Peckham, G. W. and E. G. Peckham. The sense of sight in spiders with some observations of the color sense. Trans. Wisconsin Acad. Sci. Arts Lett., *10*: 231-261, 1894.

Peckham, G. W. and E. G. Peckham. Revision of the Attidae of North America. Trans. Wisconsin Acad. Sci. Arts Lett., *16*: 355-646, 1909.

Perdeck, A. D. The isolating value of species specific song patterns in two sibling species of grasshoppers. Behaviour, *12*: 1-75, 1958.

Peters, H. M. Die Fanghandlung der Kreuzspinne (*Epeira diademata* L.). Experimentelle Analysen des Verhaltens. Z. vergl. Physiol., *15*: 693-748, 1931.

Peters, H. M. Weitere Untersuchungen über die Fanghandlung der Kreuzspinne (*Epeira diademata* Cl.). Z. vergl. Physiol., *19*: 47-67, 1933a.

Peters, H. M. Kleine Beiträge zur Biologie der Kreuzspinne *Epeira diademata* Cl. Z. Morph. Ökol. Tiere, *26*: 447-468, 1933b.

Peters, H. M. Studien am Netz der Kreuzspinne: II. Über die Herstellung des Rahmens, der Radialfäden und der Hilfsspirale. Z. Morph. Ökol. Tiere, *33*: 128-150, 1938.

Petrunkevitch, A. Spiders of the Virgin Islands. Trans. Conn. Acad. Arts and Sci., *27*: 21-78, 1926.

Pickard-Cambridge, F. O. The spiders of Dorset, with an appendix containing short descriptions of those British spiders not yet found in Dorsetshire, II. Proc. Dorset Nat. Hist., 237-625, 1881.

Pickard-Cambridge, F. P. Newly discovered stridulating organs in the genus *Scytodes.* Ann. Mag. Nat. Hist., *16*: 371-373, 1895.

Pielou, E. C. *Mathematical Ecology.* Wiley, 1977.

Platnick, N. The evolution of courtship behaviour in spiders. Bull. Brit. Arachn. Soc., *2*: 40-47, 1971.

Pocock, R. J. On a new sound producing organ in a spider. Ann. Mag. Nat. Hist., *16*: 223-230, 1895.

Pocock, R. J. On the presence of Wood-Mason organs in *Techoila zebrata.* Ann. Mag. Nat. Hist., *17*: 177-179, 1896.

Pocock, R. J. On the spiders of the suborder Mygalomorphae from the Ethiopian region contained in the collection of the British Museum. Proc. zool. Soc. London, 724-774, 1897.

Pocock, R. J. A new stridulating theraphosid spider from South America. Ann. Mag. Nat. Hist., *3*: 347-349, 1899.

Pocock, R. J. Notes on the commensalism subsisting between a gregarious spider and the moth *Batrachedra stegodyphobius.* Ent. Monthly, *2*: 167-170, 1903.

Pointing, P. J. A quantitative field study of predatory behavior by the sheet-web spider *Frontinella communis* on European pine shoot moth adults. Can. J. Zool., *44*: 256-273, 1966.

Popper, K. R. Objective knowledge: an evolutionary approach. Clarendon Press, 1974.

Precht, H. Über das angeborene Verhalten von Tieren: Versuche an Springspinnen (Salticidae). Z. Tierpsychol., *9*: 207-230, 1952.

Precht, H. and G. Freytag. Über Ermüdung und Hemmung angeborener Verhaltensweisen bei Springspinnen (Salticidae). Zugleich ein Beitrag zum Triebproblem. Behaviour, *13*: 143-211, 1958.

Prell, H. Über trommelnde Spinnen. Zool. Anz., *48*: 61-64, 1916.

Premack, D. and G. Woodruff. Does the chimpanzee have a theory of mind? Behav. Brain Sciences, *4*: 515-526, 1978.

Preston, L. J. Communication systems and social interactions in a goby-shrimp symbiosis. Anim. Behav., *26*: 791-802, 1978.

Prószyński, J. Notes on systematics of the Salticidae (Arachnida, Aranei). I-VI. Ann. Zool., *28*: 227-255, 1971.

Randall, J. B. New observations of maternal care exhibited by the green lynx spider *Peucetia viridans* Hentz (Araneae, Oxypidae). Psyche, *84*: 286-291, 1977.

Rathmayer, W. Elektrophysiologische Untersuchungen an Proprioceptoren im Bein einer Vogelspinne (*Eurypelma hentzi* Chamb.). Z. vergl. Physiol., *54*: 438-454, 1967.

Readshaw, J. L. The numerical response of predators to prey density. J. Applied Biol., *10*: 342-351, 1973.

Reed, C. F. Cues in the web-building process. Am. Zool., *9*: 211-222, 1969.

Reese, E. S. Submissive posture as an adaptation to aggressive behavior in hermit crabs. Z. Tierpsychol., *19*: 645-651, 1962.

Reiskind, J. Ant-mimicry in Panamanian clubionid and salticid spiders (Araneae: Clubionidae, Salticidae). Biotropica, *9*: 1-8, 1977.

Reissland, A. Electrophysiology of trichobothria in orb-weaving spiders (Agelenidae, Araneae). J. comp. Physiol., *123*: 71-84, 1978.

Reissland, A. and P. Görner. Mechanics of trichobothria in orb-weaving spiders (Agelenidae, Araneae). J. comp. Physiol., *123*: 59-69, 1978.

Richards, O. W. Sexual selection and allied problems in the insects. Biol. Rev., *2*: 298-364, 1927.

Richman, D. B. The relationship of epigamic display to the systematics of jumping spiders (Araneae: Salticidae). Ph.D. dissertation, University of Florida, Gainesville, 1977a.

Richman, D. B. On the relationship of sexual selection to sexual dimorphism in jumping spiders. Peckhamia, *1*: 36-39, 1977b.

Richman, D. B. and R. C. Hemenway. Field cage evaluation of predators of the soybean looper *Pseudoplusia includens* (Walker) Lepidoptera: Noctuidae. Environ. Ent. (in press).

Richter, C.J.J. Production de soie, contenant probablement un phéromone, par mâles et femelles adultes de *Pardosa amentata*. Proc. 5th Int. Arachn. Congr., Brno, 227-231, 1971.

Richter, C.J.J. and C. van der Kraan. Silk production in adult males of the wolf spider *Pardosa amentata* (Cl.) (Araneae, Lycosidae). Neth. J. Zool., *20*: 392-400, 1970.

Richter, C.J.J., C.J. Stolting, and L. Vlijm. Silk production in adult females of the wolf spider *Pardosa amentata* (Lycosidae, Araneae). J. Zool. Lond., *165*: 285-290, 1971.

Riechert. S. E. The pattern of local web distribution in a desert spider: mechanisms and seasonal variation. J. Anim. Ecol., *43*: 733-746, 1974a.

Riechert, S. E. Thoughts on the ecological significance of spiders. Bioscience, *24*: 352-356, 1974b.

Riechert, S. E. Web-site selection in the desert spider, *Agelenopsis aperta* (Gertsch). Oikos, *27*: 311-315, 1976.

Riechert. S. E. Energy-based territoriality in populations of the desert-spider, *Agelenopsis aperta* (Gertsch). Symp. zool. Soc. London, *42*: 211-222, 1978a.

Riechert, S. E. Games spiders play: behavioral variability in territorial disputes. Behav. Ecol. Sociobiol., *3*: 135-152, 1978b.

Riechert, S. E. Games spiders play: II. Resource assessment strategies. Behav. Ecol. Sociobiol., *6*: 121-128, 1979.

Riechert, S. E. The consequences of being territorial: spiders, a case study. Amer. Nat., *117*: 871-892, 1981.

Riechert, S. E., W. G. Reeder, and T. G. Allen. Patterns of spider distribution *Agelenopsis aperta* (Gertsch) in desert grassland and recent lava bed habitats, south central New Mexico. J. Anim. Ecol., *42*: 19-35, 1973.

Riechert, S. E. and C. R. Tracy. Thermal balance and prey availability: bases for a model relating web-site characteristics to spider reproductive success. Ecology, *56*: 265-285, 1975.

Risch, P. Quantitative analysis of orb web patterns in four species of spiders. Behav. Gen., *7*: 199-238, 1977.

Roberts, M. Observations on an environmental association between *Entelecara erythropus* (Linyphiidae) and *Ciniflo similis* (Dictynidae). Bull. Brit. Arachn. Soc., *1*: 63, 1969.

Robertson, D. R., A. D. Fletcher, and M. G. Cleland. Schooling as a mechanism for circumventing the territoriality of competitors. Ecology, *57*: 1208-1220, 1976.

Robinson, M. H. Predatory behavior of *Argiope argentata* (Fabricius). Am. Zool., *9*: 161-173, 1969.

Robinson, M. H. Symbiosis between insects and spiders: an association between lepidopteran larvae and the social spider *Anelosimus eximius* (Araneae: Theridiidae). Psyche, *83*: 225-232, 1977a.

Robinson, M. H. Tropical spinners. New Scientist, *1*, 1977b.

Robinson, M. H. and B. Robinson. Prey caught by a sample population of the spider *Argiope argentata* (Araneae: Araneidae) in Panama: a year's census data. Zool. J. Linn. Soc., *49*: 345-357, 1970.

Robinson, M. H. and B. Robinson. The predatory behavior of the ogre faced spider *Dinopis longipes* F. Cambridge (Araneae: Dinopidae). Am. Midl. Nat., *85*: 85-96, 1971.

Robinson, M. H. and B. Robinson, Ecology and behaviour of the giant wood spider *Nephila maculata* (Fabricius) in New Guinea. Smithson. Contrib. Zool., *149*: 1-76, 1973a.

Robinson, M. H. and B. Robinson. The stabilimenta of *Nephila clavipes* and the origins of stabilimentum-building in araneids. Psyche, *80*: 277-288, 1973b.

Robinson, M. H. and B. Robinson. A tipulid associated with spider webs in Papua New Guinea. Entomol. Mon. Mag., *112*: 1-4, 1976.

Robinson, M. H. and B. Robinson. The evolution of courtship systems in tropical araneid spiders. Symp. zool. Soc. London, *42*: 17-29, 1978.

Robinson, M. H. and H. Mirick. The predatory behavior of the golden-web spider *Nephila clavipes* (Araneae: Araneidae). Psyche, *78*: 123-139, 1971.

Robinson, M. H., H. Mirick, and O. Turner. The predatory behavior of some araneid spiders and the origin of immobilization wrapping. Psyche, *76*: 487-501, 1969.

Robinson, M. H. and J. Olazarri. Units of behavior and complex sequences in the predatory behavior of *Argiope argentata* Fabricius. Smithson. Contrib. Zool., *65*: 1-36, 1971.

Robinson, M. H. and C. E. Valerio. Attacks on large or highly defended prey by tropical salticid spiders. Psyche, *84*: 1-10, 1977.

Roelofs, W. L. and A. Comeau. Sex pheromone specificity: taxonomic and evolutionary aspects in Lepidoptera. Science, *165*: 398-400, 1969.

Romanes, G. J. *Animal Intelligence*. Kegan Paul, Trench,1882.

Rosenblatt, J. S. and D. S. Lehrman. Maternal behavior of the laboratory rat. In *Maternal Behavior in Mammals*, H. L. Rheingold (ed.), Wiley, pp. 8-57, 1963.

Ross, J. W. Evidences for territoriality in the sheetline weaving spider *Florinda coccinea* (Hentz) (Araneae: Linyphiidae). M.S. thesis, University of Tennessee, Knoxville, 1977.

Ross, K. and R. L. Smith. Aspects of the courtship behavior of the black widow spider, *Latrodectus hesperus* (Araneae, Theridiidae) with evidence for the existence of a contact sex pheromone. J. Arachnol., *7*: 69-77, 1979.

Rovner, J. S. Acoustic communication in a lycosid spider (*Lycosa rabida*) Walckenaer. Anim. Behav., *15*: 273-281, 1967.

Rovner, J. S. Territoriality in the sheet web spider *Linyphia triangularis* (Clerck) (Araneae, Linyphiidae). Z. Tierpsychol., *25*: 232-242, 1968a.

Rovner, J. S. An analysis of display in the lycosid spider *Lycosa rabida* Walckenaer. Anim. Behav., *16*: 358-369, 1968b.

Rovner, J. S. Copulation in the lycosid spider (*Lycosa rabida*) Walckenaer: a quantitative study. Anim. Behav., *20*: 133-138, 1972.

Rovner, J. S. Copulatory pattern supports generic placement of *Schizocosa avida* (Walckenaer). Psyche, *80*: 245-248, 1973.

Rovner, J. S. Copulation in the lycosid spider *Schizocosa saltatrix* (Hentz): An analysis of palpal insertion patterns. Anim. Behav., *22*: 94-99, 1974.

Rovner, J. S. Sound production by Nearctic wolf spiders: a substratum-coupled stridulatory mechanism. Science, *190*: 1309-1310, 1975.

Rovner, J. S. Adhesive hairs in spiders: Behavioral functions and hydraulically mediated movement. Symp. zool. Soc. London, *42*: 99-107, 1978.

Rovner, J. S. Morphological and ethological adaptations for prey capture in wolf spiders (Araneae: Lycosidae). J. Arachnol., *8*: 201-216, 1980a.

Rovner J. S. Vibration in *Heteropoda venatoria* (Sparassidae): a third method of sound production in spiders. J. Arachnol., *8*: 193-200, 1980b.

Rovner, J. S., G. A. Higashi, and R. F. Foelix. Maternal behavior in wolf-spiders: the role of abdominal hairs. Science, *182*: 1153-1155, 1973.

Rovner, J. S., and F. G. Barth. Vibratory communication through living plants by a tropical wandering spider. Science, *214*: 464-466, 1981.

Rypstra, A. L. Foraging flocks of spiders. A study of aggregate behavior in *Cyrtophora citricola* Forskal (Araneae: Araneidae) in West Africa. Behav. Ecol. Sociobiol., *5*: 291-300, 1979.

Salmon, M. Waving display and sound production in the courtship behavior of *Uca pugilator* with comparisons to *U. minax* and *U. pugnax*. Zoologica, *50*: 124-150, 1965.

Salpeter, M. and C. Walcott. An electron microscopical study of a vibration receptor in the spider. Exp. Neurol., *2*: 232-250, 1960.

Sarinana, F. O., J. S. Kittredge, and D. C. Lowrie. A preliminary investigation of the sex pheromone of *Pardosa ramulosa*. Notes Arachnol. Southwest, *2*: 9-11, 1971.

Savory, T. H. A theory of animal courtship. Ann. Mag. Nat. Hist., *16*: 548-551, 1925.

Savory, T. H. *The Biology of Spiders*. Sidgwick and Jackson, 1928.

Savory, T. H. *The Spiders and Allied Orders of the British Isles*. Frederick Warne, 1935.

Savory, T. H. The male spider: myth wrongly labels him few in numbers and sorry in prospects. Nat. Hist., New York, *70*: 51-54, 1961.

Scharrer, E. Zirpende Spinnen. Nat. und Mus., *62*: 267-268, 1932.

Scheuring, L. Die Augen der Arachnoiden. II. Zool. Jb. (Anat.), *37*: 369-464, 1914.

Schick, R. X. The early instars, larval feeding and the significance of larval feeding in the crab spider genus *Misumenops* (Araneida, Thomisidae). Notes Arachnol. Southwest, *3*: 12-19, 1972.

Schiffman, S. S. Physicochemical correlates of olfactory quality. Science, *185*: 112-116, 1974.

Schleidt, W. M. Tonic communication: continual effects of discrete signs in animal communication systems. J. Theor. Biol., *42*: 359-386, 1973.

Schleidt, W. M. How "fixed" is the fixed action pattern? Z. Tierpsychol., *36*: 184-211, 1974.

Schmitt, B. C. and B. W. Ache. Olfaction: responses of a decapod crustacean are enhanced by flicking. Science, *205*: 204-206, 1979.

Schneider, D. Electrophysiological investigation on the antennal receptors of the silkworm moth during chemical and mechanical stimulation. Experientia, *31*: 89-91, 1957.

Schneider, D. Insect antennae. Ann. Rev. Entomol., *9*: 103-122, 1964.

Schneider, D. Insect olfaction: deciphering system for chemical messages. Science, *163*: 1031-1037, 1969.

Schoener, T. W. Theory of feeding strategies. Ann. Rev. Ecol. Syst., *2*: 369-404, 1971.

Schopenhauer, A. *Von der Nichtigkeit des Daseins.* Deutsche Bibliotek Berlin (n.d.).

Schopenhauer, A. In *Encyclopedia Britannica, 20*: 100-102. Chicago, 1974.

Schwartz, E. Zur Lokalisation akustischer Reize von Fischen und Amphibien. Fortsch. Zool., *21*: 121-135, 1973.

Scott, H. Eight months' entomological collecting in the Seychelles Islands. Trans. Linn. Soc. Lond. (Z. Zool.), *14*: 32-33, 1910.

Scott, H. Tipulids with white tarsi hanging on spiders' webs and their dancing movements. Entomol. Mon. Mag., *94*: 247, 1958.

Sebeok, T. A. (ed.). *How Animals Communicate.* Indiana University Press, 1977.

Semichon, L. Observation sur une Araignée mexicaine transportée en France. Bull. Soc. Entom. France, 338-340, 1910.

Sengun, A. Experimente zur Sexuell-mechanischen Isolation. Rev. Fac. Sci. Istanbul(b), *9*: 239-253, 1944.

Shannon, C. and W. Weaver. *The Mathematical Theory of Communication.* University of Illinois Press, 1949.

Shear, W. A. Observations on the predatory behavior of the spider *Hypochilus gertschi* Hoffman (Hypochilidae). Psyche, *76*: 407-417, 1969.

Shear, W. A. The evolution of social phenomena in spiders. Bull. Brit. Arachn. Soc., *1*: 65-76, 1970.

Shepard, M. Spider-ant symbiosis: *Cotinusa* spp. (Araneida: Salticidae) and *Tapinoma melanocephalum* (Hymenoptera: Formicidae). Canad. Entomol., *104*: 1951-1954, 1972.

Shorey, H. H. Sex pheromones of noctuid moths: II. Mating behavior of *Trichoplusia ni* (Lepidoptera: Noctuidae) with special reference to the role of the sex pheromone. Ann. Ent. Soc. Amer., *57*: 371-377, 1964.

Shorey, H. H. Behavioral response to insect pheromones. Ann. Rev. Entomol., *18*: 349-380, 1973.

Shorey, H. H. *Animal Communication by Pheromones*. Academic Press, 1976.

Shorey, H. H., L. K. Gaston, and J. S. Roberts. Sex pheromones of noctuid moths. VI. Absence of behavioral specificity for the female sex pheromones of *Trichoplusia ni* versus *Autographa californica*, and *Heliothis zea* versus *H. virescens* (Lepidoptera: Noctuidae). Ann. Ent. Soc. Amer., *58*: 600-603, 1965.

Simon, E. Histoire naturelle des araignées (Aranéides). Libr. Encyclop. de Roret, Paris, 1864.

Simon, E. Observations biologiques sur les Arachnides. I. Araignées sociales. Ann. Soc. Ent. France, *60*: 5-14, 1891.

Simon, E. Sur un organe stridulatoire dans le genre *Sicarius*. Ann. Soc. Ent. France, *62*: 224-225, 1893.

Simon, E. Sur l'araignée *Mosquéro*. C. R. Acad. Sc. Paris, *148*: 736-737, 1909.

Simon, H. A. *Models of Man: Social and Rational:* Mathematical Essays on Rational Human Behavior in a Social Setting. Wiley, 1957.

Skudrzyk, E. *Die Grundlagen der Akustik*. Springer-Verlag, 1954.

Slater, P.J.B. and J. C. Ollason. The temporal pattern of behaviour in isolated male zebra finches: transition analysis. Behaviour, *42*: 248-269, 1972.

Smith, H. M. Size of breeding populations in relation to egg-laying and reproductive success in the eastern red-wing (*Agelaius p, phoeniceus*). Ecology, *24*: 183-207, 1943.

Smith, W. J. *The Behavior of Communicating*. Harvard University Press, 1977.

Sokal, R. R. and F. J. Rohlf. *Biometry*. W. H. Freeman, 1969.

Spencer, R. B. On the presence and structure of a stridulating organ in *Phlogius (=Phrictus) crassipes*. Horn. Scient. Exp. Centr. Austral., *2*: 412-415, 1896.

Steel, R.D.G. and J. H. Torrie. *Principles and Procedures of Statistics*. McGraw-Hill, 1960.

Stratton, G. E. and G. W. Uetz. Courtship behavior, acoustic communication and reproductive isolation in 2 sibling species of wolf spiders (Aranea: Lycosidae) (in prep.).

Struthsaker, T. T. Notes on the spiders *Uloborus mundior* (Chamberlin and Ivie) and *Nephila clavipes* (Linnaeus) in Panama. Am. Midl. Nat., *82*: 611-613, 1969.

Suter, R. B. *Cyclosa turbinata* (Araneae, Araneidae): Prey discrimination via web-borne vibrations. Behav. Ecol. Sociobiol., *3*: 283-296, 1978.

Suzuki, H., M. Watanabe, Y. Tsukahara, and K. Tasaki. Duplex system in the simple retina of a gastropod mollusc, *Limax flavus* L. J. comp. Physiol., *133*: 125-130, 1979.

Szlep, R. Change in the response of spiders to repeated web vibrations. Behaviour, *23*: 203-239, 1964.

Szlep, R. The web-spinning process and web-structure of *Latrodectus tredecimguttatus, L. pallidus* and *L. revivensis.* Proc. zool. Soc. London, *145*: 75-89, 1965.

Tautz, J. Reception of medium vibration by thoracal hairs of caterpillars of *Barathra brassicae* L. (Lepidoptera, Noctuidae). I. Mechanical properties of the receptor hairs. J. comp. Physiol., *118*: 13-31, 1977.

Tautz, J. Reception of medium vibration by thoracal hairs of caterpillars of *Barathra brassicae* L. (Lepidoptera, Noctuidae). II. Response characteristics of the sensory cell. J. comp. Physiol., *125*: 67-77, 1978.

Tavolga, W. N. Levels of interaction in animal communication. In *Development and Evolution of Behavior,* L. R. Aronson, E. Tobach, D. S. Lehrman, and J. S. Rosenblatt (eds.), W. H. Freeman, pp. 281-302, 1970.

Taylor, B. B. and W. B. Peck. A comparison of the northern and southern forms of *Phidippus audax* (Hentz) (Araneida: Salticidae). J. Arachnol., *2*: 89-99, 1974.

Thomas, M. Observations sur *Philaeus chrysops.* Bull. Ann. Soc. Entomol. Belgium, *69*: 266-270, 1929.

Thompson, H. R. The statistical study of plant distribution patterns using a grid of quadrats. Aust. J. Bot., *6*: 322-342, 1954.

Thompson, H. R. Distribution of distance to nth neighbor in a population of randomly distributed individuals. Ecology, *37*: 391-394, 1956.

Thornhill, R. Scorpionflies as kleptoparasites of web-building spiders. Nature, *258*: 709-711, 1975.

Thorpe, W. H. The definition of terms used in animal behaviour studies. Bull. Anim. Behav., *9*: 34-40, 1951.

Tietjen, W. J. Dragline-following by male lycosid spiders. Psyche, *84*: 165-178, 1977.

Tietjen, W. J. Tests for olfactory communication in four species of wolf spiders (Araneae, Lycosidae). J. Arachnol., *6*: 197-206, 1979a.

Tietjen, W. J. Is the sex pheromone of *Lycosa rabida* deposited on a substratum? J. Arachnol., *7*: 207-212, 1979b.

Tietjen, W. J. and J. S. Rovner. Physico-chemical trail-following behaviour in two species of wolf spiders: sensory and etho-ecological concomitants. Anim. Behav., *28*: 735-741, 1980.

Tinbergen, L. The natural control of insects in pine wood. Arch. Nederl. Zool., *13*: 266-336, 1960.

Tolbert, W. W. Aerial dispersal behavior of two orb-weaving spiders. Psyche, *84*: 13-27, 1977.

Tretzel, E. Zum Begegnungsverhalten von Spinnen. Zool. Anz., *163*: 194-205, 1959.

Tretzel, E. Biologie, Ökologie und Brutpflege von *Coelotes terrestris* (Wider) (Araneae: Agelenidae). I. Biologie und Ökologie. II. Brutpflege. Z. Morph. Ökol. Tiere, *49*: 658-745, 1961; *50*: 375-542, 1961.

Tretzel, E. Die Sprache bei Spinnen. Umschau, *13*: 403-407, 1963.

Trivers, R. L. Parental investment and sexual selection. In *Sexual Selection and the Descent of Man*, B. Campbell (ed.), Aldine, pp. 136-179, 1972.

Turnbull, A. L. The prey of the spider *Linyphia triangularis* (Clerck) (Araneae, Linyphiidae). Can. J. Zool., *38*: 859-873, 1960.

Turnbull, A. L. The search for prey by a web-building spider *Achaearanea tepidariorum* (C. L. Koch) (Araneae, Theridiidae). Canad. Entomol., *96*: 568-579, 1964.

Turnbull, A. L. A population of spiders and their potential prey in an overgrazed pasture in eastern Ontario. Can. J. Zool., *44*: 557-583, 1966.

Turnbull, A. L. Ecology of the true spiders (Araneomorphae). Ann. Rev. Entomol., *18*: 305-348, 1973.

Uetz, G. W. and J. W. Burgess. Habitat structure and colonial behavior in *Metepeira spinipes* (Araneae: Araneidae), an orb weaving spider from Mexico. Psyche, *86*: 79-89, 1979.

Uetz, G. W. and G. Denterlein. Courtship behavior, habitat and reproductive isolation in *Schizocosa rovneri* Uetz and Dondale. (Araneae: Lycosidae). J. Arachnol., *7*: 121-128, 1979.

Uetz, G. W., A. D. Johnson, and D. W. Schemske. Web placement, web structure, and prey capture in orb-weaving spiders. Bull. Brit. Arachn. Soc., *4*: 141-148, 1978.

Valerio, C. E. Prey capture by *Drymusa dinora* (Araneae, Scytodidae). Psyche, *81*: 284-287, 1974a.

Valerio, C. E. Feeding on eggs by spiderlings of *Achaearanea tepidariorum* (Araneae, Theridiidae) and the significance of the quiescent instar in spiders. J. Arachnol., *2*: 57-63, 1974b.

Valerio, C. E. Population structure in the spider *Achaearanea tepidariorum* (Araneae, Theridiidae). J. Arachnol. *3*: 185-194, 1975.

Valerio, C. E. and M. V. Herrero. Tendencia social en adultos de la Araña *Leucauge* sp. (Araneae, Araneidae) en Costa Rica. Brenesia, *10/11*: 69-76, 1977.

Vancassel, M. Le développement du cycle parental de *Labidura riparia*. Biologie du Comportement, *2*: 51-75, 1977.

Verdcourt, B. An odd behaviour in a species of crane-fly, *Limonia (Euglochina) connectans* Alexander (Dipt., Tipulidae). Entomol. Mon. Mag., *94*: 163, 1958.

Vlijm, L. and W. J. Borsje. Comparative research of the courtship behaviour in the genus *Pardosa* (Arachn.-Araneae). II. Some remarks about courtship behaviour in *Pardosa pullata* (Clerck). Bull. Mus. Hist. Nat. Paris, *41*: 112-116, 1969.

Vlijm, L. and H. Dijkstra. Comparative research of the courtship behaviour in the genus *Pardosa*. I. Some remarks about the courtship of *P. amentata, P. hortensis, P. nigriceps* and *P. lugubris*. Senck. Biol., *47*: 51-55, 1966.

Vlijm, L., J. Hollander, and S. E. Wenderlaar-Bonga. Locomotory activity and sexual display in *Pardosa amentata* (Lycosidae, Araneae). Neth. J. Zool., *20*: 475-484, 1970.

Vlijm, L., A. Kessler, and C.J.J. Richter. The life history of *Pardosa amentata* (Cl.) (Araneae, Lycosidae). Ent. Berichten, *23*: 75-80, 1963.

Vlijm, L. and C.J.J. Richter, Activity fluctuations of *Pardosa lugubris* (Walckenaer), Araneae: Lycosidae, during the breeding season. Ent. Berichten, *26*: 222-230, 1966.

Vlijm, L. and A. M. Kessler-Geschiere. The phenology and habitat of *Pardosa monticola, P. nigriceps* and *P. pullata* (Araneae, Lycosidae). J. Anim. Ecol., *36*: 31-56, 1967.

Vogel, H. Über die Spaltsinnesorgane der Radnetzspinnen. Jena Z. Med. Naturw., *59*: 171-208, 1923.

Vogel, B. R. Courtship of some wolf spiders. The Armadillo papers, *4*: 1-6, 1970.

Vogel, B. R. Individual interactions of *Pardosa*. The Armadillo papers, *5*: 1-12, 1971.

Vogel, B. R. Apparent niche sharing of two *Pardosa* species (Araneidae, Lycosidae). The Armadillo papers, *7*: 1-13, 1972.

Vollrath, F. Konkurrenzvermeidung bei tropischen kleptoparasitischen Haubennetzspinnen der Gattung *Argyrodes* (Arachnida: Araneae: Theridiidae). Ent. Germ., *3*: 104-108, 1976.

Vollrath, F. Behaviour of two kleptoparasitic species of the genus *Argyrodes* Simon (Araneae, Theridiidae). Symp. zool. Soc. London, *42*: 483, 1978.

Vollrath, F. Behavior of the kleptoparasitic spider *Argyrodes elevatus* (Araneae, Theridiidae). Anim. Behav., *27*: 515-521, 1979a.

Vollrath, F. Vibrations: their signal function for a spider kleptoparasite. Science, *205*: 1149-1151, 1979b.

von Uexküll, J. *Umwelt und Innenwelt der Tiere.* Springer-Verlag, 1909.

Waage, J. K. Reproductive character displacement in *Calopteryx* (Odonata: Calopterygidae). Evolution, *33*: 104-116, 1979.

Walcott, C. The effect of the web on vibration sensitivity in the spider, *Achaearanea tepidariorum* (Koch). J. exp. Biol., *40*: 595-611, 1963.

Walcott, C. A spider's vibration receptor: its anatomy and physiology. Am. Zool., *9*: 133-144, 1969.

Walcott, C. and W. G. van der Kloot. The physiology of the spider vibration receptor. J. exp. Zool., *141*: 191-244, 1959.

Walker, T. J. Factors responsible for intraspecific variation in the calling song of crickets. Evolution, *16*: 407-428, 1962.

Walker, T. J. The taxonomy and calling songs of United States tree crickets, II. Ann. Ent. Soc. Amer., *56*: 772-789, 1963.

Walker, T. J. Cryptic species among several sound producing ensiferan orthoptera (Gryllidae and Tettigoniidae). Quart. Rev. Biol., *30*: 345-355, 1964.

Walker, T. J. Character displacement and acoustic insects. Am. Zool., *14*: 1147-1150, 1974.

Wallace, A. R. *Darwinism: An Exposition of the Theory of Natural Selection with Some of its Applications.* MacMillan, 1889.

Wallace, A. R. The colours of animals. Nature, *42*: 289-291, 1890.

Waters, W. E. A quantitative measure of aggregation in insects. J. Econ. Entom., *52*: 1180-1184, 1959.

Weems, H. V. and W. H. Whitcomb. The green lynx spider, *Peucetia viridans* (Hentz) (Araneae: Oxyopidae). Entomol. Circ., *181*, 1977.

Westring, N. Om stridulationsorganet hoe *Asagena serratipes*. Schrk. Nat. Tidsskr., *4*: 349-360, 1843.

Westring, N. Bekrifning pa Stridulationsorganer hos slagtena *Pachycoris* Burm. och *Scuterella* Lamarck, af Insektordningen Hemiptera jemte oftversigt af alla de hittils bekanta olika satten for sadane ljuds framalstrande bland andra Insekter. Goteb. Kongl. Vet. Handl., *4*: 47-57, 1858.

Weygoldt, R. Communication in crustaceans and arachnids. In *How Animals Communicate,* T. A. Sebeok (ed.), Indiana University Press, pp. 303-333, 1977.

Whitcomb, W. and R. Eason. The mating behavior of *Peucetia viridans* (Araneida: Oxyopidae). Florida Entomol., *48*: 163-167, 1965.

Whitcomb, W. H., H. Exline, and M. Hite. Comparison of spider populations of ground stratum in Arkansas pasture and adjacent cultivated field. Arkansas Acad. Sci. Proc., *17*: 1-6, 1963.

Whitcomb, W. H. and M. Tardic. Araneida as predators of the fall webworm. J. Kansas Entomol. Soc., *36*: 186-190, 1963.

Wickler, W. *Mimicry in Plants and Animals.* Weidenfeld and Nicholson, 1968.

Wickler, W. Über Koloniegründung und soziale Bindung von *Stegodyphus mimosarum* Parvesi und anderen sozialen Spinnen. Z. Tierpsychol., *32*: 522-531, 1973.

Wiehle, H. Beiträge zur Biologie der Araneen, insbesondere zur Kenntnis des Radnetzbaues. Z. Morph. Ökol. Tiere, *11*: 115-151, 1928.

Wiehle, H. Neue Beiträge zur Kenntnis des Fanggewebes der Spinnen aus den Familien Argiopidae, Uloboridae und Theridiidae. Z. Morph. Ökol. Tiere, *23*: 349-400, 1931.

Wilde, J. de. Some physical properties of the spinning threads of *Aranea diademata* L. Arch. Neerl. Physiol., *27*: 117-132, 1943.

Williams, D. S. The physiological optics of a nocturnal semi-aquatic spider, *Dolomedes aquaticus* (Pisauridae). Z. Naturforsch., *34c*: 463-469, 1979.

Williams, G. C. Measurement of consociation among fishes: comments on the evolution of schooling. Pub. Mich. St. Univ. Biol. Serv., *2*: 351-383, 1964.

Williams, G. C. *Adaptation and Natural Selection.* Princeton University Press, 1966.

Wilson, C. S. and D. G. Kleiman. Eliciting play: A comparative study (*Octodon, Octidontomys, Pediolagus, Phoca, Chocropsis, Ailuropoda*). Am. Zool., *14*: 341-370, 1974.

Wilson, D. S. Structured demes and the evolution of group-advantageous traits. Am. Nat., *111*: 157-185, 1977.

Wilson, E. O. Chemical communication among workers of the fire ant

Solenopsis saevissima (Fr. Smith), 2. An information analysis of the odor trail. Anim. Behav., *10*: 148-164, 1962.

Wilson, E. O. Chemical systems. In *Animal Communication*, T. A. Sebeok (ed.), Indiana University Press, 1968.

Wilson, E. O. Chemical communication within animal species. In *Chemical Ecology*, E. Sondheimer and J. B. Simeone (eds.), Academic Press, pp. 133-155, 1970.

Wilson, E. O. *The Insect Societies*. Harvard University Press, 1971.

Wilson, E. O. *Sociobiology: The New Synthesis.* Harvard University Press, 1975.

Wilson, E. O. and W. H. Bossert, Chemical communication among animals. Recent Progr. Hormone Res., *19*: 673-716, 1963.

Witt, P. N. Ein einfaches Prinzip zur Deutung einiger Proportionen im Spinnennetz. Behaviour, *4*: 172-189, 1952.

Witt, P. N. Do we live in the best of all worlds? Spider webs suggest an answer. Perspect. Biol. Med., *8*: 475-487, 1965.

Witt, P. N. The web as a means of communication. Biosci. Commun., *1*: 7-23, 1975.

Witt, P. N., J. O. Rawlings, and C. F. Reed. Ontogeny of web-building behavior in two orb-weaving spiders. Am. Zool., *12*: 445-454, 1972.

Witt, P. N., C. F. Reed, and D. B. Peakall. *A Spider's Web. Problems in Regulatory Biology.* Springer, 1968.

Witt, P. N., M. B. Scarboro, and D. B. Peakall. Comparative feeding data in three spider species of different sociality: *Araneus diadematus* Cl., *Mallos trivittatus* (Banks) and *Mallos gregalis* (Simon). Symp. zool. Soc. London, *42*: 89-97, 1978.

Wood-Mason, J. On the gigantic stridulating spider. Ann. Mag. Nat. Hist., *16*: 96, 1876.

Work, R. W. The force-elongation behavior of web fibers and silks forcibly obtained from orb-web-spinning spiders. Textile Res. J., *46*: 485-492, 1976.

Wynne-Edwards, V. C. *Animal Dispersion in Relation to Social Behavior.* Oliver and Byrd, 1962.

Yamashita, S. and H. Tateda. Spectral sensitivities of jumping spiders' eyes. J. comp. Physiol., *105*: 29-41, 1976.

TAXONOMIC INDEX

SUBJECT INDEX

Library of Congress Cataloging in Publication Data

Main entry under title:

Spider communication.

Bibliography: p.
Includes index.
Contents: Introduction: Communication in spiders / by Peter N. Witt—The significance and complexity of communication in spiders / by Bertrand Krafft—Spiders and vibratory signals / by Friedrich G. Barth—[etc.]
1. Spiders—Behavior. 2. Animal communication. 3. Arachnida—Behavior.
I. Witt, Peter Nikolaus. II. Rovner, Jerome S.
QL458.4.S64 595.4'40459 81-47164
ISBN 0-691-08291-X AACR2